OT 47
Operator Theory: Advances and Applications
Vol. 47

Birkhäuser Verlag
Basel · Boston · Berlin

Extension and Interpolation of Linear Operators and Matrix Functions

Edited by

I. Gohberg

1990

Birkhäuser Verlag
Basel · Boston · Berlin

Editor's address:

Prof. I. Gohberg
Raymond and Beverly Sackler Faculty of Exact Sciences
School of Mathematical Sciences
Tel Aviv University
69978 Tel Aviv, Israel

CIP-Titelaufnahme der Deutschen Bibliothek

Extension and Interpolation of Linear Operators and Matrix Functions /
ed. by I. Gohberg. – Basel ; Boston ; Berlin :
Birkhäuser, 1990
(Operator theory ; Vol. 47)
ISBN 3-7643-2233-0
NE: Gohberg, Izrael' [Hrsg.]; GT

Printed in Germany
ISBN 3-7643-2530-5
ISBN 0-8176-2530-5

Table of Contents

EDITORIAL INTRODUCTION

This volume consists of five papers which develop further the theory of extension and interpolation of linear operators and matrix functions.

The first paper, "Realization and factorization for rational matrix functions with symmetries," of D. Alpay, J.E. Ball, I. Gohberg, and L. Rodman, is related to the problem of homogeneous interpolation for rational matrix functions with symmetries. It contains a description of properties of realization of a rational matrix function with given symmetries and the solution of the inverse problem which is the interpolation problem of constructing a symmetric rational matrix function from its null pole data. Factorization is also one of the themes of this paper. The paper concludes with an analysis of different types of symmetries.

The second paper, "Lossless inverse scattering and reproducing kernels for upper triangular operators," of D. Alpay, P. Dewilde, and H. Dym, is concerned with lossless inverse scattering for nonstationary systems. The theory of point evaluations and Blaschke products is developed for the nonstationary case. These tools will be used in future publications on nonstationary generalizations of Nevanlinna Pick interpolation problems, where both the interpolation points and the values consigned to these points will be block diagonal operators.

The third paper, "Zero-pole structure of nonregular rational matrix functions," of J.A. Ball and M. Rakowski, is concerend with homogeneous interpolation problems for nonregular rational matrix functions. Here is described the null pole structure of such matrix functions. The inverse problem, which is an interpolation problem of constructing a rational matrix function from its null pole data, is also solved.

In the fourth paper, "Structured interpolation theory," of H. Bercovici, C. Foias and A. Tannenbaum, is solved a Nevanlinna Pick type interpolation problem for matrix valued functions. The novelty here is that instead of putting up restrictions on the norm of the matrix function which is its singular value, a more sophisticated structured singular value from systems theory is used. The latter is a generalization of the norm and the spectral radius. The paper contains a new version of the commutant lifting theorem.

The fifth paper, "Extension theorems for contraction operators on Krein spaces," of D.A. Dritschel and J. Rovnyak, starts with an introduction to operator theory on Krein spaces and continues with general theorems on extensions of contractive operators. The main theme is the theory of commutant lifting for contraction operators in Krein spaces.

One of the most eminent mathematicans of our time, Mark Grigorievich Krein passed away in Odessa on October 17, 1989. Almost all of the authors submitted their papers with a dedication to the memory of M.G. Krein, who made extremely important contributions in the topics of this volume. The editor feels it very appropriate to follow this spontaneous action and is joining the authors in dedicating the whole volume to the memory of M.G. Krein as an expression of respect and admiration.

Operator Theory
Advances and Applications, Vol. 47

REALIZATION AND FACTORIZATION FOR RATIONAL MATRIX FUNCTIONS WITH SYMMETRIES

Daniel Alpay, Joseph A. Ball[1], Israel Gohberg and Leiba Rodman[1]

IN MEMORY OF MARK GRIGORIEVICH KREIN

In this paper we study rational matrix valued functions which are symmetric in various senses. The problems of realization and minimal factorization are the main themes discussed for these classes of functions.

0. INTRODUCTION.

Let F be a field which will be considered fixed thoughout the paper. By $\mathcal{R}$ we denote the quotient field $F(z)$ of the polynomial ring $F[z]$ of F in the indeterminate z; more concretely one can view $\mathcal{R}$ as consisting of formal Laurent series $\sum_{j=-\infty}^{q} f_j z^j$ which have a presentation as $p(z) \cdot q(z)^{-1}$ with p and q polynomials. When the field is infinite, one can identify elements of $\mathcal{R}$ as functions from F into F; for a finite field this point of view has the draw back that a nonzero element of $\mathcal{R}$ may correspond to the zero function. Let $GL_m(\mathcal{R})$ be the multiplicative group of all regular (i.e. with determinant not equal to the zero element of $\mathcal{R}$) $m \times m$ matrices over $\mathcal{R}$; elements of $GL_m(\mathcal{R})$ thus more concretely may be thought of as formal Laurent series with coefficients equal to matrices over F or, if F is infinite, as matrix (over F)-valued functions on F. Given an automorphism $\mathcal{A}$ of this group $GL_m(\mathcal{R})$, we say that the rational matrix $W \in GL_m(\mathcal{R})$ is $\mathcal{A}$-symmetric if $\mathcal{A}(W) = W$. For example, let $F = \mathbb{C}$ and let $\mathcal{A}$ be the automorphism on $GL_m(\mathcal{R})$ defined by

$$\mathcal{A}(W)(z) = W(-\bar{z})^{*-1}.$$

Then the $\mathcal{A}$-symmetric rational matrices are exactly those rational matrix functions with unitary values on the imaginary line. In general, our main aim is to describe all $\mathcal{A}$-symmetric rational matrices and secondly, to describe all factorizations of $\mathcal{A}$-symmetric rational matrices with factors which are also $\mathcal{A}$-symmetric. We are also interested in the problem of characterizing the elementary $\mathcal{A}$-symmetric matrices, i.e. those $\mathcal{A}$-symmetric matrices with no nontrivial representation as a product of $\mathcal{A}$-symmetric matrices. These three problems can also be considered for rational matrices with respect to several symmetries.

In this paper we will be especially interested in the case when $F = \mathbb{C}$ and the automor-

[1]Research of these authors was partially supported by grants from the National Science Foundation.

phism $\mathcal{A}$ has one of the following four forms:

$$\text{(Ia)} \qquad \mathcal{A}(W)(z) = M^{-1}W \circ \varphi(z)M$$

$$\text{(Ib)} \qquad \mathcal{A}(W)(z) = M^{-1}\overline{W \circ \varphi(\bar{z})}M$$

$$\text{(IIa)} \qquad \mathcal{A}(W)(z) = M^{-1}[W \circ \varphi(z)]^{-1T}M$$

$$\text{(IIb)} \qquad \mathcal{A}(W)(z) = M^{-1}[W \circ \varphi(\bar{z})]^{-1*}M$$

where M is a fixed invertible matrix and φ is a Möbius transformation. Our main approach to solving this problem is the representation of proper rational matrices in the form

$$W(z) = D + C(zI - A)^{-1}B \tag{1.1}$$

where A, B, C and D are finite matrices of appropriate sizes. Such a representation for $W(z)$ we refer to as a *realization* of $W(z)$.

All four types of automorphisms mentioned above belong to the class of automorphisms which have the following property: there exists a map $(A, B, C, D) \to (A_{\mathcal{A}}, B_{\mathcal{A}}, C_{\mathcal{A}}, D_{\mathcal{A}})$ such that

$$\mathcal{A}(W)(z) = D_{\mathcal{A}} + C_{\mathcal{A}}(zI - A_{\mathcal{A}})^{-1}B_{\mathcal{A}}$$

if

$$W(z) = D + C(zI - A)^{-1}B$$

whenever both $W(z)$ and $\mathcal{A}(W)(z)$ are analytic and invertible at infinity. Here we work only with minimal realizations, i.e., we assume that the (square) size of A is as small as possible among all presentations of $W(z)$ in the form (1.1). We also assume that the map $(A, B, C, D) \to (A_{\mathcal{A}}, B_{\mathcal{A}}, C_{\mathcal{A}}, D_{\mathcal{A}})$ satisfies some additional properties which hold for the types of automorphisms listed above.

Namely, we demand that

$$\left(\begin{bmatrix} A_1 & B_1C_2 \\ 0 & A_2 \end{bmatrix}_{\mathcal{A}}, \begin{bmatrix} B_1D_2 \\ B_2 \end{bmatrix}_{\mathcal{A}}, [C_1 \quad D_1C_2]_{\mathcal{A}}, (D_1D_2)_{\mathcal{A}}\right) = \\ \left(\begin{bmatrix} A_{1\mathcal{A}} & B_{1\mathcal{A}}C_{2\mathcal{A}} \\ 0 & A_{2\mathcal{A}} \end{bmatrix}, \begin{bmatrix} B_{1\mathcal{A}}D_{2\mathcal{A}} \\ B_{2\mathcal{A}} \end{bmatrix}, [C_{1\mathcal{A}} \quad D_{1\mathcal{A}}C_{2\mathcal{A}}], (D_{1\mathcal{A}}D_{2\mathcal{A}})\right).$$

and, for $T \in \mathbb{C}^{n\times n}$ there exists a map $T \to T_{\mathcal{A}}$ (called the *associated map* to $\mathcal{A}$) such that

$$((TAT^{-1})_{\mathcal{A}}, (TB)_{\mathcal{A}}, (CT^{-1})_{\mathcal{A}}, D_{\mathcal{A}}) = (T_{\mathcal{A}}A_{\mathcal{A}}T_{\mathcal{A}}^{-1}, T_{\mathcal{A}}B_{\mathcal{A}}, C_{\mathcal{A}}T_{\mathcal{A}}^{-1}, D_{\mathcal{A}}).$$

In each of the four types of examples listed above these four properties are easily verified. For any automorphism $\mathcal{A}$ which satisfies these properties, we classify the $\mathcal{A}$-symmetric matrices in terms of their realizations and describe all minimal factorizations of $\mathcal{A}$–symmetric matrices with $\mathcal{A}$–symmetric factors. Examples of elementary $\mathcal{A}$–symmetric matrices are given in all four concrete cases. The basis for this analysis is the geometric state space characterization of minimal factorization for biproper rational matrix functions from [BGK]. This approach has already been used extensively to analyze Wiener–Hopf and spectral factorization and various analogues of spectral factorization associated with other types of antiautomorphisms (see [R],[RR1,2,3]). The theory we present here can be viewed as a direct generalization of [AG] to a more general setting.

Section 2 presents the axiomatic framework, preliminary analysis of the examples and the general theory. Section 3 deals with the group of $\mathcal{A}$–symmetric functions, their realizations and minimal factorizations of them into $\mathcal{A}$–symmetric factors of lower McMillan degree. The last four sections contain a detailed analysis of the four main types of examples mentioned above.

In future work we plan to apply the same approach as in this paper to analyze functions satisfying a combination of symmetry conditions with applications to various kinds of interpolation problems. A number of interesting results in this direction have already been obtained by Efimov–Potapov [EP] and Dyukarev–Katsnelson [DK].

In the last two sections of the paper we given an intrinsic description of degree preserving automorphisms of regular rational matrix functions and of the four specific classes of automorphisms analyzed in this paper. The starting point here is the characterization due to Dieudonné of automorphisms of the general linear group over an arbitrary field (see [D]). Indeed, the four classes of automorphisms discussed here illustrate Dieudonné's theory for the case where the field is the set of scalar rational functions over a field.

Although most applications are concerned with rational matrix functions over the complex or real field, other fields appear as well. For example, see [OVW], where rational matrix functions over the field of formal power series are used in the study of certain perturbed system. We develop the theory in Sections 1–3 in the framework of rational matrices over an arbitrary field.

Notation which we shall need is as follows:

$\mathbb{C}$	The field of complex numbers
$\mathbb{R}$	The field of real numbers
A^T	Transpose of the matrix A
A^*	Conjugate transpose of A ($F = \mathbb{C}$)
$\sigma(M)$	The set of eigenvalues of the matrix M
A^{-T}	$(A^T)^{-1}$
A^{-*}	$(A^*)^{-1}$
$\mathcal{R} = F(z)$	The field of scalar rational Laurent series with coefficients in a field F
$GL_m(\mathcal{R})$	Invertible $m \times m$ matrices over $\mathcal{R}$ (i.e. matrices over $\mathcal{R}$ with determinant

not the zero element of $\mathcal{R}$)

$Im\ A$ The image of the matrix A (where A is viewed as a linear transformation

$Ker\ A$ The kernel of the matrix A

We denote by $\sum_{m,n}$ the set of $\begin{pmatrix} A & B \\ C & D \end{pmatrix}$ where $A \in F^{n\times n}$, $B \in F^{n\times m}$, $C \in F^{m\times n}$ and $D \in F^{m\times m}$, and by $*$ system multiplication

$$\begin{pmatrix} A_1 & B_1 \\ C_1 & D_1 \end{pmatrix} * \begin{pmatrix} A_2 & B_2 \\ C_2 & D_2 \end{pmatrix} = \begin{pmatrix} A_1 & B_1C_2 & \vdots & B_1D_2 \\ 0 & A_2 & \vdots & B_2 \\ \cdots & \cdots\cdots & \vdots & \cdots\cdots \\ C_2 & D_1C_2 & \vdots & D_1D_2 \end{pmatrix}$$

This operation is connected with multiplication of rational matrices as follows: if $W_i(z) = D_i + C_i(zI - A_i)^{-1}B_i$ for $i = 1,2$ then $W(z) = W_1(z)W_2(z)$ has the realization $W(z) = D + C(zI - A)^{-1}B$ where $\begin{pmatrix} A & B \\ C & D \end{pmatrix} = \begin{pmatrix} A_1 & B_1 \\ C_1 & D_1 \end{pmatrix} * \begin{pmatrix} A_2 & B_2 \\ C_2 & D_2 \end{pmatrix}$.

1. PRELIMINARIES.

Let F be a (commutative) field. We consider matrices W of the form

$$W(z) = [w_{ij}(z)]_{i,j=1}^{m}, \qquad z \in F, \tag{1.1}$$

where $w_{ij}(z)$ is a Laurent series of the form $\frac{p_{ij}(z)}{q_{ij}(z)}$, where $p_{ij}(z)$ and $q_{ij}(z)$ are polynomials with coefficients in F. Without loss of generality we assume also that each $q_{ij}(z)$ is a *monic* polynomial, i.e., with leading coefficient 1.

Functions of the form (1.1) will be called *rational matrices* (r.m.). A matrix Laurent series $W(z)$ will be called *proper* if $W(z)$ has the form $\sum_{j=-\infty}^{0} W_j z^j$. For a proper Laurent series $W(z)$ its value at infinity $W(\infty)$ is defined to be the leading coefficient W_0 in its Laurent series representation. A *realization* of $W(z)$ is a representation in the form

$$W(z) = D + C(zI - A)^{-1}B, \tag{1.2}$$

where A, B, C and D are matrices (of appropriate sizes) with entries in F.

If $W(z)$ has a realization as in (1.2) it is easily seen that W is rational and that the Laurent series for W is given by

$$W(z) = D + \sum_{j=1}^{\infty} CA^{j-1}Bz^j$$

from which it follows that $W(z)$ is proper with $W(\infty) = D$. We now state the converse.

PROPOSITION 1.1. *A Laurent series $W(z) = \sum_{j=-\infty}^{\infty} W_j z^j$ admits a realization (1.2) if and only if W is proper and the Hankel matrix H_W associated with W*

$$H_W = \begin{bmatrix} W_0 & W_1 & W_2\cdots \\ W_1 & W_2 & W_3\cdots \\ W_2 & W_3 & W_4\cdots \\ \vdots & \vdots & \vdots \end{bmatrix}$$

has finite rank, or equivalently, if and only if W is proper and rational.

This result is well known; for the proof we refer to [BBV] where it is done more generally over certain types of rings.

We now consider minimal realizations. A realization (A, B, C) of a proper rational matrix function $W(z)$ is called *minimal* if the size of the matrix A is minimal possible among all realizations of $W(z)$. To characterize minimal realizations we need some preliminaries. (In what follows, an $m \times n$ matrix G with entries in F will be considered also as a linear transformation $F^n \to F^m$ and the restriction of G to a subspace $\mathcal{M} \subset F^n$ will be denoted $G|_{\mathcal{M}}$.) A pair of matrices C, A (with entries in F) of sizes $n \times m$ and $m \times m$ respectively will be called a *null kernel pair* if $\bigcap_{j=0}^{\infty} Ker(CA^j) = \{0\}$. A pair of matrices A, B of sizes $m \times m$ and $m \times n$, respectively, will be called a *full range pair* if

$$\sum_{j=0}^{\infty} Im(A^j B) = F^m.$$

Let now (A, B, C) be a realization of a proper rational function $W(z)$, and let $m \times m$ be the size of A. Given a direct sum decomposition $F^m = \mathcal{L} \dot{+} \mathcal{M} \dot{+} \mathcal{N}$, where $\mathcal{L}, \mathcal{M}, \mathcal{N}$ are subspaces such that $\mathcal{L}$ and $\mathcal{L} \dot{+} \mathcal{M}$ are A–invariant, $C|_{\mathcal{L}} = 0$ and $Im B \subset \mathcal{L} + \mathcal{M}$, a realization $(P_{\mathcal{M}} A|_{\mathcal{M}}, P_{\mathcal{M}} B, C|_{\mathcal{M}})$, where $P_{\mathcal{M}} : F^m \to \mathcal{M}$ is a projector of $\mathcal{M}$ with $Ker\ P_{\mathcal{M}} \supset \mathcal{L}$, is called a *reduction* of (A, B, C). Note that $(P_{\mathcal{M}} A|_{\mathcal{M}}, P_{\mathcal{M}} B, C|_{\mathcal{M}})$ is again a realization for the same $W(z)$. We shall also say that (A, B, C) is a *dilation* of $(P_{\mathcal{M}} A|_{\mathcal{M}}, P_{\mathcal{M}} B, C|_{\mathcal{M}})$.

THEOREM 1.2. *Any realization (A, B, C) of $W(z)$ is the dilation of a realization (A_0, B_0, C_0) of $W(z)$ with null kernel pair (C_0, A_0) and full range pair (A_0, B_0).*

The proof is the same as that of Theorem 7.1.3 in [GLR].

We need further some observations concerning one–sided invertibility of matrices. Let (C, A) be a null kernel pair and (A, B) be a full range pair. Then we have

$$Ker\ col[CA^j]_{j=0}^{p-1} = \{0\}, \qquad Im[B, AB, \ldots, A^{p-1}B] = F^m$$

where p is any integer not smaller than the degree of the minimal polynomial for A. Hence there exists a *left inverse* $[col[CA^j]_{j=0}^{p-1}]^{-L}$. Thus

$$[col[CA^j]_{j=0}^{p-1}]^{-L}\ col[CA^j]_{j=0}^{p-1} = I$$

Also, there exists a *right inverse* $[B, AB, \ldots, A^{p-1}B]^{-R}$:

$$[B, AB, \ldots, A^{p-1}B][B, AB, \ldots, A^{p-1}B]^{-R} = I$$

Note that in general the left and right inverses involved are not unique.

THEOREM 1.3. *Let (A_1, B_1, C_1) and (A_2, B_2, C_2) be realizations for a proper rational matrix function $W(z)$ for which (C_1, A_1) and (C_2, A_2) are null kernel pairs and (A_1, B_1), (A_2, B_2) are full range pairs. Then the sizes of A_1 and A_2 coincide, and there exists a nonsingular matrix S such that*

$$A_1 = S^{-1}A_2S, \qquad B_1 = S^{-1}B_2, \qquad C_1 = C_2S. \tag{1.4}$$

Moreover, the matrix S is unique and is given by

$$\begin{aligned} S &= [col[C_2A_2^j]_{j=0}^{p-1}]^{-L}[col[C_1A_1^j]_{j=0}^{p-1}] \\ &= [B_2, A_2B_2, \ldots, A_2^{p-1}B_2][B_1, A_1B_1, \ldots, A_1^{p-1}B_1]^{-R}. \end{aligned} \tag{1.5}$$

Here p is any integer greater than or equal to the maximum of the degrees of minimal polynomials for A_1 and A_2 and the superscript $-L$ (resp. $-R$) indicates left (resp. right) inverse.

PROOF: We have

$$\sum_{j=1}^{\infty} C_1A_1^jB_1z^j = \sum_{j=1}^{\infty} C_2A_2^jB_2z^j \tag{1.6}$$

The proof of Theorem 1.3 can be done in the same way as the proof of Theorem 7.1.4 in [GLR]. □

Theorems 1.2 and 1.3 allow us to interpret minimality of a realization (A, B, C) as the null kernel property of (C, A) and the full range property of (A, B) holding simultaneously.

A closely related concept that will be used throughout the paper is the degree of a rational matrix function. This concept is well–known in the theory of linear systems.

Let $W(z)$ be $n \times n$ rational matrix function with coefficients in F and such that $det\, W(z) \not\equiv 0$. We can write

$$W(z) = W_1(z) + W_2(z) \tag{1.10}$$

where $W_1(z)$ is proper and $W_2(z)$ is a polynomial. By adjusting the value of $W_1(z)$ at ∞ between $W_1(z)$ and $W_2(z)$ we can assume that both $W_1(\infty)$ and $W_2(0)$ are nonsingular matrices. Now $W_1(z)$ admits a minimal realization

$$W_1(z) = D + C(zI - A)^{-1}B,$$

which is unique up to similarity. The rational matrix $\tilde{W}_2(z) = W_2(z^{-1})$ is proper and, as a polynomial in z^{-1}, is rational, and hence admits a minimal realization as well:

$$\tilde{W}_2(z) = \tilde{D} + \tilde{C}(zI - \tilde{A})^{-1}\tilde{B}.$$

(again unique up to similarity). We take the sum of the sizes of A and $\tilde{A}$ to be, by definition, the *McMillan degree*, or simply the *degree* of $W(z)$ (notation: deg(W)). It is immediately seen that the degree is well defined, i.e. is independent of the choice of the minimal realization of $W_1(z)$ and $\tilde{W}_2(z)$ and also independent of the choice of the splitting (1.10) of $W(z)$ with the required properties.

In particular, the notion of degree applies to scalar rational series. It is not difficult to see that if $w(z) = p(z)/q(z)$ is a scalar rational Laurent series over F and the polynomials $p(z)$ and $q(z)$ have no common factor, then $\deg w = max(\deg p, \deg q)$.

We consider next the minimal factorization of rational matrices. Let $W(z)$ be an $m \times m$ rational matrix with coefficients in F. A factorization (or product)

$$W(z) = W_1(z)W_2(z)$$

where $W_i(z)$ are also $m \times m$ rational matrices, is called *minimal* if

$$\deg(W) = \deg(W_1(z)) + \deg(W_2(z)).$$

In the sequel we will apply this notion in case that $W(z)$ biproper (i.e. both W and W^{-1} are proper). For a comprehensive treatment of minimal factorizations, see [BGK] (see also [GLR]) for the case $F = \mathbb{C}$.

We will need a description of minimal factorizations in terms of supporting invariant decomposition for $W(z)$ as follows. Let $W(z) = I + C(zI - A)^{-1}B$ be a minimal realization of W (it is assumed here that $W(\infty) = I$), where A is of size $n \times n$. We say that

$$F^n = \mathcal{M} \dot{+} \mathcal{N} \tag{1.11}$$

is a *supporting invariant decomposition* for $W(z)$ if $\mathcal{M}$ is A–invariant and $\mathcal{N}$ is $A^\times$–invariant, where $A^\times = A - BC$. The projection $\pi : F^n \to F^n$ which projects on $\mathcal{N}$ along $\mathcal{M}$ is called the *supporting projection* associated with the supporting invariant decomposition (1.11).

The role of supporting invariant decompositions in the minimal factorization problem is revealed in the next theorem (proved in [BGK] (see also [GLR]) for the case when $F = \mathbb{C}$).

THEOREM 1.4. *Let (1.11) be a supporting invariant decomposition for $W(z)$. Then $W(z)$ admits a minimal factorization*

$$\begin{aligned} W(z) =& [I + C\pi_{\mathcal{M}}(zI - A)^{-1}\pi_{\mathcal{M}}B] \times [I + C\pi_{\mathcal{N}}(zI - A)^{-1}\pi_{\mathcal{N}}B] \\ =& [I + C(zI - A)^{-1}\pi_{\mathcal{M}}B] \times [I + C\pi_{\mathcal{N}}(zI - A)^{-1}B] \end{aligned} \tag{1.12}$$

where $\pi_{\mathcal{M}}$ is the projector on $\mathcal{M}$ along $\mathcal{N}$ and $\pi_{\mathcal{N}}$ is the projector on $\mathcal{N}$ along $\mathcal{M}$.

Conversely, for every minimal factorization $W(z) = W_1(z)W_2(z)$ where the factors are rational matrices with value I at infinity, there exists a unique supporting invariant decomposition $F^n = \mathcal{M} \dot{+} \mathcal{N}$ such that

$$\begin{aligned} W_1(z) =& I + C\pi_{\mathcal{M}}(zI - A)^{-1}\pi_{\mathcal{M}}B \\ W_2(z) =& I + C\pi_{\mathcal{N}}(zI - A)^{-1}\pi_{\mathcal{N}}B. \end{aligned} \tag{1.13}$$

Note that the second equality in (1.12) follows from the relations $\pi_{\mathcal{M}} A \pi_{\mathcal{M}} = A \pi_{\mathcal{M}}$ and $\pi_{\mathcal{N}} A \pi_{\mathcal{N}} = \pi_{\mathcal{N}} A$, which express the A–invariance of $\mathcal{M}$.

The proof goes exactly as in the proof of Theorem 7.3.1 in [GLR] and therefore is omitted.

The results presented in this section will be used repeatedly throughout the paper.

2. AXIOMATIC THEORY.

This section is divided into two subsections. In 2.1 we introduce the family of automorphisms considered in the present paper while 2.2 is devoted to the general theory of these automorphisms. Recall that $\mathcal{R}$ stands for the field of scalar rational Laurent series over the field F.

2.1 Admissible Automorphisms.

We begin with:

DEFINITION 2.1. *The automorphism $W \to \mathcal{A}(W)$ of $GL_m(\mathcal{R})$ is called admissible if it is degree preserving: for every W in $GL_m(\mathcal{R})$,*

$$\deg W = \deg \mathcal{A}(W) \tag{2.1}$$

and if there exists a (in general defined everywhere but on an algebraic set) map on systems

$$\begin{pmatrix} A & B \\ C & D \end{pmatrix} \to \begin{pmatrix} A & B \\ C & D \end{pmatrix}_{\mathcal{A}} = \begin{pmatrix} A_{\mathcal{A}} & B_{\mathcal{A}} \\ C_{\mathcal{A}} & D_{\mathcal{A}} \end{pmatrix} \tag{2.2}$$

where $\begin{pmatrix} A & B \\ C & D \end{pmatrix}$ and $\begin{pmatrix} A & B \\ C & D \end{pmatrix}_{\mathcal{A}}$ belong to $\sum_{m,n}$, m and $n \geq 1$, with the following properties:

(i) If both W and $\mathcal{A}(W)$ are proper and if $W(z) = D + C(zI_n - A)^{-1}B$ is a minimal realization of W, then $\mathcal{A}(W)(z) = D_{\mathcal{A}} + C_{\mathcal{A}}(zI_n - A_{\mathcal{A}})^{-1}B_{\mathcal{A}}$ is a minimal realization of $\mathcal{A}(W)$.

(ii) If W_1 and W_2 are elements in $GL_m(\mathcal{R})$ with minimal realizations $W_i(z) = D_i + C_i(zI_{n_i} - A_i)^{-1}B_i$, $i = 1,2$, and if moreover the product $W = W_1W_2$ is minimal and $\mathcal{A}(W_1)$ and $\mathcal{A}(W_2)$ are also proper, then

$$\left(\begin{pmatrix} A_1 & B_1 \\ C_1 & D_1 \end{pmatrix} * \begin{pmatrix} A_2 & B_2 \\ C_2 & D_2 \end{pmatrix}\right)_{\mathcal{A}} = \begin{pmatrix} A_1 & B_1 \\ C_1 & D_1 \end{pmatrix}_{\mathcal{A}} * \begin{pmatrix} A_2 & B_2 \\ C_2 & D_2 \end{pmatrix}_{\mathcal{A}} \tag{2.3}$$

(iii) For every $n \geq 1$ and every invertible matrix T in $F^{n\times n}$ there exist a unique invertible matrix $T_{\mathcal{A}}$ in $F^{n\times n}$ such that, for every minimal system $\begin{pmatrix} A & B \\ C & D \end{pmatrix}$ for which $\begin{pmatrix} A & B \\ C & D \end{pmatrix}_{\mathcal{A}}$ exists ($D \in F^{m\times m}$) the equality

(2.4)
$$\left(\begin{pmatrix} T & 0 \\ 0 & I_m \end{pmatrix}\begin{pmatrix} A & B \\ C & D \end{pmatrix}\begin{pmatrix} T & 0 \\ 0 & I_m \end{pmatrix}^{-1}\right)_{\mathcal{A}} = \begin{pmatrix} T_{\mathcal{A}} & 0 \\ 0 & I_m \end{pmatrix}\begin{pmatrix} A & B \\ C & D \end{pmatrix}_{\mathcal{A}}\begin{pmatrix} T_{\mathcal{A}} & 0 \\ 0 & I_m \end{pmatrix}^{-1}$$

holds.

Roughly speaking, condition (2.2) means that we define and study these automorphisms from the state space point of view while (2.3) expresses that the automorphism property is translated into the system framework.

The map (2.2) is not uniquely defined, but in every example which we will consider, there is a particular natural map, called the *associated system map*, which defines the automorphism.

We note that the properties of degree conservation and (i) together "almost imply" (ii) and (iii). Indeed, since both sides of (2.3) define a minimal realization for $\mathcal{A}(W)$, (2.3) will always hold, up to a similarity, that is

$$\begin{pmatrix} T & 0 \\ 0 & I \end{pmatrix}\left(\begin{pmatrix} A_1 & B_1 \\ C_1 & D_1 \end{pmatrix} * \begin{pmatrix} A_2 & B_2 \\ C_2 & D_2 \end{pmatrix}\right)_{\mathcal{A}} = \left(\begin{pmatrix} A_1 & B_1 \\ C_1 & D_1 \end{pmatrix}_{\mathcal{A}} * \begin{pmatrix} A_2 & B_2 \\ C_2 & D_2 \end{pmatrix}_{\mathcal{A}}\right)\begin{pmatrix} T & 0 \\ 0 & I \end{pmatrix}$$

for some invertible element T in $F^{n \times n}$, $n = n_1 + n_2$.

Similarly, if both $\begin{pmatrix} A & B \\ C & D \end{pmatrix}$ and $\begin{pmatrix} T & 0 \\ 0 & I \end{pmatrix}\begin{pmatrix} A & B \\ C & D \end{pmatrix}\begin{pmatrix} T & 0 \\ 0 & I \end{pmatrix}^{-1}$ are minimal systems with transfer function $W(z)$ and if $\mathcal{A}(W)$ is proper, then (i) implies that

$$\begin{pmatrix} A & B \\ C & D \end{pmatrix}_{\mathcal{A}} \text{ and } \left(\begin{pmatrix} T & 0 \\ 0 & I \end{pmatrix}\begin{pmatrix} A & B \\ C & D \end{pmatrix}\begin{pmatrix} T & 0 \\ 0 & I \end{pmatrix}^{-1}\right)_{\mathcal{A}}$$

are minimal systems with transfer function $\mathcal{A}(W)$, and thus there exists a matrix $T_{\mathcal{A}}$ such that (2.4) holds. A priori, $T_{\mathcal{A}}$ may depend on A, B, C and D. Condition (iii) expresses that $T \to T_{\mathcal{A}}$ is well defined and not multivalued.

DEFINITION 2.2. *The map $T \to T_{\mathcal{A}}$ will be called the associated similarity map.*

We now present four families of admissible automorphisms. In the following, M denotes an invertible $n \times n$ matrix and φ a degree one Möbius transform, i.e.

$$\varphi(z) = (kz + l)/(rz + s)$$

with $ks - lr \neq 0$, where k, s, l, r are elements of F.

EXAMPLE 2.1. *Type Ia Symmetries.*

These are the automorphisms defined by

$$\mathcal{A}(W)(z) = M^{-1}W(\varphi(z))M \tag{2.5}$$

The map $W \to \mathcal{A}(W)$ is clearly an automorphism. It is degree preserving since neither the Möbius transform nor multiplication by M or M^{-1} change the McMillan degree of W.

When F is infinite and we may identify $\mathcal{R}$ with functions from a generic subset of F into F, then the rational matrix $\mathcal{A}(W)$ is proper if and only if W is defined at $\varphi(\infty)$ (i.e., the denominators of the entries in W take non-zero value at $\varphi(\infty)$). In general, $\mathcal{A}(W)$ is proper if $W(z)$ has a proper series expansion in powers of $(z-\varphi(\infty))$:

$$W(z) = P(z)q(z)^{-1} = \tilde{P}(z-\varphi(\infty))\tilde{q}(z-\varphi(\infty))^{-1} = \sum_{j=-\infty}^{0} \tilde{W}_j(z-\varphi(\infty))^j$$

where P and $\tilde{P}$ are polynomial matrices and q and $\tilde{q}$ are scalar polynomials. We have

$$\varphi(\infty) = \begin{cases} k/r & if\ k \neq 0\ and\ r \neq 0 \\ 0 & if\ k = 0\ (and\ then\ r \neq 0\ since\ ks - lr \neq 0) \\ \infty & if\ r = 0\ (and\ then\ k \neq 0). \end{cases}$$

Thus if W is itself proper and if $W(z) = D + C(zI_n - A)^{-1}B$ is a minimal realization of W, the function $\mathcal{A}(W)$ is proper if and only if $kI_n - rA$ is invertible. A minimal realization of $W(\varphi(z))$ is then $W(\varphi(z)) = D_\varphi + C_\varphi(zI_n - A_\varphi)^{-1}B_\varphi$ where

$$A_\varphi = (kI_n - rA)^{-1}(sA - lI_n) \tag{2.6a}$$

$$B_\varphi = (kI_n - rA)^{-1}B \tag{2.6b}$$

$$C_\varphi = C(kI_n - rA)^{-1}(ks - lr) \tag{2.6c}$$

and

$$D_\varphi = D + rC(kI_n - rA)^{-1}B \tag{2.6d}$$

(see [BGK] or Chapter 5 in [BGR]), and we may take as the associated system map (2.2)

$$\begin{pmatrix} A & B \\ C & D \end{pmatrix}_{\mathcal{A}} = \begin{pmatrix} I & 0 \\ 0 & M^{-1} \end{pmatrix} \begin{pmatrix} A & B \\ C & D \end{pmatrix}_{\varphi} \begin{pmatrix} I & 0 \\ 0 & M \end{pmatrix} \tag{2.7}$$

(where we define $\begin{pmatrix} A & B \\ C & D \end{pmatrix}_\varphi$ to be $\begin{pmatrix} A_\varphi & B_\varphi \\ C_\varphi & D_\varphi \end{pmatrix}$).

To check that property (2.3) is met, let W_1 and W_2 be as in definition 2.1 and let A, B, C and D be defined by

$$A = \begin{pmatrix} A_1 & B_1C_2 \\ 0 & A_2 \end{pmatrix},\ B = \begin{pmatrix} B_1D_2 \\ B_2 \end{pmatrix},\ C = (C_1,\ D_1C_2)\ and\ D = D_1D_2. \tag{2.8}$$

By the minimality hypothesis both $kI_{n_1} - A_1$ and $kI_{n_2} - A_2$ are invertible and, with $n = n_1 + n_2$,

$$(kI_n - A)^{-1} = \begin{pmatrix} (kI_n - A_1)^{-1} & r(kI_{n_1} - rA_1)^{-1}B_1C_2(kI_{n_2} - rA_2)^{-1} \\ 0 & (kI_{n_2} - A_2)^{-1} \end{pmatrix} \tag{2.9}$$

so that

$$\begin{aligned} &(kI_n - A)^{-1}(sA - lI_n) \\ =&\begin{pmatrix} (kI_{n_1} - A_1)^{-1} & r(kI_{n_1} - rA_1)^{-1}B_1C_2(kI_{n_2} - rA_2)^{-1} \\ 0 & (kI_{n_2} - A_2)^{-1} \end{pmatrix} \begin{pmatrix} sA_1 - lI_{n_1} & sB_1C_2 \\ 0 & sA_2 - lI_{n_2} \end{pmatrix} \end{aligned}$$

The (1,2) block of this product is equal to

$$\begin{aligned}&s(kI_{n_1}-A_1)^{-1}B_1C_2+r(kI_{n_1}-rA_1)^{-1}B_1C_2(kI_{n_2}-rA_2)^{-1}(sA_2-lI_{n_2})\\=&B_{1\mathcal{A}}C_2(kI_{n_2}-rA_2)^{-1}\{s(kI_{n_2}-rA_2)+r(sA_2-lI_2)\}\\=&B_{1\mathcal{A}}C_{2\mathcal{A}}\end{aligned}$$

and so,

$$A_{\mathcal{A}}=\begin{pmatrix}A_{1\mathcal{A}} & B_{1\mathcal{A}}C_{2\mathcal{A}}\\ 0 & A_{2\mathcal{A}}\end{pmatrix}$$

which is one of the four equalities needed to prove (2.3). We next prove that

$$B_{\mathcal{A}}=\begin{pmatrix}B_{1\mathcal{A}}D_{2\mathcal{A}}\\ B_{2\mathcal{A}}\end{pmatrix}$$

Indeed,

$$B_{\mathcal{A}}=(kI_n-rA)^{-1}BM$$

and, using (2.9) and the definition of B, we obtain

$$B_{\mathcal{A}}=\begin{pmatrix}(kI_{n_1}-A_1)^{-1} & r(kI_{n_1}-rA_1)^{-1}B_1C_2(kI_{n_2}-rA_2)^{-1}\\ 0 & (kI_{n_2}-A_2)^{-1}\end{pmatrix}\begin{pmatrix}B_1D_2M\\ B_2M\end{pmatrix}$$

The (2,1) block of this equality is equal to $B_{2\mathcal{A}}$. As for the (1,1) block, it is equal to

$$\begin{aligned}&(kI_{n_1}-A_1)^{-1}B_1D_2M+r(kI_{n_1}-rA_1)^{-1}B_1C_2(kI_{n_2}-rA_2)^{-1}B_2M\\=&B_{1\mathcal{A}}\{M^{-1}D_2M+rM^{-1}C_2(kI_{n_2}-rA_2)^{-1}B_2M\}\\=&B_{1\mathcal{A}}D_{2\mathcal{A}}\end{aligned}$$

Similarly, one proves that

$$C_{\mathcal{A}}=(C_{1\mathcal{A}},D_{1\mathcal{A}}C_{2\mathcal{A}})$$

and the equality $D_{\mathcal{A}}=D_{1\mathcal{A}}\cdot D_{2\mathcal{A}}$ drops from taking $z=\infty$ in the identity $\mathcal{A}(W)=\mathcal{A}(W_1)\mathcal{A}(W_2)$. The equality (2.3) is thus satisfied.

Similar computations would appear in the next three families of automorphisms to be defined, but we will omit them.

To prove that $\mathcal{A}(W)$ defines an admissible automorphism, it remains to show that (iii) of definition (2.1) is satisfied. If we replace A, B and C by respectively TAT^{-1}, TB and CT^{-1} in (2.6), D_φ stays the same and A_φ, B_φ and C_φ are replaced by $TA_\varphi T^{-1}$, TB_φ and $C_\varphi T^{-1}$ respectively, i.e.

$$\left[\begin{pmatrix}T&0\\0&I\end{pmatrix}\begin{pmatrix}A&B\\C&D\end{pmatrix}\begin{pmatrix}T&0\\0&I\end{pmatrix}^{-1}\right]_\varphi=\begin{pmatrix}T&0\\0&I\end{pmatrix}\begin{pmatrix}A&B\\C&D\end{pmatrix}_\varphi\begin{pmatrix}T&0\\0&I\end{pmatrix}^{-1}\tag{2.10}$$

In combination with (2.7) we obtain as associated similarity map

(2.11) $$T_{\mathcal{A}} = T.$$

EXAMPLE 2.2. *Type Ib Symmetries.*

Here $F = \mathbb{C}$. They are defined by

(2.12) $$\mathcal{A}(W)(z) = M^{-1}\overline{W(\overline{\varphi(z)})}M,$$

where the upper bar denotes complex conjugation. These are clearly degree preserving automorphisms of $GL_m(\mathcal{R})$. The map (2.2) can be chosen to be:

(2.13) $$\begin{pmatrix} A & B \\ C & D \end{pmatrix}_{\varphi} = \begin{pmatrix} I & 0 \\ 0 & M^{-1} \end{pmatrix} \begin{pmatrix} \bar{A} & \bar{B} \\ \bar{C} & \bar{D} \end{pmatrix}_{\varphi} \begin{pmatrix} I & 0 \\ 0 & M \end{pmatrix}$$

where

$$\begin{pmatrix} \bar{A} & \bar{B} \\ \bar{C} & \bar{D} \end{pmatrix}_{\varphi} = \begin{pmatrix} (\bar{A})_{\varphi} & (\bar{B})_{\varphi} \\ (\bar{C})_{\varphi} & (\bar{D})_{\varphi} \end{pmatrix}$$

and $(\bar{A})_{\varphi}, \ldots, (\bar{D})_{\varphi}$ are defined by (2.6) with $\bar{A}, \ldots, \bar{D}$ instead of $A, \ldots, D$ and the associated similarity map is

(2.14) $$T_{\mathcal{A}} = \bar{T}$$

More generally, type Ib symmetries can be defined (using the same formula (2.12)) for a field F (instead of $\mathbb{C}$) and its nontrivial field automorphism μ (instead of complex conjugation). We will consider the type Ib symmetries only for the case of $F = \mathbb{C}$ and μ=complex conjugation.

EXAMPLE 2.3. *Type IIa Symmetries.*

There symmetries are defined by

(2.15) $$\mathcal{A}(W)(z) = M(W(\varphi(z))^{-1})^T M^{-1}$$

The associated system map (2.2) is defined now by:

(2.16a) $$A_{\mathcal{A}} = (A_{\varphi} - B_{\varphi}D_{\varphi}^{-1}C_{\varphi})^T$$

(2.16b) $$B_{\mathcal{A}} = C_{\varphi}^T(D_{\varphi}^{-1})^T M^{-1}$$

(2.16c) $$C_{\mathcal{A}} = -M(D_{\varphi}^{-1})^T B_{\varphi}^T$$

and

(2.16d) $$D_{\mathcal{A}} = M(D_{\varphi}^{-1})M^{-1}$$

which can be written in the form

$$\begin{pmatrix} A & B \\ C & D \end{pmatrix}_{\mathcal{A}} = \begin{pmatrix} I & 0 \\ 0 & M \end{pmatrix} \left(\begin{pmatrix} A & B \\ C & D \end{pmatrix}_{\varphi} \right)^{\times T} \begin{pmatrix} I & 0 \\ 0 & M^{-1} \end{pmatrix} \tag{2.17}$$

with the convention that

$$\begin{pmatrix} A & B \\ C & D \end{pmatrix}^{\times} = \begin{pmatrix} A - BD^{-1}C & -BD^{-1} \\ D^{-1}C & D^{-1} \end{pmatrix} \tag{2.18}$$

If we replace in (2.16) A, B and C by TAT^{-1}, TB and CT^{-1}, then D stays unchanged and $A_{\mathcal{A}}$, $B_{\mathcal{A}}$ and $C_{\mathcal{A}}$ are respectively replaced by $(T^T)^{-1}A_{\mathcal{A}}T^T$, $C_{\mathcal{A}}T^T$ and $(T^T)^{-1}B_{\mathcal{A}}$ and so the associated similarity map is

$$T_{\mathcal{A}} = (T^T)^{-1}$$

EXAMPLE 2.4. *Type IIb Symmetries.*

Here again $F = \mathbb{C}$, as in Type Ib symmetries. These are the automorphisms of $GL_m(\mathcal{R})$ defined by

$$\mathcal{A}(W)(z) = M(W(\overline{\varphi(z)})^*)^{-1}M^{-1} \tag{2.19}$$

and the map (2.2) is chosen to be

$$A_{\mathcal{A}} = ((\overline{A})_{\varphi} - (\overline{B})_{\varphi}(\overline{D})_{\varphi}^{-1}(\overline{C})_{\varphi})^T \tag{2.20a}$$

$$B_{\mathcal{A}} = (\overline{C})_{\varphi}^T((\overline{D})_{\varphi}^{-1})^T M^{-1} \tag{2.20b}$$

$$C_{\mathcal{A}} = -\,M((\overline{D})_{\varphi}^{-1})^T(\overline{B})_{\varphi}^T \tag{2.20c}$$

and

$$D_{\mathcal{A}} = M((\overline{D})_{\varphi}^{-1})M^{-1} \tag{2.20d}$$

and the associated similarity map is

$$T_{\mathcal{A}} = (T^*)^{-1} \tag{2.21}$$

Even in the scalar case and, for example, with $F = \mathbb{C}$ these four families of automorphisms exhaust but a small part of the set of all possible automorphisms. Indeed, any element in $GL_1(\mathcal{R})$ can be written in a unique way as

$$w(z) = c \cdot \prod_{\alpha \in \mathbb{C}} (w_\alpha(z))^{p_\alpha}$$

where $c \in \mathbb{C}$, $w_\alpha(z) = z - \alpha$ and $p_\alpha \in \mathbb{Z}$ for each α and $p_\alpha = 0$ for all but finitely many α.

If $\alpha \rightarrow z_\alpha$ is a one–to–one map from $\mathbb{C}$ onto $\mathbb{C}$ and k is some integer, then the map $w \rightarrow \mathcal{A}(w)$ defined by

$$\mathcal{A}(w)(z) = c^k \prod_{\alpha \in \mathbb{C}} (z - z_\alpha)^{p_\alpha}$$

is an automorphism of $GL_1(\mathcal{R})$ which, as is easily checked, is generally not of the form (2.5), (2.12), (2.15) or (2.19).

One can also replace in the examples 2.1–2.4 M by $M(z)$, a non–constant element of $GL_m(\mathcal{R})$. The corresponding automorphisms will not be degree preserving (see Theorem 8.2), as is illustrated by the example

$$M(z) = \begin{pmatrix} 0 & b(z) \\ b^{-1}(z) & 0 \end{pmatrix}, \qquad b(z) = (z - i)/(z + i).$$

Then,

$$\mathcal{A}(W)(z) = M^{-1}(z)W(z)M(z)$$

is an automorphism of $GL_2(\mathcal{R})$. The image of $W(z) = \begin{pmatrix} 0 & 1 \\ 1 & 0 \end{pmatrix}$ is

$$\mathcal{A}(W) = \begin{pmatrix} 0 & b^2 \\ b^{-2} & 0 \end{pmatrix}$$

and $\deg \mathcal{A}(W) \neq \deg W$.

More generally, it is possible to determine all the automorphisms of $GL_m(\mathcal{R})$ using the general theory of [D] and to give a characterization of degree perserving automorphisms. Details will be presented in Section 8.

2.2 General Theory.

Let $W \rightarrow \mathcal{A}(W)$ be an admissible automorphism and let $W = W_1W_2$ be a minimal factorization of $W \in GL_m(\mathcal{R})$ into two elements of $GL_m(\mathcal{R})$. Since the automorphism $\mathcal{A}$ is degree preserving, $\mathcal{A}(W) = \mathcal{A}(W_1)\mathcal{A}(W_2)$ is a minimal factorization of $\mathcal{A}(W)$ and so, up to multiplicative constants, minimal factorizations of W and of $\mathcal{A}(W)$ are in a one–to–one correspondence. When both W and $\mathcal{A}(W)$ are biproper, this one–to–one correspondence induces a one–to–one correspondence $\pi \rightarrow \pi_{\mathcal{A}}$ between the supporting projections of A and of $A_{\mathcal{A}}$.

DEFINITION 2.2. *The map $\pi \rightarrow \pi_{\mathcal{A}}$ will be called the associated projection map.*

It is possible to give a formula for the associated projection map. For that purpose, let W, W_1, and W_2 be as above, proper and invertible at infinity and let $W_i(z) = D_i + C_i(zI_{n_i} - A_i)^{-1}B_i$, $i = 1, 2$, be a minimal realization of W_i, $i = 1, 2$. Then a minimal realization for W is given by A, B, C and D as in (2.8). If $W(z) = \tilde{D} + \tilde{C}(zI_n - \tilde{A})^{-1}\tilde{B}$ is another minimal realization of W then there exists a unique similarity matrix T such that

$$\begin{pmatrix} T & 0 \\ 0 & I \end{pmatrix} \begin{pmatrix} \tilde{A} & \tilde{B} \\ \tilde{C} & \tilde{D} \end{pmatrix} = \begin{pmatrix} A & B \\ C & D \end{pmatrix} \begin{pmatrix} T & 0 \\ 0 & I \end{pmatrix} \tag{2.22}$$

With respect to the realization (2.8) the supporting projection corresponding to the factorization $W = W_1W_2$ is $\begin{pmatrix} 0 & 0 \\ 0 & I_{n_2} \end{pmatrix}$. The supporting projection π corresponding to the same factorization but with respect to the realization $W(z) = \tilde{D} + \tilde{C}(zI_n - \tilde{A})^{-1}\tilde{B}$ is

$$\pi = T^{-1}\begin{pmatrix} 0 & 0 \\ 0 & I_{n_2} \end{pmatrix}T. \tag{2.23}$$

When $\mathcal{A}(W_1)$ and $\mathcal{A}(W_2)$ are biproper, (2.22) leads to

$$\begin{pmatrix} T_{\mathcal{A}} & 0 \\ 0 & I \end{pmatrix}\begin{pmatrix} \tilde{A} & \tilde{B} \\ \tilde{C} & \tilde{D} \end{pmatrix}_{\mathcal{A}} = \begin{pmatrix} A & B \\ C & D \end{pmatrix}_{\mathcal{A}}\begin{pmatrix} T_{\mathcal{A}} & 0 \\ 0 & I \end{pmatrix}.$$

Using (iii) of definition 2.1, we see that $\begin{pmatrix} A & B \\ C & D \end{pmatrix}_{\mathcal{A}}$ is defined by (2.8) where all the letters have now an index $\mathcal{A}$,

$$A_{\mathcal{A}} = \begin{pmatrix} A_{1\mathcal{A}} & B_{1\mathcal{A}}C_{2\mathcal{A}} \\ 0 & A_{2\mathcal{A}} \end{pmatrix}, \; B_{\mathcal{A}} = \begin{pmatrix} B_{1\mathcal{A}}D_{2\mathcal{A}} \\ B_{2\mathcal{A}} \end{pmatrix} \tag{2.24a}$$

$$C_{\mathcal{A}} = (C_{1\mathcal{A}}, D_{1\mathcal{A}}C_{2\mathcal{A}}) \tag{2.24b}$$

and

$$D_{\mathcal{A}} = D_{1\mathcal{A}}D_{2\mathcal{A}}. \tag{2.24c}$$

The supporting projection corresponding to the minimal factorization $\mathcal{A}(W) = \mathcal{A}(W_1)\mathcal{A}(W_2)$ is equal to $\begin{pmatrix} 0 & 0 \\ 0 & I_{n_2} \end{pmatrix}$ when we consider the minimal realization (2.24) of $\mathcal{A}(W)$ and is equal to

$$\pi_{\mathcal{A}} = T_{\mathcal{A}}^{-1}\begin{pmatrix} 0 & 0 \\ 0 & I_{n_2} \end{pmatrix}T_{\mathcal{A}} \tag{2.25}$$

when we consider the minimal realization $\begin{pmatrix} \tilde{A} & \tilde{B} \\ \tilde{C} & \tilde{D} \end{pmatrix}_{\mathcal{A}}$ of $\mathcal{A}(W)$.

Comparing (2.23) and (2.25), we obtain the formula

$$\pi_{\mathcal{A}} = (T_{\mathcal{A}}^{-1}T)\pi(T_{\mathcal{A}}^{-1}T)^{-1}. \tag{2.26}$$

The map $\pi \to \pi_{\mathcal{A}}$ takes a nice form in each of the previous families of automorphisms. Indeed we have:

PROPOSITION 2.1. *The following hold:*

(1) for type **Ia** *symmetries,*

$$\pi_{\mathcal{A}} = \pi.$$

(2) for type **Ib** *symmetries,*

$$\pi_{\mathcal{A}} = \overline{\pi}.$$

(3) *for type* **IIa** *symmetries,*

$$\pi_{\mathcal{A}} = \pi^T.$$

(4) *for type* **IIb** *symmetries,*

$$\pi_{\mathcal{A}} = \pi^*(= \bar{\pi}^T).$$

PROOF: In case (1), $T_{\mathcal{A}} = T$ and formulas (2.23) and (2.25) lead to $\pi_{\mathcal{A}} = \pi$ and similarly $T_{\mathcal{A}} = \overline{T}$ and $\pi_{\mathcal{A}} = \overline{\pi}$ for type Ib symmetries. For type IIa symmetries, $T_{\mathcal{A}} = (T^T)^{-1}$ and so,

$$\pi_{\mathcal{A}} = T^T \begin{pmatrix} 0 & 0 \\ 0 & I_{n_2} \end{pmatrix} (T^T)^{-1} = \pi^T$$

and the case of type IIb symmetries is treated in the same way: $T_{\mathcal{A}} = (T^*)^{-1}$ and thus

$$\pi_{\mathcal{A}} = T^* \begin{pmatrix} 0 & 0 \\ 0 & I_{n_2} \end{pmatrix} (T^*)^{-1} = \pi^*.$$

3. THE ASSOCIATED GROUP.

Let $W \to \mathcal{A}(W)$ be an automorphism of $GL_m(\mathcal{R})$. An element of $GL_m(\mathcal{R})$ satisfying $W = \mathcal{A}(W)$ will be called *$\mathcal{A}$-symmetric.* The set of *$\mathcal{A}$-symmetric* elements of $GL_m(\mathcal{R})$ form a subgroup of $GL_m(\mathcal{R})$ which we will call the *associated group* (to the given automorphism $\mathcal{A}$). It will be denoted by $G_{\mathcal{A}}$. For instance if $F = \mathbb{C}$ and $\mathcal{A}(W)(z) = \overline{W(\overline{z})}$, then $G_{\mathcal{A}}$ consists of real elements of $GL_m(\mathcal{R})$. The purpose of this section is to study the associated group when $\mathcal{A}$ is an admissible automorphism. In section 3.1 we study the minimal realizations of elements of $G_{\mathcal{A}}$ while in 3.2 we study factorizations within the group.

3.1 Realization Theorems, the associated matrix, and examples.

The first result is

THEOREM 3.1. *Let $\mathcal{A}$ be an admissible automorphism of $GL_m(\mathcal{R})$ and let W be in $GL_m(\mathcal{R})$ with minimal realization $W(z) = D + C(zI_n - A)^{-1}B$ (in particular W is proper.) Then, W belongs to the associated group $G_{\mathcal{A}}$ if and only if the following conditions hold:*

(1) *The rational matrix function $\mathcal{A}(W)$ is proper.*

(2) *There exists an invertible $n \times n$ matrix S such that*

$$\begin{pmatrix} S & 0 \\ 0 & I_m \end{pmatrix} \begin{pmatrix} A & B \\ C & D \end{pmatrix} = \begin{pmatrix} A_{\mathcal{A}} & B_{\mathcal{A}} \\ C_{\mathcal{A}} & D_{\mathcal{A}} \end{pmatrix} \begin{pmatrix} S & 0 \\ 0 & I_m \end{pmatrix} \tag{3.1}$$

PROOF: If $W = \mathcal{A}(W)$ then clearly $\mathcal{A}(W)$ is also proper. Using (i) of definition 2.1 we obtain that $W = \mathcal{A}(W)$ if and only if

$$D + C(zI_n - A)^{-1}B = D_{\mathcal{A}} + C_{\mathcal{A}}(zI_n - A_{\mathcal{A}})^{-1}B_{\mathcal{A}} \tag{3.2}$$

which is an equality between two minimal realizations of a given proper rational function. Hence, by Theorem 1.3 there exists a unique invertible matrix S such that (3.2) holds. This proves one half of the theorem. The converse is proved in much the same way. □

The matrix S satisfying (3.2) is uniquely defined from A, B, C, D and $A_{\mathcal{A}}, B_{\mathcal{A}}, C_{\mathcal{A}}$, as seen in the proof of Theorem 3.1. It is given by the formulas

$$\begin{aligned} S &= \left[\mathrm{col}(B_{\mathcal{A}}A_{\mathcal{A}}^j)_0^{n-1}\right]^+ \left[(\mathrm{col}CA^j)_0^{n-1}\right] \\ &= \left[\mathrm{row}(A_{\mathcal{A}}^jC_{\mathcal{A}})_0^{n-1}\right] \ \mathrm{row}\left[(A^jB)_0^{n-1}\right]^\dagger \end{aligned}$$

(see Theorem 1.3), where $+$ denotes a left inverse and $\dagger$ a right inverse, and it will be called the *associated matrix* (to the given realization of W). The next two theorems study the main properties of the associated matrix.

THEOREM 3.2. *Let $\mathcal{A}$ be an admissible automorphism of $GL_m(\mathcal{R})$ and let W_1 and W_2 two elements in the associated group $G_{\mathcal{A}}$, such that the product $W = W_1W_2$ is minimal. Let $W_i(z) = D_i + C_i(zI_n - A_i)^{-1}B_i$ be a minimal realization of W_i with associated matrix $S_i, i = 1, 2$. Then, (2.8) defines a minimal realization of W with associated matrix*

$$S = \begin{pmatrix} S_1 & 0 \\ 0 & S_2 \end{pmatrix}. \tag{3.3}$$

PROOF: It suffices to write (with A, B, C, D as in (2.8))

$$\begin{aligned} \begin{pmatrix} S & 0 \\ 0 & I_m \end{pmatrix}\begin{pmatrix} A & B \\ C & D \end{pmatrix} &= \begin{pmatrix} S_1 & 0 & 0 \\ 0 & S_2 & 0 \\ 0 & 0 & I_m \end{pmatrix}\begin{pmatrix} A_1 & 0 & B_1 \\ 0 & I_{n_2} & 0 \\ C_1 & 0 & D_1 \end{pmatrix}\begin{pmatrix} I_{n_1} & 0 & 0 \\ 0 & A_2 & B_2 \\ 0 & C_2 & D_2 \end{pmatrix} \\ &= \begin{pmatrix} S_1 & 0 & 0 \\ 0 & I_{n_2} & 0 \\ 0 & 0 & I_m \end{pmatrix}\begin{pmatrix} A_1 & 0 & B_1 \\ 0 & I_{n_2} & 0 \\ C_1 & 0 & D_1 \end{pmatrix}\begin{pmatrix} I_{n_1} & 0 & 0 \\ 0 & S_2 & 0 \\ 0 & 0 & I_m \end{pmatrix}\begin{pmatrix} I_{n_1} & 0 & 0 \\ 0 & A_2 & B_2 \\ 0 & C_2 & D_2 \end{pmatrix} \\ &= \begin{pmatrix} A_{1\mathcal{A}} & 0 & B_{1\mathcal{A}} \\ 0 & I_{n_2} & 0 \\ C_{1\mathcal{A}} & 0 & D_{1\mathcal{A}} \end{pmatrix}\begin{pmatrix} S_1 & 0 & 0 \\ 0 & I_{n_2} & 0 \\ 0 & 0 & I_m \end{pmatrix}\begin{pmatrix} I_{n1} & 0 & 0 \\ 0 & A_{2\mathcal{A}} & B_{2\mathcal{A}} \\ 0 & C_{2\mathcal{A}} & D_{2\mathcal{A}} \end{pmatrix}\begin{pmatrix} I_{n_1} & 0 & 0 \\ 0 & S_2 & 0 \\ 0 & 0 & I_m \end{pmatrix} \\ &= \begin{pmatrix} A_{1\mathcal{A}} & B_{1\mathcal{A}}C_{1\mathcal{A}} & B_{1\mathcal{A}}D_{2\mathcal{A}} \\ 0 & A_{2\mathcal{A}} & B_{2\mathcal{A}} \\ C_{1\mathcal{A}} & D_{1\mathcal{A}}C_{2\mathcal{A}} & D_{1\mathcal{A}}D_{2\mathcal{A}} \end{pmatrix}\begin{pmatrix} S_1 & 0 & 0 \\ 0 & S_2 & 0 \\ 0 & 0 & I_m \end{pmatrix} \end{aligned}$$

That is, by using hypothesis (ii) of definition 2.1, we get

$$\begin{pmatrix} S & 0 \\ 0 & I_m \end{pmatrix}\begin{pmatrix} A & B \\ C & D \end{pmatrix} = \begin{pmatrix} A_{\mathcal{A}} & B_{\mathcal{A}} \\ C_{\mathcal{A}} & D_{\mathcal{A}} \end{pmatrix}\begin{pmatrix} S & 0 \\ 0 & I_m \end{pmatrix}.$$

This expresses S as the associated matrix to the realization (2.8) of W.

THEOREM 3.3. *Let $\mathcal{A}$ be an admissible automorphism of $GL_m(\mathcal{R})$ and let W be an element of the associated group $G_{\mathcal{A}}$. Let W be proper and $W(z) = D_i + C_i(zI_n - A_i)^{-1}B_i, i = 1, 2$ two minimal* *realizations of W with associated matrices $S_i, i = 1,2$. Then the two realizations are similar: there exists a unique invertible matrix T such that*

$$\begin{pmatrix} T & 0 \\ 0 & I_m \end{pmatrix}\begin{pmatrix} A_1 & B_1 \\ C_1 & D_1 \end{pmatrix} = \begin{pmatrix} A_2 & B_2 \\ C_2 & D_2 \end{pmatrix}\begin{pmatrix} T & 0 \\ 0 & I_m \end{pmatrix} \tag{3.4}$$

and S_1 and S_2 are related by

$$S_2 = T_{\mathcal{A}} S_1 T^{-1} \tag{3.5}$$

PROOF: The existence of the similarity matrix T is well known; see Theorem 1.3. The matrices S_1 and S_2 satisfy respectively the equations

$$\begin{pmatrix} S_1 & 0 \\ 0 & I_m \end{pmatrix}\begin{pmatrix} A_1 & B_1 \\ C_1 & D_1 \end{pmatrix} = \begin{pmatrix} A_1 & B_1 \\ C_1 & D_1 \end{pmatrix}_{\mathcal{A}}\begin{pmatrix} S_1 & 0 \\ 0 & I_m \end{pmatrix} \tag{3.6}$$

and

$$\begin{pmatrix} S_2 & 0 \\ 0 & I_m \end{pmatrix}\begin{pmatrix} A_2 & B_2 \\ C_2 & D_2 \end{pmatrix} = \begin{pmatrix} A_2 & B_2 \\ C_2 & D_2 \end{pmatrix}_{\mathcal{A}}\begin{pmatrix} S_2 & 0 \\ 0 & I_m \end{pmatrix} \tag{3.7}$$

while (2.4) implies that

$$\begin{pmatrix} A_2 & B_2 \\ C_2 & D_2 \end{pmatrix}_{\mathcal{A}} = \begin{pmatrix} T_{\mathcal{A}} & 0 \\ 0 & I_m \end{pmatrix}\begin{pmatrix} A_1 & B_1 \\ C_1 & D_1 \end{pmatrix}_{\mathcal{A}}\begin{pmatrix} T_{\mathcal{A}} & 0 \\ 0 & I_m \end{pmatrix}^{-1} \tag{3.8}$$

Combining equations (3.5), (3.6) and (3.7) we obtain

$$\begin{aligned}
&\begin{pmatrix} S_2 & 0 \\ 0 & I_m \end{pmatrix}\begin{pmatrix} A_2 & B_2 \\ C_2 & D_2 \end{pmatrix} \\
&\quad = \begin{pmatrix} T_{\mathcal{A}} & 0 \\ 0 & I_m \end{pmatrix}\begin{pmatrix} A_1 & B_1 \\ C_1 & D_1 \end{pmatrix}_{\mathcal{A}}\begin{pmatrix} T_{\mathcal{A}} & 0 \\ 0 & I_m \end{pmatrix}^{-1}\begin{pmatrix} S_2 & 0 \\ 0 & I_m \end{pmatrix} \\
&\quad = \begin{pmatrix} T_{\mathcal{A}} & 0 \\ 0 & I_m \end{pmatrix}\begin{pmatrix} S_1 & 0 \\ 0 & I_m \end{pmatrix}\begin{pmatrix} A_1 & B_1 \\ C_1 & D_1 \end{pmatrix}\begin{pmatrix} S_1 & 0 \\ 0 & I_m \end{pmatrix}^{-1}\begin{pmatrix} T_{\mathcal{A}} & 0 \\ 0 & I_m \end{pmatrix}^{-1}\begin{pmatrix} S_2 & 0 \\ 0 & I_m \end{pmatrix}
\end{aligned}$$

and hence

$$\begin{pmatrix} S_2^{-1}T_{\mathcal{A}}S_1 & 0 \\ 0 & I_m \end{pmatrix}\begin{pmatrix} A_1 & B_1 \\ C_1 & D_1 \end{pmatrix} = \begin{pmatrix} A_2 & B_2 \\ C_2 & D_2 \end{pmatrix}\begin{pmatrix} S_2^{-1}T_{\mathcal{A}}S_1 & 0 \\ 0 & I_m \end{pmatrix}$$

By the uniqueness of the similarity matrix, we obtain

$$T = S_2^{-1}T_{\mathcal{A}}S_1$$

from which (3.5) follows. □

We now make explicit the form of the associated group for the four families of admissible automorphisms introduced in section 2.

EXAMPLE 3.1. *The associated group for type Ia symmetries.*

An element W of $GL_m(\mathcal{R})$ is $\mathcal{A}$ symmetric if it satisfies

$$MW(z) = W(\varphi(z))M. \tag{3.9}$$

Such rational matrices will be studied in detail in section 4, while the next family will be studied in section 5.

EXAMPLE 3.2.. *The associated group for type Ib symmetries ($F = \mathbb{C}$).*

The matrix function W in $G_{\mathcal{A}}$ satisfies then the equation

$$MW(z) = \overline{W(\overline{\varphi(z)})}M \tag{3.10}$$

EXAMPLE 3.3.. *The associated group for type IIa symmetries.*

The equation characterizing an element of the associated group is now

$$W(\varphi(z))MW(z) = M \tag{3.11}$$

The special case $\varphi(z) = z$ and $M = \begin{pmatrix} 0 & I_p \\ I_p & 0 \end{pmatrix}$ is of special interest. The values of W are then symplectic matrices.

EXAMPLE 3.4.. *The associated group for type IIb symmetries ($F = \mathbb{C}$).*

The elements of $G_{\mathcal{A}}$ satisfy

$$W(\varphi(z))^* MW(z) = M \tag{3.12}$$

The cases where $\varphi(z) = \dfrac{1}{z}, \varphi(z) = z$ and $M = M^* = M^{-1}$ were studied in detail in [AG].

Section 6 and section 7 deal with associated groups with type IIa and type IIb symmetries respectively.

3.2 Factorization within the associated group.

Let $\mathcal{A}$ be an admissible automorphism and W be a $\mathcal{A}$-symmetric element of $GL_m(\mathcal{R})$. The purpose of this section is to study minimal factorizations of W within $G_{\mathcal{A}}$, that is minimal factorizations $W = W_1W_2$ where both W_1 and W_2 belong to $G_{\mathcal{A}}$. As already known e.g. from the real case or the J-unitary case [AG], such factorizations need not always exist (even if the ground field F is algebraically closed such as $\mathbb{C}$). An element W in $G_{\mathcal{A}}$ may admit nontrivial minimal factorizations in $GL_m(\mathcal{R})$ and none within $G_{\mathcal{A}}$.

THEOREM 3.4. *Let $\mathcal{A}$ be an admissible automorphism of $GL_m(\mathcal{R})$ and let W be a proper rational matrix function in $G_{\mathcal{A}}$ and assume that $W(\infty)$ is invertible. Let $W(z) = D + C(zI_n - A)^{-1}B$ be* *a minimal realization of* W, *with the associated matrix* S. *Let* π *be a supporting projection of* A *and let* W_1, W_2 *be defined by*

$$W_1(z) = D_1 + C(zI_n - A)^{-1}(1_n - \pi)BD_2^{-1} \tag{3.13}$$

$$W_2(z) = D_2 + D_1^{-1}C\pi(zI_n - A)^{-1}B \tag{3.14}$$

and suppose that

$$W_1(\infty) = (\mathcal{A}(W_1))(\infty) \tag{3.15}$$

$$\pi_{\mathcal{A}} = S\pi S^{-1}. \tag{3.16}$$

Then, $W = W_1W_2$ *is a minimal factorization of* W *within* $G_{\mathcal{A}}$. *Conversely, any minimal factorization of* W *in* $G_{\mathcal{A}}$ *is obtained in such a way.*

We note that equation (3.14) is in fact an equation which D_1 must satisfy. This point will be made clearer in the proof. Before turning to the proof of Theorem 3.4 we first need the following preliminary result.

LEMMA 3.1. *With the notation of Theorem 3.4, let* $n_1 = \dim \operatorname{Ker}\pi$ *and let* $W_1(z) = D_1 + C_1(zI_{n_1} - A_1)^{-1}B_1$ *be a minimal realization of* W_1, *where* $A_1 \in F^{n_1 \times n_1}, C_1 \in F^{n \times n_1}$ *and* $B_1 \in F^{n_1 \times n}$. *Let* $\mathcal{M}_{\mathcal{A}} = \operatorname{Ker}\pi_{\mathcal{A}}$. *Then, there exists a similarity matrix* $\sigma_{\mathcal{A}}$ *from* $\mathcal{M}_{\mathcal{A}}$ *onto* F^{n_1} *such that*

$$\begin{pmatrix} \sigma_{\mathcal{A}} & 0 \\ 0 & I_n \end{pmatrix}\begin{pmatrix} A_{\mathcal{A}|\mathcal{M}_{\mathcal{A}}} & (I - \pi_{\mathcal{A}})B_{\mathcal{A}}D_{\mathcal{A}}^{-1}D_{1\mathcal{A}} \\ C_{\mathcal{A}|\mathcal{M}_{\mathcal{A}}} & D_{1\mathcal{A}} \end{pmatrix} = \begin{pmatrix} A_{1\mathcal{A}} & B_{1\mathcal{A}} \\ C_{1\mathcal{A}} & D_{1\mathcal{A}} \end{pmatrix}\begin{pmatrix} \sigma_{\mathcal{A}} & 0 \\ 0 & I_n \end{pmatrix} \tag{3.17}$$

where $A_{\mathcal{A}|\mathcal{M}_{\mathcal{A}}}$ *denotes the restriction of* $A_{\mathcal{A}}$ *to* $\mathcal{M}_{\mathcal{A}}$.

PROOF: Let W_2 be as in (3.14). Then by Theorem 1.4, $W = W_1W_2$ is a minimal factorization of W and thus $\mathcal{A}(W) = \mathcal{A}(W_1)\mathcal{A}(W_2)$ is a minimal factorization of $\mathcal{A}(W)$. By the definition of an admissible automorphism, a minimal realization of $\mathcal{A}(W_1)$ is given by

$$\mathcal{A}(W_1)(z) = D_{1\mathcal{A}} + C_{1\mathcal{A}}(zI_n - A_{1\mathcal{A}})^{-1}B_{1\mathcal{A}}$$

On the other hand, by the description of the minimal factorizations of $\mathcal{A}(W)$ given in Theorem 1.4 and the definition of $\pi_{\mathcal{A}}$, there exists a decomposition $D_{1\mathcal{A}} = M_1M_2$ of $D_{1\mathcal{A}}$ into two $F^{n \times n}$ matrices such that

$$\mathcal{A}(W_1)(z) = M_1 + C_{\mathcal{A}}(zI_n - A_{\mathcal{A}})^{-1}(I_n - \pi_{\mathcal{A}})B_{\mathcal{A}}M_2^{-1}$$

Thus, by the uniqueness up to similarity of minimal realizations, there exists a unique linear transformation $\sigma_{\mathcal{A}} : \mathcal{M}_{\mathcal{A}} \to \mathbb{C}^{n_1}$ such that

$$\begin{pmatrix} \sigma_{\mathcal{A}} & 0 \\ 0 & I_n \end{pmatrix}\begin{pmatrix} A_{\mathcal{A}|\mathcal{M}_{\mathcal{A}}} & (I - \pi_{\mathcal{A}})B_{\mathcal{A}}M_2^{-1} \\ C_{\mathcal{A}|\mathcal{M}_{\mathcal{A}}} & M_1 \end{pmatrix} = \begin{pmatrix} A_{1\mathcal{A}} & B_{1\mathcal{A}} \\ C_{1\mathcal{A}} & D_{1\mathcal{A}} \end{pmatrix}\begin{pmatrix} \sigma_{\mathcal{A}} & 0 \\ 0 & I_n \end{pmatrix}$$

from which it follows that $M_1 = D_{1\mathcal{A}}$ and we obtain (3.17). □

We now turn to the proof of Theorem 3.4.

PROOF OF THEOREM 3.4: We first prove that (3.15) and (3.16) are sufficient to insure factorization within the associated group $G_{\mathcal{A}}$ and proceed in a number of steps.

STEP 1. *The linear transformation $S_{\mathcal{A}} = (I_n - \pi_{\mathcal{A}})S|_{Ker\pi}$ is invertible, as a mapping from $Ker\pi$ to $Ker\pi_{\mathcal{A}}$.*

PROOF OF STEP 1: Let m be in $Ker\pi$ and such that $S_{\mathcal{A}}m = 0$. Then,

$$Sm = \pi_{\mathcal{A}}Sm$$

Using (3.15) we obtain

$$Sm = S\pi m$$

and thus, since $\pi m = 0$,

$$Sm = 0$$

which implies that $m = 0$.

STEP 2. *consists in proving that*

$$\begin{pmatrix} S_{\mathcal{A}} & 0 \\ 0 & I_m \end{pmatrix}\begin{pmatrix} A|_{\mathcal{M}} & (I_n - \pi)B \\ C|_{\mathcal{M}} & D \end{pmatrix} = \begin{pmatrix} A_{\mathcal{A}|\mathcal{M}_{\mathcal{A}}} & (I_n - \pi_{\mathcal{A}})B_{\mathcal{A}} \\ C_{\mathcal{A}|\mathcal{M}_{\mathcal{A}}} & D_{\mathcal{A}} \end{pmatrix}\begin{pmatrix} S_{\mathcal{A}} & 0 \\ 0 & I_n \end{pmatrix} \tag{3.18}$$

where $M = Ker\pi$.

PROOF OF STEP 2: By definition of the associated matrix,

$$\begin{pmatrix} S & 0 \\ 0 & I_n \end{pmatrix}\begin{pmatrix} A & B \\ C & D \end{pmatrix} = \begin{pmatrix} A_A & B_A \\ C_A & D_A \end{pmatrix}\begin{pmatrix} S & 0 \\ 0 & I \end{pmatrix}$$

Hence,

$$\begin{pmatrix} (I_n - \pi_{\mathcal{A}}) & 0 \\ 0 & I_m \end{pmatrix}\begin{pmatrix} S & 0 \\ 0 & I_m \end{pmatrix}\begin{pmatrix} A & B \\ C & D \end{pmatrix} = \begin{pmatrix} I_m - \pi_{\mathcal{A}} & 0 \\ 0 & I_m \end{pmatrix}\begin{pmatrix} A_{\mathcal{A}} & B_{\mathcal{A}} \\ C_{\mathcal{A}} & D_{\mathcal{A}} \end{pmatrix}\begin{pmatrix} S & 0 \\ 0 & I_m \end{pmatrix}$$

and we obtain

$$(I_n - \pi_{\mathcal{A}})SA = (I_n - \pi_{\mathcal{A}})A_{\mathcal{A}}S \tag{3.19}$$

$$(I_n - \pi_{\mathcal{A}})SB = (I_n - \pi_{\mathcal{A}})B_{\mathcal{A}} \tag{3.20}$$

and

$$C = C_{\mathcal{A}}S \tag{3.21}$$

Taking into account that $\pi_{\mathcal{A}}S = S\pi$ and that π and $\pi_{\mathcal{A}}$ are supporting projections for A and $A_{\mathcal{A}}$ respectively, we obtain from (3.19) that

$$(I_n - \pi_{\mathcal{A}})SA(I - \pi) = (I_n - \pi_{\mathcal{A}})A_{\mathcal{A}}S(I - \pi)$$

Therefore,

$$(I_n - \pi_{\mathcal{A}})S(I - \pi)A(I - \pi) = (I - \pi_{\mathcal{A}})A_{\mathcal{A}}(I - \pi_{\mathcal{A}})S(I - \pi)$$

from which it follows that

$$S_{\mathcal{A}}A|_{\mathcal{M}} = A_{\mathcal{A}|\mathcal{M}_{\mathcal{A}}} \cdot S_{\mathcal{A}} \tag{3.22}$$

Similarly we obtain from (3.19) that

$$S_{\mathcal{A}}(I-\pi)B = (I-\pi_{\mathcal{A}})B_{\mathcal{A}} \tag{3.23}$$

and from (3.21)

$$C_{|\mathcal{M}} = C_{\mathcal{A}|\mathcal{M}_{\mathcal{A}}} \cdot S_{\mathcal{A}} \tag{3.24}$$

and $D = D_{\mathcal{A}}$. This in conjunction with (3.22), (3.23), (3.24) leads to (3.18).

STEP 3. *consists in concluding the proof of the sufficiency of (3.15) and (3.16).*

PROOF OF STEP 3: We first note that there exists an invertible linear transformation σ from $Ker\pi = \mathcal{M}$ onto $\mathbb{C}^{n_1}$ such that

$$\begin{aligned}&\begin{pmatrix}\sigma & 0\\ 0 & I_m\end{pmatrix}\begin{pmatrix}A|_{\mathcal{M}} & (I-\pi)BD_2^{-1}\\ C|_{\mathcal{M}} & D_1\end{pmatrix}\\ &=\begin{pmatrix}A_1 & B_1\\ C_1 & D_1\end{pmatrix}\begin{pmatrix}\sigma & 0\\ 0 & I_m\end{pmatrix}\end{aligned} \tag{3.25}$$

This equation, together with (3.18), leads to

$$\begin{aligned}&\begin{pmatrix}\sigma & 0\\ 0 & I_m\end{pmatrix}\begin{pmatrix}S_{\mathcal{A}}^{-1} & 0\\ 0 & I_m\end{pmatrix}\begin{pmatrix}A_{\mathcal{A}|\mathcal{M}_{\mathcal{A}}} & (I_n-\pi_{\mathcal{A}})B_{\mathcal{A}}\\ C_{\mathcal{A}|\mathcal{M}_{\mathcal{A}}} & D_{\mathcal{A}}\end{pmatrix}\begin{pmatrix}I_n & 0\\ 0 & D_2^{-1}\end{pmatrix}\\ &=\begin{pmatrix}A_1 & B_1\\ C_1 & D_1\end{pmatrix}\begin{pmatrix}\sigma & 0\\ 0 & I_m\end{pmatrix}\begin{pmatrix}S_{\mathcal{A}}^{-1} & 0\\ 0 & I_m\end{pmatrix}\end{aligned} \tag{3.26}$$

We note that $D = D_{\mathcal{A}}$ since $W \in G_{\mathcal{A}}$ and $D_{1\mathcal{A}} = D_1$ by equation (3.15). Hence $D_2 = D_{2\mathcal{A}}$ and (3.17) and (3.26) imply that

$$\begin{pmatrix}X & 0\\ 0 & I_m\end{pmatrix}\begin{pmatrix}A_{1\mathcal{A}} & B_{1\mathcal{A}}\\ C_{1\mathcal{A}} & D_{1\mathcal{A}}\end{pmatrix} = \begin{pmatrix}A_1 & B_1\\ C_1 & D_1\end{pmatrix}\begin{pmatrix}X & 0\\ 0 & I_m\end{pmatrix} \tag{3.27}$$

where

$$X = \sigma S_{\mathcal{A}}^{-1}\sigma_{\mathcal{A}}^{-1}$$

is an invertible $n_1 \times n_1$ matrix. Thus, $W_1 = \mathcal{A}(W_1)$, and the factorization $W = W_1W_2$ is indeed within $G_{\mathcal{A}}$.

We now turn to the proof of the converse statement. Let $W = W_1W_2$ be a minimal factorization of W with W, W_1 and W_2 in $G_{\mathcal{A}}$ and let $W_i(z) = D_i + C_i(zI_n - A_i)^{-1}B_i$ be a minimal realization of W_i with the associated matrix S_i, for $i = 1,2$. Then, with respect to the minimal realization (2.8) of W the associated matrix is given by (3.3) and

$$\pi = \pi_{\mathcal{A}} = \begin{pmatrix}0 & 0\\ 0 & I_{n_2}\end{pmatrix}$$

from which (3.16) follows clearly.

As for (3.15), it follows from equating $W_1 = \mathcal{A}(W_1)$ at infinity. □

4. TYPE Ia SYMMETRIES.

In this section we specialize the theory developed in section 3 to matrix valued functions which are $\mathcal{A}$–symmetric where $\mathcal{A}$ is a type Ia symmetry and thus satisfy

$$MW(z) = W(\varphi(z))M$$

where M is a $F^{n\times n}$ invertible matrix and φ is a Möbius transform.

The section is divided into three parts which deal respectively with realization theorems and examples, inverse problems and factorization problems.

4.1 Realization Theorem and Examples.

The associated map $\begin{pmatrix} A & B \\ C & D \end{pmatrix} \to \begin{pmatrix} A & B \\ C & D \end{pmatrix}_{\mathcal{A}}$ was defined in (2.7) and it was remarked in the discussion of example 2.1 that $\mathcal{A}(W)$ is proper if and only if $kI_n - rA$ is invertible. Thus, the characterization of $\mathcal{A}$-symmetric rational matrices given in Theorem 3.1 takes the following form in the present framework:

THEOREM 4.1. *Let M be an invertible element of $F^{m\times m}$ and let $\varphi(z) = (kz+l)(rz+s)^{-1}$ be a Möbius transform ($ks - lr \neq 0$). Let W be in $GL_m(\mathcal{R})$ and proper and let $W(z) = D + C(zI_n - A)^{-1}B$ be a minimal realization of W. Then, W satisfies the equality*

$$MW(z) = W(\varphi(z))M \tag{4.1}$$

if and only if the following conditions hold:

(1) The matrix $kI_n - rA$ is invertible.

(2) There exists an invertible element S of $F^{m\times m}$ such that

$$\begin{aligned} &\begin{pmatrix} S & 0 \\ 0 & M \end{pmatrix}\begin{pmatrix} A & B \\ C & D \end{pmatrix} \\ &= \begin{pmatrix} (kI_n - rA)^{-1}(sA - lI_n) & (kI_n - rA)^{-1}B \\ C(kI_n - rA)^{-1}(ks - lr) & D + rC(kI_n - rA)^{-1}B \end{pmatrix}\begin{pmatrix} S & 0 \\ 0 & M \end{pmatrix}. \end{aligned} \tag{4.2}$$

As follows from Theorem 3.1 the matrix S, i.e. the associated matrix, is uniquely defined by the given minimal realization of W. For special choices of φ the properties of M and S are related as follows:

THEOREM 4.2. *Let W be a proper rational matrix valued function invertible at infinity and satisfying the symmetry condition (4.1), and let $W(z) = D + C(zI_n - A)^{-1}B$ be a minimal realization of W with associated matrix S. Let P be a scalar polynomial in z. Then, if $\varphi(z) = z$ or $\varphi(z) = \frac{1}{z}$,*

$$P(M) = 0 \implies P(S) = 0 \tag{4.3}$$

while, if $\varphi(z) = -z$ or $\varphi(z) = -\frac{1}{z}$

$$P(M^2) = 0 \implies P(S^2) = 0. \tag{4.4}$$

PROOF: When $\varphi(z) = z$, equation (4.2) leads to

$$SA = AS \quad and \quad MC = CS$$

so that, for every choice of positive integer k and l,

$$M^k C A^l = C A^l S^k$$

Hence

$$\begin{bmatrix} M^k & & & 0 \\ & M^k & & \\ & & \ddots & \\ 0 & & & M^k \end{bmatrix} \begin{bmatrix} C \\ CA \\ \vdots \\ CA^m \end{bmatrix} = \begin{bmatrix} C \\ CA \\ \vdots \\ CA^m \end{bmatrix} S^k \tag{4.5}$$

from which (4.3) follows since the pair (C, A) is observable.

When $\varphi(z) = -z$ we obtain from (4.2) the equations

$$SA = -AS \quad and \quad MC = CS$$

and so, for every choice of positive integers k and l,

$$M^k C A^l = C A^l S^k (-1)^{lk}.$$

When k is even, $(-1)^{lk} = 1$ and (4.5) holds for such k, from which we get (4.4).

The case $\varphi(z) = \frac{1}{z}$ is done as follows. Equation (4.2) now implies that

$$SA = A^{-1}S \quad and \quad MC = CA^{-1}S$$

and hence

$$M^k C A^l = C A^{-(l+1)} S^k$$

for positive integers k and l. To conclude as above, it suffices to remark that the pair (C, A^{-1}) is observable when the pair (C, A) is observable. The case $\varphi(z) = -\frac{1}{z}$ is treated in a similar way.□

In particular, by taking $P(z)$ to be the minimal polynomial of M with coefficients in the algebraic closure of F, we obtain that every eigenvalue of S is also an eigenvalue of M (if $\varphi(z) = z$ or $\varphi(z) = \frac{1}{z}$). Analogously, in cases $\varphi(z) = -z$ or $\varphi(z) = -\frac{1}{z}$, every eigenvalue of S^2 is also an eigenvalue of M^2.

It is easy to verify that a type Ia symmetry $\mathcal{A}(W)$ (defined by M and φ as in (2.5)) is an involution if and only if $\varphi(\varphi(z)) = z$ and $M^2 = \alpha I_n$ for some $\alpha \in F \setminus \{0\}$. We have $\varphi(\varphi(z)) = z$ if and only if $\varphi(z) = \pm z$ or $\varphi(z) = \pm\frac{1}{z}$, and the preceding theorem is thus of interest for involutions. In particular, if $M^2 = I_m$, then $S^2 = I_n$.

We now present examples of $\mathcal{A}$–symmetric functions, for $\mathcal{A}$ a type Ia symmetry. To keep computations under control we focus on the case $\varphi(z) = z$.

EXAMPLE 4.3. *Degree one rational series.*

For degree one series, A and S are scalars and B and C are vectors. All these quantities will be represented by lower cases letters. Since $\varphi(z) = z$, equation (4.2) leads to

$$sb = bM, \qquad Mc = sc \tag{4.6}$$

and

$$MD = DM. \tag{4.7}$$

Hence, b and c are left and right eigenvectors, respectively, of M with eigenvalue s, and we obtain that any degree one $\mathcal{A}$ symmetric function W is of the form

$$W(z) = D + \frac{cb}{z-a} \tag{4.8}$$

with $a \in F$ and c, b and D satisfying (4.6) and (4.7).

In the next example, we give an $\mathcal{A}$–symmetric function W of McMillan degree two, which admits degree one minimal factors and no nontrivial minimal factorizations within the associated group (these latter assertions will be proved after the factorization theorem, Theorem 4.5).

EXAMPLE 4.4. *Let M be an invertible matrix and let $s_1, s_2 \in F$ be two different points in $\sigma(M)$. Let c_1, c_2, b_1, b_2 be such that*

$$Mc_1 = s_1c_1 \quad Mc_2 = s_2c_2 \quad b_1M = s_1b_1 \quad b_2M = s_2b_2$$

(with $b_i^T, c_i \in F^n$, $i = 1, 2$).

Then for every choice of elements a and b in F

$$W(z) = I + \frac{c_1b_1}{z-a} + \frac{c_2b_2}{z-b} \tag{4.9}$$

is $\mathcal{A}$–symmetric of McMillan degree two.

If $a = b$ and $b_1c_2 \neq 0$, $b_2c_1 \neq 0$, W admits no minimal factorization in the associated group.

It is easy to see that W is $\mathcal{A}$–symmetric. Moreover, we have

$$W(z) = I + [c_1\ c_2](zI_2 - A)^{-1}\begin{pmatrix} b_1 \\ b_2 \end{pmatrix} \tag{4.10}$$

which is a minimal realization of W, with the associated matrix S,

$$S = \begin{pmatrix} s_1 & 0 \\ 0 & s_2 \end{pmatrix}. \tag{4.11}$$

We note that in (4.8) and (4.9) we can replace $\frac{1}{z-a}$ and $\frac{1}{z-b}$ by any scalar rational functions $r_1(z)$ and $r_2(z)$. The resulting W is still $\mathcal{A}$–symmetric (but of course of degree depending on r_1 and r_2).

We now turn to the version of the factorization Theorem 3.4 for type Ia symmetries. The associated projection map $\pi \to \pi_{\mathcal{A}}$ was computed in section 3 to be $\pi_{\mathcal{A}} = \pi$ and thus equation (3.16) becomes

$$S\pi = \pi S$$

which is equivalent to

$$S\mathcal{N} = \mathcal{N}$$
$$S\mathcal{M} = \mathcal{M}.$$

THEOREM 4.5. *Let $\mathcal{A}$ be a type Ia symmetry and let W be a $\mathcal{A}$–symmetric element of $GL_m(\mathcal{R})$, proper and invertible at infinity. Let $W(z) = D + C(zI_n - A)^{-1}B$ be a minimal realization of W with the associated matrix S and let $\mathcal{M}$ and $\mathcal{N}$ be a pair of subspaces of F^n such that*

$$A\mathcal{M} \subset \mathcal{M}, \quad A^{\times}\mathcal{N} \subset \mathcal{N} \tag{4.12}$$

$$S\mathcal{M} = \mathcal{M}, \quad S\mathcal{N} = \mathcal{N} \tag{4.13}$$

and

$$\mathcal{M} \dot{+} \mathcal{N} = F^n$$

Define the projection π by

$$Ker\,\pi = \mathcal{M}, \quad Im\pi = \mathcal{N}. \tag{4.14}$$

Finally, let $D = D_1D_2$ where D_1 and D_2 are in $F^{m\times m}$ and such that

$$MD_1 = D_1M + rC(kI_n - rA)^{-1}(I_n - \pi)BD_2^{-1}M. \tag{4.15}$$

Then, the factorization $W = W_1W_2$ where

$$W_1(z) = D_1 + C(zI_n - A)^{-1}(I_n - \pi)BD_2^{-1} \tag{4.16}$$

$$W_2(z) = D_2 + D_1^{-1}C\pi(zI_n - A)^{-1}B \tag{4.17}$$

is a minimal factorization of W into two $\mathcal{A}$–symmetric factors W_1 and W_2.

Conversely, any minimal factorization of W within the associated group $G_{\mathcal{A}}$ is obtained in such a way.

We note that (4.15) is equation (3.15) in the present framework. When $r = 0$, it takes the simpler form

$$MD_1 = D_1M.$$

We now turn back to example 4.4 and show that the given function W (with appropriate b_1 and c_1) has no minimal factorizations within the associated group. Indeed, from (4.11) we get that (with $s_1 \neq s_2$) the only possible choices for $\mathcal{N}$ and $\mathcal{M}$ in Theorem 4.5 are:

$$\mathcal{M} = span\left\{\begin{pmatrix} 0 \\ 1 \end{pmatrix}\right\}, \quad \mathcal{N} = span\left\{\begin{pmatrix} 1 \\ 0 \end{pmatrix}\right\}$$

or

$$\mathcal{M} = span\left\{\begin{pmatrix} 1 \\ 0 \end{pmatrix}\right\}, \quad \mathcal{N} = span\left\{\begin{pmatrix} 0 \\ 1 \end{pmatrix}\right\}.$$

When $a = b$, $A = I_2$ and in either case we do have $A\mathcal{M} \subset \mathcal{M}$. But the matrix $A^\times$ is equal to

$$A^\times = A - BD^{-1}C = \begin{pmatrix} 1 - b_1c_1 & -b_1c_2 \\ -b_2c_1 & 1 - b_2c_2 \end{pmatrix}.$$

When b_1c_2 and b_2c_1 are both nonzero, we will not have $A^\times\mathcal{N} \subset \mathcal{N}$ for $\mathcal{N}$ as above.

On the other hand, W admits a minimal factorization into degree one factors (which will not be $\mathcal{A}$–symmetric) at least when F is algebraically closed. To exhibit such a factorization it suffices to take any $\mathcal{N}$, $\dim\mathcal{N} = 1$, such that $A^\times\mathcal{N} \subset \mathcal{N}$. Since $A = I_2$, one can always find a $\mathcal{M}$, $A\mathcal{M} \subset \mathcal{M}$, $\mathcal{M}\dot{+}\mathcal{N} = F^2$. The corresponding W_1 and W_2 defined by (4.16) and (4.17) (with arbitrary decomposition $D = D_1D_2$) is a minimal factorization of W, as follows from the general theory of [BGK] (see also Theorem 1.4).

5. TYPE Ib SYMMETRIES.

In this section we assume that $F = \mathbb{C}$.

Type 1b symmetries were defined by (2.12) and a rational matrix valued function W is $\mathcal{A}$–symmetric if

$$MW(z) = \overline{W(\overline{\varphi(z)})}M \tag{5.1}$$

for z points of analyticity of W.

The associated map was defined in (2.13) and the present version of Theorem 3.1 is:

THEOREM 5.1. *Let M be an invertible element of $\mathbb{C}^{m\times m}$ and let $\varphi(z) = \frac{kz+l}{rz+s}$, $ks - lr \neq 0$, be a degree one Möbius transform. Let W be a proper element of $GL_m(\mathcal{R})$ and let $W(z) = D + C(zI_n - A)^{-1}B$ be a minimal realization of W. Then, W satisfies (5.1) if and only if the following conditions hold:*

(1) The matrix $kI_n - r\overline{A}$ is invertible.

(2) There exists an invertible element S of $\mathbb{C}^{n\times n}$ such that

$$\begin{aligned} &\begin{pmatrix} S & 0 \\ 0 & M \end{pmatrix}\begin{pmatrix} A & B \\ C & D \end{pmatrix} \\ &= \begin{pmatrix} (kI_n - r\overline{A})^{-1}(s\overline{A} - lI_n) & (kI_n - r\overline{A})^{-1}\overline{B} \\ \overline{C}(kI_n - r\overline{A})^{-1}(ks - lr) & \overline{D} + r\overline{C}(kI_n - r\overline{A})^{-1}\overline{B} \end{pmatrix}\begin{pmatrix} S & 0 \\ 0 & M \end{pmatrix}. \end{aligned} \tag{5.2}$$

To obtain this result from Theorem 3.1 we remark that $\overline{W(\overline{\varphi(z)})}$ is proper if and only if $kI_n - r\overline{A}$ is invertible.

We now relate properties of M and S.

THEOREM 5.2. *Let φ be $\varphi(z) = z$, $\varphi(z) = -z$, $\varphi(z) = \frac{1}{z}$ or $\varphi(z) = -\frac{1}{z}$. Let W be a proper $\mathbb{C}^{m\times m}$ valued rational function satisfying (5.1). Let $W(z) = D + C(zI_n - A)^{-1}B$ be a minimal realization of W with associated matrix S. Then, for a polynomial P with coefficients in $\mathbb{C}$,*

$$P(\overline{M}M) = 0 \implies P(\overline{S}S) = 0 \tag{5.3}$$

or, equivalently

$$P(M\overline{M}) = 0 \implies P(S\overline{S}) = 0. \tag{5.4}$$

PROOF: When $\varphi(z) = z$, equation (5.2) leads to

$$SA = \overline{A}S \quad \text{and} \quad MC = \overline{C}S$$

from which we obtain

$$\overline{M}MC = C\overline{S}S \tag{5.5}$$

and

$$\overline{S}SA = A\overline{S}S \tag{5.6}$$

So

$$(\overline{M}M)^kCA^l = CA^l(\overline{S}S)^k; \qquad l,k = 0,\ldots$$

and (5.3) is now derived in a similar way as (4.3).

The other cases are treated similarly. For instance, when $\varphi(z) = \frac{1}{z}$, (5.5) and (5.6) are now replaced by

$$\overline{M}MCA^{-1} = CA^{-1}\overline{S}S$$

and

$$A^{-1}\overline{S}S = \overline{S}SA^{-1}.$$

Moreover, it is clear that (5.3) and (5.4) are equivalent. Indeed, from

$$M(\overline{M}M)^n = (M\overline{M})^nM$$

we obtain

$$MP(\overline{M}M) = P(M\overline{M})M$$

and so,

$$P(\overline{M}M) = 0 \iff P(M\overline{M}) = 0. \quad \square$$

Of particular interest are the cases where $M\overline{M} = \alpha I_m$. Then, the type Ib symmetry (2.12) is an involution and $S\overline{S} = \alpha Im$. Examples of $\mathcal{A}$–symmetric functions are easily adapted from section 4 and we have:

EXAMPLE 5.3. *Degree one function.*

We take M a nonsingular element of $\mathbb{C}^{m\times m}$ and s in $\mathbb{C}\setminus\{0\}$. Let b and c be such that

(5.6) $$Mc = s\overline{c}$$

(5.7) $$\overline{b}M = \overline{s}b$$

(Note that $\overline{M}Mc = |s|^2c$.)

Then, for every real number a, the function

(5.8) $$W(z) = I + \frac{c\overline{b}}{z-a}$$

satisfies (5.1) with $\varphi(z) = z$, and, as follows from Theorem 5.1, (5.8) is in fact the general form of a degree one $\mathcal{A}$–symmetric function when $\varphi(z) = z$.

To get a specific example of M, c, b and s satisfying (5.6) and (5.7), let $M = \begin{pmatrix} 0 & I_p \\ I_p & 0 \end{pmatrix}$ and $c = \begin{pmatrix} x \\ \overline{x} \end{pmatrix}$, $b = (y, \overline{y})$, x and y in $\mathbb{C}^{p\times 1}$ and $\mathbb{C}^{1\times p}$ respectively. Then

$$Mc = \overline{c}$$

and

$$\overline{b}M = b$$

EXAMPLE 5.4.

Let s_1 and s_2 be two nonzero complex numbers of different modulus and let c_1, c_2, b_1 and b_2 be such that

(5.9) $$Mc_1 = s_1\overline{c}_1 \qquad \overline{b}_1M = \overline{s}_1b_1$$

and

(5.10) $$Mc_2 = s_2\overline{c}_2 = b \qquad \overline{b}_2M = \overline{s}_2b_2.$$

For instance, $M = \begin{pmatrix} \begin{pmatrix} 0 & I \\ I & 0 \end{pmatrix} & 0 \\ 0 & \frac{1}{2}\begin{pmatrix} 0 & I \\ I & 0 \end{pmatrix} \end{pmatrix}$, $c_1 = \begin{pmatrix} x \\ \overline{x} \\ 0 \\ 0 \end{pmatrix}$, $c_2 = \begin{pmatrix} 0 \\ 0 \\ x \\ \overline{x} \end{pmatrix}$, $b_1 = (y, \overline{y}, 0, 0)$ and $b_2 = (0, 0, y, \overline{y})$.

Then, for any $a, b \in \mathbb{R}$,

$$W(z) = I + \frac{c_1 \overline{b}_1}{z - a} + \frac{c_2 \overline{b}_2}{z - b} \tag{5.11}$$

satisfies (5.1).

When $a = b$ and $\overline{b}_1 c_2$ and $\overline{b}_2 c_1$ are nonzero we will see that W admits no nontrivial minimal factorizations within the associated group although it admits minimal factorizations into degree one factors in $GL_m(\mathcal{R})$.

To get a wider class of examples, we could replace $\frac{1}{z-a}$ and $\frac{1}{z-b}$ by rational functions with real coefficients in (5.11).

We now turn to the factorization theorem. As seen in Section 3, the associated projection map is given by $\pi_{\mathcal{A}} = \overline{\pi}$ for $\mathcal{A}$ a Ib symmetry and hence (3.16) becomes

$$\overline{\pi} S = S \pi \tag{5.12}$$

which is equivalent to

$$S\mathcal{N} = \overline{\mathcal{N}} \tag{5.13}$$

and

$$S\mathcal{M} = \overline{\mathcal{M}} \tag{5.14}$$

(Indeed, if $x \in \mathcal{M}$, $\pi x = 0$ and $\overline{\pi} S x = 0$ so that $Sx \in \overline{\mathcal{M}}$. Hence $S\mathcal{M} \subset \overline{\mathcal{M}}$ and, moreover, the equality follows by dimension count. A similar argument gives $SM = \overline{\mathcal{N}}$). Hence Theorem 3.4 becomes:

THEOREM 5.6. *Let W be a proper $\mathbb{C}^{m \times m}$ valued rational function invertible at infinity and suppose W is $\mathcal{A}$–symmetric where $\mathcal{A}$ is a Ib automorphism. Let $W(z) = D + C(zI_n - A)^{-1}B$ be a minimal realization of W with associated matrix S, and let $\mathcal{M}, \mathcal{N}$ be a pair of subspaces of $\mathbb{C}^n$ such that*

$$A\mathcal{M} \subset \mathcal{M}, \qquad A^{\times}\mathcal{N} \subset \mathcal{N} \tag{5.15}$$

$$S\mathcal{M} = \overline{\mathcal{M}}, \qquad S\mathcal{N} = \overline{\mathcal{N}} \tag{5.16}$$

and

$$\mathcal{M} \dot{+} \mathcal{N} = \mathbb{C}^n$$

and let π be the supporting projection defined by

$$Ker\,\pi = \mathcal{M}, \qquad Im\pi = \mathcal{N}.$$

Finally, let $D = D_1D_2$ where D_1 and D_2 are $\mathbb{C}^{m\times m}$ matrices such that

$$MD_1 = (\overline{D}_1 + r\overline{C}_1(k - r\overline{A})^{-1}(I_n - \pi)BD_2^{-1})M. \tag{5.17}$$

Then, the factorization $W = W_1W_2$, where

$$\begin{aligned} W_1(z) =& D_1 + C(zI_n - A)^{-1}(I_n - \pi)BD_2^{-1} \\ W_2(z) =& D_2 + D_1^{-1}C\pi(zI_n - A)^{-1}B, \end{aligned}$$

is a minimal factorization of W into two $\mathcal{A}$–symmetric factors.

Conversely, any minimal factorization of W within the associated group is obtained in such a way.

We note that (5.17) is the present form of (3.15). When $r = 0$, it takes the simpler form

$$MD_1 = D_1M.$$

We now show that the function $W(z)$ defined in (5.11) has no nontrivial minimal factorizations within the associated group. We suppose $a = b$ and $|s_1| \neq |s_2|$. A minimal realization of W is given by

$$W(z) = I + [c_1, c_2]\left(zI_2 - \begin{pmatrix} a & 0 \\ 0 & a \end{pmatrix}\right)^{-1}\begin{pmatrix} \overline{b}_1 \\ \overline{b}_2 \end{pmatrix}$$

and its associated matrix is

$$S = \begin{pmatrix} s_1 & 0 \\ 0 & s_2 \end{pmatrix}.$$

Since S satisfies

$$SA = \overline{A}S,$$

we have

$$MC = [s_1\overline{c}_1, s_2\overline{c}_2] = [\overline{c}_1, \overline{c}_2]S$$

and

$$SB - M\overline{B}.$$

We look for $\mathcal{N} \subset \mathbb{C}^2$ of dimension 1 such that $S\mathcal{N} = \overline{\mathcal{N}}$. If $\begin{pmatrix} a \\ b \end{pmatrix}$ is a basis of $\mathcal{N}$, there exists $k \in \mathbb{C}$ such that

$$\begin{aligned} s_1a =& k\overline{a} \\ s_2b =& k\overline{b}. \end{aligned}$$

Suppose $b \neq 0$. Then $\frac{s_1}{s_2}\frac{a}{b} = \frac{\overline{a}}{\overline{b}}$ so that $|\frac{s_1}{s_2}| = 1$, which is impossible since $|s_1| \neq |s_2|$. Hence,

$$\mathcal{N} = span\left\{\begin{pmatrix} 0 \\ 1 \end{pmatrix}\right\} \quad or \quad \mathcal{N} = span\left\{\begin{pmatrix} 1 \\ 0 \end{pmatrix}\right\}$$

and then,

$$\mathcal{M} = span\left\{\begin{pmatrix} 1 \\ 0 \end{pmatrix}\right\} \quad or \quad \mathcal{M} = span\left\{\begin{pmatrix} 0 \\ 1 \end{pmatrix}\right\}.$$

On the other hand,

$$A^\times = A - BD^{-1}C = \begin{pmatrix} a - \overline{b}_1 a & -\overline{b}_1 c_2 \\ -\overline{b}_2 c_1 & a - \overline{b}_2 c_2 \end{pmatrix}$$

and, if $\overline{b}_1 c_2 \neq 0$ and $\overline{b}_2 c_1 \neq 0$, we will not have $A^\times \mathcal{N} \subset \mathcal{N}$ for $\mathcal{N}$ as required.

On the other hand, W admits minimal factorizations within $GL_m(\mathcal{R})$ by Theorem 1.4 since $A = aI_2$.

6. TYPE IIa SYMMETRIES.

We now study rational matrices which are $\mathcal{A}$-symmetric when $\mathcal{A}$ is a type IIa symmetry. Such matrices W hence satisfy

$$W(z)MW(\varphi(z))^T = M$$

where $M \in F^{m\times m}$ is invertible and φ is a degree one scalar rational function, i.e., Möbius transform (as before, F is an arbitrary infinite field).

The case where $F = \mathbb{R}$, $\varphi(z) = z$ and $M = \begin{pmatrix} 0 & -I_p \\ I_p & 0 \end{pmatrix}$ is of importance: the values of W are then symplectic for real z which are not poles of W.

Our first result is the version of Theorem 3.1 in the present framework.

THEOREM 6.1. *Let M be an invertible matrix in $F^{m\times m}$ and let $\varphi(z) = (kz+l)/(rz+s)$ be a Möbius transform. Let W be in $GL_m(\mathcal{R})$ proper and invertible at infinity and let $W(z) = D + C(zI_n - A)^{-1}B$ be a minimal realization of W. Then, W satisfies*

$$W(z)MW(\varphi(z))^T = M \tag{6.1}$$

if and only if the following conditions hold:

(1) The matrix $kI_n - rA$ is invertible.

(2) There exists an invertible matrix S in $F^{m\times m}$ such that

$$\begin{pmatrix} S & 0 \\ 0 & M^{-1} \end{pmatrix}\begin{pmatrix} A & B \\ C & D \end{pmatrix} = \begin{pmatrix} A_\varphi - B_\varphi D_\varphi^{-1} C_\varphi & -B_\varphi D_\varphi^{-1} \\ D_\varphi^{-1} C_\varphi & D_\varphi^{-1} \end{pmatrix}^T \begin{pmatrix} S & 0 \\ 0 & M^{-1} \end{pmatrix} \tag{6.2}$$

(where A_φ, B_φ, C_φ and D_φ are defined in (2.6).)

To obtain Theorem 6.1 from Theorem 3.1 we first observe that $\mathcal{A}(W)$ is proper if and only if $\mathcal{A}(W)^{-1}$ is proper (as follows from (6.1) and from the fact that W is proper and invertible at infinity). Hence, condition (1) in the statement of Theorem 3.1 leads to condition (1) here. To obtain (6.2) we note that the associated map is given by formula (2.16).

We note that (6.2) may be rewritten as

$$SA = (A_\varphi - B_\varphi D_\varphi^{-1} C_\varphi)^T S \tag{6.3a}$$

$$SB = C_\varphi^T D_\varphi^{-T} M^{-1} \tag{6.3b}$$

$$M^{-1}C = -D_\varphi^{-T} B_\varphi^T S \tag{6.3c}$$

and

$$M^{-1}D = D_\varphi^{-T} M^{-1}. \tag{6.3d}$$

In particular, S and A are related by

$$SA - A_\varphi^T S = -C_\varphi^T M^{-1} C$$

which can also be rewritten as

$$-rA^T SA + kSA - sA^T S + lS = -(ks - ls) C^T M^{-1} C$$

which, for special cases of $\varphi(z) = \frac{kz+l}{rz+s}$ becomes a Lyapunov or a Stein equation.

The special cases $\varphi(z) = \frac{1}{z}$, $\varphi(z) = z$ are of interest and we now study them in greater details. In particular one can then relate properties of the associated matrix S and of M.

THEOREM 6.2. *(The case $\varphi(z) = z$.) Let W be a proper $F^{m\times m}$ valued function invertible at infinity and let $W(z) = D + C(zI_n - A)^{-1}B$ be a minimal realization of W. Then,*

$$W(z)MW(z)^T = M \tag{6.5}$$

if and only if the following properties hold:

(1)

$$DMD^T = M \tag{6.6}$$

(2) There exists an invertible matrix S such that

$$SA - A^T S = -C^T M^{-1} C \tag{6.7}$$

$$C = DMB^T S \tag{6.8}$$

$$C^T = SBMD^T \tag{6.9}$$

If, furthermore, $M^T = \epsilon M$, $\epsilon \in \{-1, 1\}$, *then (6.8) and (6.9) are equivalent and* $S^T = -\epsilon S$.

Proof: The Lyapunov equation (6.7) is (6.5) for $\varphi(z) = z$, and (6.6), (6.8) and (6.9) follow also from (6.3). Since S is uniquely determined from the given minimal realization, the last claim is proved by checking that both S and $-\epsilon S^T$ satisfy (6.7)–(6.9) and it follows then that (6.8) and (6.9) are equivalent. □

We note that (6.7) may be replaced by

$$-S^{-1}A^T + AS^{-1} = -BMB^T. \tag{6.10}$$

Similarly, one proves that:

Theorem 6.3. *(The case* $\varphi(z) = \frac{1}{z}$.*) Let* W *be a* $F^{m\times m}$ *valued rational function proper and invertible at infinity and let* $W(z) = D + C(zI_n - A)^{-1}B$ *be a minimal realization of* W. *Then*

$$W(z)MW(\frac{1}{z})^T = M$$

if and only if there exists an invertible matrix S *such that*

$$\begin{pmatrix} A & B \\ C & D \end{pmatrix}^T \begin{pmatrix} S & 0 \\ 0 & -M^{-1} \end{pmatrix} \begin{pmatrix} A & B \\ C & D \end{pmatrix} = \begin{pmatrix} S & 0 \\ 0 & -M^{-1} \end{pmatrix}.$$

Moreover, if $M^T = \epsilon M$, *where* $\epsilon = \pm 1$, *then* $S^T = \epsilon S$.

We note that S satisfies the Stein equation

$$S - ASA^T = BMB^T. \tag{6.12}$$

Similar theorems hold for the case $\varphi(z) = -\frac{1}{z}$ and $\varphi(z) = -z$ but we will not write them down.

We now present examples of $\mathcal{A}$–symmetric functions for type IIa symmetries.

Example 6.1.

Let M be such that $M + M^T = 0$, $M^2 + I = 0$. Let $u \neq 0$ be such that $u^T M u = 0$. (For example, if n is even, the existence of such an M and u is assured: $M = \begin{bmatrix} 0 & 1 \\ -1 & 0 \end{bmatrix} \oplus \cdots \oplus \begin{bmatrix} 0 & 1 \\ -1 & 0 \end{bmatrix}$ ($\frac{n}{2}$ times), $u = [1, 0, \ldots, 0]^T$.) Then, for every rational scalar function $r(z)$ the function

$$W(z) = I_m + uu^T M r(z)$$

satisfies

$$W(z)MW(z)^T = M.$$

Indeed,

$$\begin{aligned} W(z)MW(z)^T &= (M + uu^TM^2r(z))(I_m - Muu^Tr(z)) \\ &= M + (uu^TM^2 - M^2uu^T)r(z) - uu^TM^3uu^Tr(z) \\ &= M \end{aligned}$$

EXAMPLE 6.2. *Let M be as in example 6.1 and let u_1, u_2 be vectors such that*

$$u_1^TMu_1 = u_2^TMu_2 = 0$$

$$u_1^TMu_2 \neq 0.$$

Let, for $i \neq j$

$$W_{ij} = u_i(u_j^TMu_i)^{-1}u_j^TM$$

and r_1, r_2 be two scalar rational functions such that

$$r_1 + r_2 + r_1r_2 = 0.$$

Then

$$W(z) = I_m + r_1W_{12} + r_2W_{21}$$

satisfies

$$W(z)MW(z)^T = M. \tag{6.13}$$

To verify (6.13), we first remark that $W_{12}MW_{12}^T = W_{21}MW_{21}^T = 0$, and

$$\begin{aligned} W_{12}MW_{21}^T &= -\frac{u_1u_2^T}{u_2^TMu_1} = W_{12}M = MW_{21}^T \\ W_{21}MW_{12}^T &= -\frac{u_2u_1^T}{u_1^TMu_2} = W_{21}M = MW_{12}^T. \end{aligned}$$

Hence,

$$\begin{aligned} W(z)MW(z)^T &= (M + r_1W_{12}M + r_2W_{21}M)(I + r_1W_{12}^T + r_2W_{21}^T) \\ &= M + r_1r_2(W_{12}MW_{21}^T + W_{21}MW_{12}^T) \\ &\quad + r_1W_{12}M + r_2W_{21}M + r_1MW_{12}^T + r_2MW_{21}^T \\ &= M - \frac{u_1u_2^T}{u_2^TMu_1}(r_1 + r_2 + r_1r_2) \\ &\quad - \frac{u_2u_1^T}{u_1^TMu_2}(r_1 + r_2 + r_1r_2) \\ &= M \end{aligned}$$

In particular, take

$$r_2 = \frac{z-\lambda_2}{z-\lambda_1} - 1; \qquad r_1 = \frac{z-\lambda_1}{z-\lambda_2} - 1,$$

where $\lambda_1, \lambda_2 \in F$ are distinct. In this case, W becomes

$$W(z) = I + \left(\frac{z-\lambda_2}{z-\lambda_1} - 1\right) W_{12} + \left(\frac{z-\lambda_1}{z-\lambda_2} - 1\right) W_{21}. \tag{6.14}$$

As explained later in the section this function W has degree two, admits minimal factorizations with degree one factors but has no (nontrivial) minimal factorization within the associated group of functions satisfying (6.11).

We now turn to the factorization theorem in the present framework. We focus on the two special cases $\varphi(z) = z$ and $\varphi(z) = \frac{1}{z}$ and $M^T = \pm M$. Then, as already noted, $S^T = \pm S$. We endow the state space F^n with the bilinear form

$$[x, y]_S = y^T S x. \tag{6.16}$$

Two vectors x and y will be said to be S–orthogonal with respect to $[,]_S$ if $[x, y]_S = 0$, and for $\mathcal{M} \subset F^n$, $\mathcal{M}^{[\perp]}$ will denote the subspace of all vectors S–orthogonal to $\mathcal{M}$

$$\mathcal{M} = \{y \in F^n, \ [y, m]_S = 0 \qquad \textit{for all } m \in \mathcal{M}\}$$

A subspace $\mathcal{M}$ will be called nondegenerate if $\mathcal{M} \cap \mathcal{M}^{[\perp]} = \{0\}$. Then

$$\mathcal{M}[\dot{+}]\mathcal{M}^{[\perp]} = F^n$$

where $[\dot{+}]$ denotes an orthogonal direct sum.

For A in $F^{n \times n}$, let

$$A^{[T]} = S^{-1} A^T S.$$

The equality

$$[Ax, y]_S = [x, A^{[T]} y]_S \tag{6.17}$$

expresses that $A^{[T]}$ is the adjoint of A with respect to $[,]_S$.

The form $[,]_S$ will be called the associated bilinear form and the following two theorems are the version of Theorem 3.4 in the present framework.

THEOREM 6.4. *Let M be an invertible element of $F^{m \times m}$ satisfying $M = \pm M^T$ and let W be an $F^{m \times m}$ valued rational function proper and invertible at infinity and such that $W(z) M W(z)^T = M$. Let $W(z) = D + C(zI_n - A)^{-1} B$ be a minimal realization of W, with the associated matrix S and bilinear form $[,]_S$. Finally let $\mathcal{M}$ be an invariant subspace of A nondegenerate in the bilinear form $[\ ,\ ]_S$, let π be the projection defined by*

$$Ker\,\pi = \mathcal{M}, \qquad Im\pi = \mathcal{M}^{[\perp]} \tag{6.18}$$

and let $D = D_1D_2$ be a factorization of D into two factors such that $D_iMD_i^T = D_i$, $i = 1,2$. Then, the factorization $W = W_1W_2$, where

$$(6.19)\qquad W_1(z) = D_1 + C(zI_n - A)^{-1}(I_n - \pi)BD_2^{-1}$$

$$(6.20)\qquad W_2(z) = D_2 + D_1^{-1}C\pi(zI_n - A)^{-1}B$$

is a minimal factorization of W into two factors satisfying (6.5).

Conversely, any such minimal factorization of W can be obtained in this way, and, with a fixed decomposition $D = D_1D_2$, the correspondence between minimal factorizations of W into factors satisfying (6.5) and nondegenerate invariant subspaces of A is one to one.

PROOF: Condition (3.15) reduces here to $D_1MD_1^T = D_1$ while the equation (3.16) leads to

$$(6.21)\qquad \pi^{[T]} = \pi$$

since the associated projection map for type IIa symmetries is $\pi_{\mathcal{A}} = \pi^T$.

Hence it follows from Theorem 3.1 that for a given $D = D_1D_2$ the minimal factorizations of W within the associated groups are in a one–to–one correspondence with supporting projections that are S–orthogonal. To conclude the proof of Theorem 6.1 we have to check that S–orthogonal projections are in one–to–one correspondence with S–nondegenerate $\mathcal{A}$-invariant subspaces of F^n.

Equation (6.3) with $\varphi(z) = z$ leads to

$$(6.22)\qquad A^{\times} = A^{[T]}$$

so that

$$(6.23)\qquad A\mathcal{M} \subset \mathcal{M} \Longrightarrow A^{\times}\mathcal{M}^{[\perp]} \subset \mathcal{M}^{[\perp]}.$$

Hence, an S–nondegenerate subspace $\mathcal{M}$ leads to a supporting projection π such that

$$Ker\,\pi = \mathcal{M}, \qquad Im\pi = \mathcal{M}^{[\perp]}$$

and this projection is clearly S–orthogonal.

Conversely, if π is a S–orthogonal supporting projection, $Ker\,\pi$ and $Im\pi$ are S–orthogonal and $\mathcal{M} = Ker\,\pi$ is an S–nondegenerate A–invariant subspace. □

For the case $\varphi(z) = \frac{1}{z}$ we have:

THEOREM 6.5. *Let M be an invertible element of $F^{m\times m}$ satisfying $M = \pm M^T$ and let W be a $F^{m\times m}$ valued rational function proper and invertible at infinity and such that*

$$(6.24)\qquad W(z)MW(\frac{1}{z})^T = M.$$

Let $W(z) = D + C(zI_n - A)^{-1}B$ be a minimal realization of W with the associated matrix S and bilinear form $[\ ,\ ]_S$. Let $\mathcal{M}$ be an invariant subspace of A, nondegenerate in the form $[\ ,\ ]_S$, and let π be the projection defined by

$$Ker\,\pi = \mathcal{M}, \qquad Im\pi = \mathcal{M}^{[\perp]}.$$

Finally, assume $I - A$ is invertible.

Then,

$$\begin{aligned} W_1(z) =& [I_m + C(zI_n - A)^{-1}(I_n - \pi)BD^{-1}]D_1 \\ W_2(z) =& D_2[I_m + D^{-1}C\pi(zI_n - A)^{-1}B] \end{aligned}$$

with

$$\begin{aligned} D =& D_1D_2 \\ D_1 =& I_m + C_1S_1^{-1}(I - A_1^T)^{-1}C_1^TM \end{aligned} \tag{6.25}$$

where

$$C_1 = C|_{\mathcal{M}}, \quad A_1 = A|_{\mathcal{M}}, \quad S_1 = (I - \pi)S|_{\mathcal{M}}, \quad B_1 = (I - \pi)B$$

define a minimal factorization $W = W_1W_2$ into two factors W_1 and W_2 satisfying (6.24).

Conversely, any such minimal factorization of W is obtained in this way, and the correspondence between minimal factorizations of W into factors satisfying (6.24) with $W_1(\infty) = D_1$ and S–nondegenerate invariant subspaces of A is one to one.

Proof: Equation (6.3a) becomes (when $\varphi(z) = \frac{1}{z}$)

$$SA = (A^{-1} + A^{-1}B(D - CA^{-1}B)^{-1}CA^{-1})^TS.$$

But

$$\begin{aligned} A^{-1} + A^{-1}B(D - CA^{-1}B)^{-1}CA^{-1} &= A^{-1}\left(I + B(D - CA^{-1}B)^{-1}CA^{-1}\right) \\ &= A^{-1}\left(I + BD^{-1}(I - D^{-1}CA^{-1}B)^{-1}CA^{-1}\right) \\ &= A^{-1}\left(I - A^{-1}BD^{-1}C\right)^{-1} \\ &= A^{\times -1}, \end{aligned}$$

hence, we obtain

$$SA = (A^{\times})^{-T}S$$

i.e., since $S^T = \pm S$,

$$A^T S = S A^{X-1}$$

so that

$$A^\times \cdot A^{[T]} = I_n.$$

Hence, if $A\mathcal{M} \subset \mathcal{M}$ we have $A^\times \mathcal{M}^{[\perp]} \subset \mathcal{M}^{[\perp]}$ and as in Theorem 6.4, we have a one–to–one correspondence between S–nondegenerate invariant subspaces of A and S–orthogonal supporting projections.

Condition (3.15) becomes here

$$D_1 M D_1^T \left(I_m - CA^{-1}(I_n - \pi)BD^{-1}\right) = M. \tag{6.26}$$

To check that D_1 given by (6.25) is a solution of (6.26) is a tedious computation which will be omitted.

The converse statement of Theorem 6.5 follows upon applying the converse statement of Theorem 3.4.□

To conclude this section we check that the function (6.14) admits no minimal factorizations with degree one factors within the associated group. The proof follows [AG], p. 191, but the reader should be aware that the reasoning is different towards the end.

We first note that $W = W_1 W_2$ where

$$\begin{aligned} W_1(z) =& I_m + \left(\frac{z-\lambda_2}{z-\lambda_1} - 1\right) W_{12} \\ W_2(z) =& I_m + \left(\frac{z-\lambda_1}{z-\lambda_2} - 1\right) W_{21}. \end{aligned}$$

Hence, W admits minimal factorizations with degree one factors.

A minimal realization of W_i is given by

$$W_i(z) = I_m + C_i(zI_n - A_i)^{-1} B_i, \qquad i = 1, 2$$

where

$$\begin{aligned} A_1 f &= \lambda_1 f \qquad f \in F \\ A_2 f &= \lambda_2 f \\ B_1 &= \frac{u_2^T M}{u_2^T M u_1} \qquad B_2 f = \frac{u_1^T M}{u_1^T M u_2} \\ C_1 &= (\lambda_1 - \lambda_2) u_1 \qquad C_2 f = (\lambda_2 - \lambda_1) u_2 \end{aligned}$$

so that W admits the minimal realization $W(z) = I + C(zIm - A)^{-1}B$ with

$$A = \begin{pmatrix} A_1 & B_1C_2 \\ 0 & A_2 \end{pmatrix} = \begin{pmatrix} \lambda_1 & 0 \\ 0 & \lambda_2 \end{pmatrix} \tag{6.27a}$$

$$B = \begin{pmatrix} B_1 \\ B_2 \end{pmatrix} = \begin{pmatrix} u_2^T M / u_2^T M u_1 \\ u_1^T M / u_1^T M u_2 \end{pmatrix} \tag{6.27b}$$

and

$$C = (C_1, C_2) = (\lambda_1 - \lambda_2)(u_1, -u_2) \tag{6.27c}$$

and so $A^\times = \begin{pmatrix} \lambda_2 & 0 \\ 0 & \lambda_1 \end{pmatrix}$.

It follows that the only minimal factorizations of W are (up to constants) $W = W_1W_2$ or $W = W_2W_1$. Now, if for some constant matrix D, W_1D was to satisfy (6.5), then the associated matrix of W with respect to the realization (6.27) would be of the form

$$S = \begin{pmatrix} s_1 & 0 \\ 0 & s_2 \end{pmatrix}, \qquad s_1 \in F$$

and equations (6.27) lead to

$$C^T M^{-1} C = 0$$

i.e. since $M^2 = -I$,

$$\begin{pmatrix} u_1^T \\ u_2^T \end{pmatrix} M\,(u_1, -u_2) = 0$$

which cannot be since $u_1^T M u_2 \neq 0$.

7. TYPE IIb SYMMETRIES.

We now focus on the case of IIb symmetries, assuming $F = \mathbb{C}$. So we study $\mathbb{C}^{m\times m}$ valued rational functions W such that $W = \mathcal{A}(W)$ where

$$\mathcal{A}(W)(z) = MW(\overline{\varphi(z)})^{-*}M^{-1}$$

with $M \in \mathbb{C}^{m\times m}$, $\det M \neq 0$ and $\varphi(z) = \dfrac{kz+l}{rz+s}$, $ks - rl \neq 0$.

The cases where $\varphi(z) = iz$ or $\varphi(z) = \dfrac{1}{z}$ and $M = M^* = M^{-1}$ were treated in detail in [AG].

We note that $\mathcal{A}(W)$ is an involution if and only if $\overline{\varphi(\overline{\varphi(z)})} = z$ and

$$MM^{-*} = \alpha \cdot I_n$$

for some $\alpha \in \mathbb{C}, \alpha \neq 0$; the case where $A(W)$ is an involution will be of importance, as explained in the sequel.

We first study the realization theorem, Theorem 3.1, in the present setting:

THEOREM 7.1. *Let $W \in GL_m(\mathcal{R})$ analytic and invertible at infinity and let $W(z) = C(zI_n - A)^{-1}B$ be a minimal realization of W. Then W satisfies*

$$W(z)MW(\overline{\varphi(z)})^* = M \qquad (z \textit{ point of analyticity of } W) \tag{7.1}$$

if and only if the following conditions hold:

1. *The matrix $(kI_n - rA)$ is invertible.*
2. *There exists an invertible matrix S of $\mathbb{C}^{m\times m}$ such that*

(7.2)
$$\begin{pmatrix} S & 0 \\ 0 & M^{-1} \end{pmatrix}\begin{pmatrix} A & B \\ C & D \end{pmatrix} = \begin{pmatrix} (\overline{A})_\varphi - (\overline{B})_\varphi(\overline{D})_\varphi^{-1}(\overline{C})_\varphi & -(\overline{B})_\varphi(\overline{D})_\varphi^{-1} \\ (\overline{D})_\varphi^{-1}(\overline{C})_\varphi & (\overline{D})_\varphi^{-1} \end{pmatrix}\begin{pmatrix} S & 0 \\ 0 & M^{-1} \end{pmatrix}$$

(where $(\overline{A})_\varphi, \ldots$ are obtained via (2.6) with $\overline{A}$ instead of $A, \ldots$).

To obtain this theorem from Theorem 3.1 we note the following: if $W = \mathcal{A}(W)$, the function $\mathcal{A}(W)$ is analytic at infinity (condition (1) of Theorem 3.1) if and only if $kI_n - r\mathcal{A}$ is invertible.

Condition (7.2) is obtained from (3.1) since the associated map is given by formula (2.20).

Equation (7.2) can be rewritten as:

$$SA = ((\overline{A})_\varphi - (\overline{B})_\varphi(\overline{D})_\varphi^{-1}(\overline{C})_\varphi)^T S \tag{7.3a}$$

$$SB = (\overline{C})_\varphi^T(\overline{D})_\varphi^{-T}M^{-1} \tag{7.3b}$$

$$M^{-1}C = -(\overline{D})_\varphi^{-T}(\overline{B})_\varphi^T S \tag{7.3c}$$

and

(7.3d) $$M^{-1}D = (\overline{D})_{\varphi}^{-T} M^{-1}$$

and in particular, A and S are related by the equation

(7.4) $$SA - (\overline{A})_{\varphi}^{T} S = (\overline{C})_{\varphi}^{T} M^{-1} C$$

i.e.

$$-rA^*SA + kSA - sA^*S + lS = (ks - lr)C^*M^{-1}C.$$

We now study in greater details the cases where $\varphi(z) = z$ or $\varphi(z) = \frac{1}{z}$.

THEOREM 7.2. *(The case $\varphi(z) = z$). Let W be a proper $\mathbb{C}^{m\times m}$ valued rational function invertible at infinity and let $W(z) = D + C(zI_n - A)^{-1}B$ be a minimal realization of W. Then*

(7.5) $$W(z)MW(\overline{z})^* = M$$

if and only if (1) and (2) below hold:

(1)

(7.6) $$DMD^* = M$$

(2) There exists an invertible matrix S such that

(7.7) $$SA - A^*S = C^*M^{-1}C$$

(7.8) $$C = -DMB^*S$$

(7.9) $$C^* = SBMD^*$$

If $M^* = M$, then (7.8) and (7.9) are equivalent and $S^* = -S$.

The proof is easily deduced from Theorem 7.1

THEOREM 7.3. *(The case $\varphi(z) = \frac{1}{z}$). Let W be a proper $\mathbb{C}^{m\times m}$ valued rational function invertible at infinity and let $W(z) = D + C(zI_n - A)^{-1}B$ be a minimal realization of W. Then*

$$W(z)MW\left(\frac{1}{\overline{z}}\right)^* = M$$

if and only if there exists an invertible matrix S such that

(7.10) $$\begin{pmatrix} A & B \\ C & D \end{pmatrix}^* \begin{pmatrix} S & 0 \\ 0 & -M^{-1} \end{pmatrix} \begin{pmatrix} A & B \\ C & D \end{pmatrix} = \begin{pmatrix} S & 0 \\ 0 & -M^{-1} \end{pmatrix}$$

Moreover, if $M^* = M$, then $S^* = S$.

When $\varphi(z) = \frac{1}{z}$, equation (7.4) becomes

$$S - A^*SA = -C^*M^{-1}C$$

which is one of the four equalities needed to prove (7.10). The others are proved in a similar way from (7.3).

These last two theorems are proved in [AG] when $M^* = M^{-1} = M$, to which we refer the reader for various examples of $\mathcal{A}$-symmetric functions for type IIb symmetries.

We now turn to the version of Theorem 3.4, the factorization theorem, in the present setting. The reasoning is very close to the one done for type Ia symmetries. We focus on the case $M^* = M^{-1}$ and either $\varphi(z) = z$ or $\varphi(z) = \frac{1}{z}$. Then, if $W(z) = D + C(zI_n - A)^{-1}B$ is a minimal realization of a $\mathcal{A}$-symmetric function W which is proper and invertible at infinity, and if S is the associated matrix, we endow the state space $\mathbb{C}^n$ with the hermitian inner product (called the associated hermitian form):

$$[x,y]_S = \begin{cases} x^*Sy & (\text{if } \varphi(z) = \frac{1}{z}) \\ ix^*Sy & (\text{if } \varphi(z) = z) \end{cases}$$

The associated projection map $\pi \to \pi_A$ is given for type IIb symmetries, by $\pi_A = \pi^*$ (see section 2, Example 2.4). Hence condition (3.15) becomes

$$\pi = \pi^{[*]}$$

where $\pi^{[*]}$ denotes the adjoint of π in the associated hermitian form.

We denote by $[\perp]$ orthogonality with respect to $[,]_S$.

THEOREM 7.4. *Let M be in $\mathbb{C}^{m\times m}, M = M^*, \det M \neq 0$ and let W be a proper $\mathbb{C}^{m\times m}$ valued rational function invertible at infinity and such that $W(z)MW(\overline{z})^* = M$. Let $W(z) = D + C(zI_n - A)^{-1}B$ be a minimal realization of W, with the associated matrix S and associated hermitian form $[\ ,\ ]_S$. Finally let $\mathcal{M}$ be an invariant subspace of A, nondegenerate with respect to $[\ ,\ ]_S$, let π be the projection be defined by:*

$$Ker\pi = \mathcal{M}, \quad Im\pi = \mathcal{M}^{[\perp]}$$

and let $D = D_1D_2$ be a multiplicative decomposition of D into two factors satisfying $D_iMD_i^* = M, i = 1,2$.

Then, the factorization $W = W_1W_2$, where W_1 and W_2 are as in (6.19), (6.20), is a minimal factorization of W into two factors satisfying (7.1).

Conversely, any such minimal factorization of W can be obtained in such a way and, for a fixed decomposition $D = D_1D_2$, the correspondence between minimal factorization of W into factors satisfying (7.1) and nondegenerate invariant subspaces of A is one to one.

This theorem is obtained from Theorem 3.4 in the same way that Theorem 6.4 is obtained from Theorem 3.4. Note that (3.15) becomes here $D_1MD_1^* = M$.

A similar result holds for the case $\varphi(z) = \frac{1}{z}$. Condition (3.15) is then more involved and becomes

$$D_1 M D_1^*(I - CA^{-1}(1-\pi)BD^{-1}) = M$$

and, following [AG], we take for D_1

$$D_1 = I_m + C_1 S_1^{-1}(I - \alpha A_1^*)^{-1} C_1^* M$$

where A_1, C_1 and S_1 are as in Theorem 6.5 (with $*$ instead of T). Then, one obtains a theorem which is word by word Theorem 6.5 when replacing T by $*$.

8. DEGREE PRESERVING AUTOMORPHISMS OF $GL_n(\mathcal{R})$.

The choice of the four types of symmetries studied in Sections 4–7 is motivated by the special properties of these symmetries within the class of all symmetries. An axiomatic description and analysis of the special properties are given in this and the next section. In this section we focus on the algebraic properties (which turn out to be degree preservation). The continuity properties will be addressed in the next section.

We start with the following result (this is a general theorem by Dieudonné [D] applied to this particular field).

THEOREM 8.1. *Let F be a field and let $\mathcal{R}$ be the field $F(z)$ of scalar rational functions in one variable z with coefficients in F. Then any group automorphism $\kappa : GL_n(\mathcal{R}) \to GL_n(\mathcal{R})$ is of the following form: There exist a group homomorphism χ of $GL_n(\mathcal{R})$ into its center such that $\chi(u) = u^{-1}$ implies $u = 1$, a matrix $M \in GL_n(\mathcal{R})$ and a field automorphism $r \mapsto r^J$ of $F(z)$ such that either*

$$\kappa(W) = \chi(W) M W^J M^{-1}, \qquad \textit{for all } W \in GL_n(\mathcal{R})$$

or

$$\kappa(W) = \chi(W) M((W^J)^T)^{-1} M^{-1}, \qquad \textit{for all } W \in GL_n(\mathcal{R})$$

where $W^J = (w_{ij}^J)$ if $W = (w_{ij})$.

We now state the main result of this section, a description of the set of all degree preserving group automorphisms of $GL_n(F(z))$ (the definition of degree is given in Section 1). To start with, consider algebraically closed fields.

THEOREM 8.2. *Let F be an algebraically closed field of characteristic different from 2. Let $\mathcal{R} = F(z)$ and $\kappa : GL_n(\mathcal{R}) \to GL_n(\mathcal{R})$ be a group automorphism (here $n \geq 2$). Assume that*

$$\deg W = 1 \Longrightarrow \deg \kappa(W) = 1. \tag{8.1}$$

Then there exists a group homomorphism $\xi : F(z)\setminus\{0\} \to F\setminus\{0\}$, a constant invertible matrix M, a field automorphism $x \mapsto x^E$ of F, and a linear fractional (Möbius) map $\psi(z)$ such that either

$$(\kappa(W))(z) = \xi(\det W) M W^E(\psi(z)) M^{-1}, \qquad W \in GL_n(\mathcal{R}) \tag{8.2}$$

or

$$(\kappa(W))(z) = \xi(\det W) M \left(W^E(\psi(z))\right)^{T-1} M^{-1}, \qquad W \in GL_n(\mathcal{R}). \tag{8.3}$$

Here for a rational matrix function $W(z) = [w_{ij}(z)]_{i,j=n}^n$, we define

$$W^E(\psi(z)) = [w_{ij}^E(\psi(z))]_{i,j=1}^n,$$

where for a scalar rational function

$$w(z) = \Sigma a_i z^i / \Sigma b_j z^j$$

we define

$$w^E(\psi(z)) = \Sigma a_i^E \psi(z)^i / \Sigma b_j^E \psi(z)^j.$$

Conversely, if $\kappa : GL_n(\mathcal{R}) \to GL_n(\mathcal{R})$ is given by (8.2) or (8.3), then κ is a group automorphism and

$$\deg W = k \Longleftrightarrow \deg \kappa(W) = k$$

for $k = 1, 2, \ldots$.

The rest of this section will be devoted to the proof of Theorem 8.2. With the exception of a remark at the end, it will be assumed for the rest of this section that F is algebraically closed.

First of all, we need the description of the field automorphisms of $F(z)$.

LEMMA 8.3. *Let $r \to r^J$ be a field automorphism of $F(z)$. Then $F^J = F$, and there exists a linear fractional (Möbius) map φ such that if*

$$r(z) = \left(\sum_{i=0}^{n} a_i z^i\right) / \left(\sum_{i=0}^{m} b_j z^j\right),$$

then

$$(r^J)(z) = \left(\sum_{i=0}^{n} a_i^J \varphi(z)^i\right) / \left(\sum_{i=0}^{m} b_j^J \varphi(z)^j\right).$$

PROOF: It is clear that $0^J = 0$ and $1^J = 1$. Let now z_0 be a nonzero element in F and, for any $n \geq 1$, let w_n be a nth root of z_0, that is $z_0 = w_n^n$. Then, for any $n \geq 1$,

$$z_0^J = (w_n^J)^n.$$

If for some n_0, $w_{n_0}^J$ is in F we thus get that $z_0^J = (w_{n_0}^J)^{n_0}$ is in F (and, in particular, w_n^J is in F for every $n \geq 1$). Suppose now that, for every $n \geq 1$, w_n^J is in $F(z)$, of strictly positive degree. Then, z_0^J would be of arbitarily large degree, a contradiction; hence, $r \to r^J$ restricted to F is a field automorphism of F.

Let $\varphi(z)$ be the image of the polynomial $p(z) = z$ under the automorphism $r \to r^J$. Then, clearly r^J is given as asserted in the Lemma for an arbitrary element $r \in F(z)$. Suppose that $\deg \varphi > 1$; then $r \to r^J$ would not be onto. Indeed, the equation

$$z = \left(\sum_0^n a_i^J(\varphi(z))^i\right) / \left(\sum_0^m b_j^J(\varphi(z))^j\right)$$

cannot be satisfied for any choice of n, m, a_i, b_j if $\deg \varphi > 1$. Thus φ is a Möbius function of degree one. The proof of the lemma is now easily completed.□

PROOF OF THE DIRECT STATEMENT OF THEOREM 8.2: Let $\kappa : GL_n(\mathcal{R}) \to GL_n(\mathcal{R})$ be a group automorphism such that

$$\deg W = 1 \Longrightarrow \deg \kappa(W) = 1.$$

By Theorem 8.1 we have either

$$\kappa(W) = \chi(W) M W^J M^{-1} \tag{8.4}$$

or

$$\kappa(W) = \chi(W) M((W^J)^T)^{-1} M^{-1}, \tag{8.5}$$

where χ, M and J are as in Theorem 8.1. By Lemma 8.3 we know that there is a linear fractional map $\psi(z)$ such that if $r(z) = \Sigma a_i z^i / \Sigma b_j z^j$, then

$$(r^J)(z) = \Sigma a_i^E \psi(z)^i / \Sigma b_j^E \psi(z)^j,$$

where $\kappa \mapsto \kappa^E$ is a field automorphism of F.

We assume that the nonzero entries of M have no common zeros and no common poles in F (if they do, we can factor out the common zeros and the common poles, producing the equality

$$MWM^{-1} = \tilde{M}W\tilde{M}^{-1}$$

for any rational matrix function W, where the non–zero entries of M have no common zeros and have no common poles in F).

We break the proof up into several steps. The trace of a matrix P will be denoted $tr\, P$.

STEP 1. *If $W(z) = I + (z - \lambda_0)^{-1} P$, where P is a constant matrix of rank 1, then*

(i) if $tr\ P = 0$, then $\chi(W)$ is a constant

(ii) if tr $P \neq 0$, then either: $\chi(W)$ is a constant or $n = 2$ and $\deg \chi(W) = 1$.

PROOF OF STEP 1: We use the notion of a local degree. Let $W \in GL_n(\mathcal{R})$, and let

$$W(z) = \sum_{j=-k}^{\infty} (z - \lambda_0)^j W_j$$

be the Laurent series of $W(z)$ at $\lambda_0 \in F \cup \{\infty\}$. The local degree of $W(z)$ at λ_0 (even for the case of an arbitrary algebraically closed field) is defined by

$$\deg\ loc(W; \lambda_0) = rank \begin{bmatrix} W_{-1} & W_{-2} & \cdot & \cdot & \cdot & W_{-k} \\ \cdot & \cdot & & & \cdot & 0 \\ \cdot & \cdot & & \cdot & & \cdot \\ \cdot & \cdot & \cdot & & & \cdot \\ W_{-k+1} & W_{-k} & & & & \cdot \\ W_{-k} & 0 & \cdot & \cdot & \cdot & 0 \end{bmatrix}$$

It follows then (see [BGK] for the case $F = \mathbb{C}$) that

$$\deg W = \sum_{\lambda_0 \in F \cup \{\infty\}} \deg\ loc(W; \lambda_0).$$

Assume that κ has the form (8.4) (if κ has the form (8.5) the proof is analogous). Let

$$W(z) = I + (z - \lambda_0)^{-1} P, \qquad \lambda_0 \in F \tag{8.6}$$

be a degree 1 proper rational function with $W(\infty) = 1$ (then $rank P = 1$). Suppose first that $tr\ P = 0$. Then

$$W^J(z) = I + (\psi(z) - \lambda_0^E)^{-1} P^E$$

and

$$\kappa(W) = \chi(W)\ (I + (\psi(z) - \lambda_0^E)^{-1} M P^E M^{-1}).$$

Because of the hypothesis we know that $\deg \kappa(W) = 1$. Let $\lambda_1 \in F \cup \{\infty\}$ be a pole of $\chi(W)$. In what follows we assume $\lambda_1 \in F$ (if $\lambda_1 = \infty$, the changes in the subsequent proof are obvious). Then λ_1 cannot be a pole of $I + (\psi(z) - \lambda_0^E)^{-1} M P^E M^{-1}$, because otherwise the local degree of $\chi(W)(I + (\psi(z) - \lambda_0^E)^{-1} M P^E M^{-1})$ at λ_1 would be at least 2, a contradiction. On the other hand, $rank\ [(\psi(z) - \lambda_0^E)^{-1} M(z) P^E M^{-1}(z)] \leq 1$ for every point z which is not a pole of this function. So for the Laurent series at λ_1 we have

$$(\psi(z) - \lambda_0^E)^{-1} M(z) P^E M^{-1}(z) = \sum_{j=0}^{\infty} (z - \lambda_1)^j Q_j,$$

where $rank Q_0 \leq 1$. It follows that

$$\chi(W)(I + (\psi(z) - \lambda_0^E)^{-1} M P^E M^{-1}) = \sum_{j=-k}^{\infty} (z - \lambda_1)^j \tilde{Q}_j,$$

where k is the multiplicity of λ_1 as a pole of $\chi(W)$, and $\tilde{Q}_{-k} = I + Q_0$ (in particular, $rank\tilde{Q}_{-k} \geq n-1$). Now $tr\ Q_0 = tr\ [(\psi(z) - \lambda_0^E)^{-1} M(z) P^E M^{-1}(z)]_{z=\lambda_1} = [(\psi(z) - \lambda_0^E)^{-n} tr P^E]_{z=\lambda_1} = 0$. Thus the sum of the eigenvalues of Q_0 is 0. Since also we have $rankQ_0 \leq 1$, Q_0 can have at most one nonzero eigenvalue. Hence all eigenvalues of Q_0 are 0. So $I + Q_0$ is invertible and

$$\deg\ loc(\chi(W)(I + (\psi(z) - \lambda_0^E)^{-1} M P^E M^{-1}; \lambda_1) = kn.$$

On the other hand, this local degree cannot exceed 1. We obtain immediately a contradiction (here we use the standing assumption that $n \geq 2$). If we start with $W(z)$ as in (8.6) but $tr\ P \neq 0$, then

$$\deg\ loc(\chi(W)(I + (\psi(z) - \lambda_0^E)^{-1} M P^E M^{-1}; \lambda_1) \geq k(n-1),$$

and we obtain a contradiction again if $n \geq 3$. Also, we obtain a contradiction with

$$\deg(\chi(W)(I + (\psi(z) - \lambda_0^E)^{-1} M P^E M^{-1}) = 1$$

if $\chi(W)$ has more that 1 pole (counted with multiplicities). □

STEP 2. *The rational matrix function $M(z)$ is actually a constant.*

PROOF OF STEP 2: Denote by E_{ij} the matrix with (i,j) entry equal to 1 and all other entries zeros. Again, we assume that κ has the form (8.4) (if κ has the form (8.5), the proof is analogous). Consider

$$W(z) = I + E_{ij}(z - \lambda_0)^{-1}, \quad \lambda_0 \in F$$

where $i \neq j$. By Step 1, we know that $\chi(W)$ is a constant (call this constant matrix K) and hence

$$\kappa(W) = K(I + M E_{ij}(\psi(z) - \lambda_0^E)^{-1} M^{-1}).$$

By the hypothesis of the theorem the degree of $\kappa(W)$ is 1. Using the concept of a local degree given in the proof of Step 1, we see that the product $M E_{ij} M^{-1}$ must be analytic at the point $\mu = \psi^{-1}(\lambda_0^E)$. Now observe that

$$M E_{ij} M^{-1} = (i^{\text{th}} \text{ column of } M) \cdot (j^{\text{th}} \text{ row of } M^{-1}).$$

Assume that some entry in the jth row of M^{-1} has a pole at μ. Then all the entries in the ith column of M must have a zero at μ. Since i can be chosen arbitrarily (subject to $i \neq j$) it follows that all entries in M, except possibly the entries in the jth column of M, must have a zero at μ. If there is an entry with a pole at μ in a row different from the jth row of M^{-1}, then arguing as before we obtain that all entries of M must have a zero at μ, a contradiction with the assumption stated at the beginning of the proof that the entries of $M(z)$ have no common factors. So we can assume that the jth row of M^{-1} is the only row containing entries with a pole at μ. Say, the $(j,k)^{\text{th}}$ entry of M^{-1} has a pole at μ. Then, by looking at the product

$$M \cdot (k^{\text{th}} \text{ column of } M^{-1}) = \begin{bmatrix} 0 \\ \vdots \\ 0 \\ 1 \\ 0 \\ \vdots \\ 0 \end{bmatrix}$$

and by using the already established properties of the entries of M and of M^{-1}, we see that all the entries in the jth column of M must also have a zero at μ. Again, we obtain a contradiction with the assumption that the entries of M have no common factor.

We have proved that no entry of M^{-1} may have a pole at μ. Analogously one proves that no entry of M^{-1} may have a zero at μ. Since $\lambda_0 \in F$ was completely arbitrary, we obtain that the only possible point in $F \cup \{\infty\}$ where the entries of M^{-1} can have a pole or a zero is at $\mu_0 := \psi^{-1}(\infty)$. But then the entries of M^{-1} must be constants, and we are done. □

STEP 3. *If $W \in GL_n(\mathcal{R})$ is a constant, then $\chi(W)$ is also a constant.*

PROOF OF STEP 3: Let $W_0 \in GL_n(\mathcal{R})$ be a constant. Assume first that the field F is of characteristic zero. Then for $p = 1, 2, \ldots$ there is a pth root $Z_p \in GL_n(F)$ of $W_0 : Z_p^p = W_0$. Since χ is a group homomorphism, we have $\chi(W_0) = (\chi(Z_p))^p$. As the range of χ consists of scalar rational functions multiplied by I, the degree of the pth power of such a function is equal to the pth multiple of the degree of the function, so we have

$$\deg \chi(W_0) = p \ \deg \chi(Z_p).$$

Thus, the integer $\deg \chi(W_0)$ is divisible by any positive integer p. So $\deg \chi(W_0) = 0$, as required. If F has prime characteristic q, then the proof is more subtle, although it uses the same idea as in the case of characteristic 0. It will suffice to prove the following statement: every constant $W_0 \in GL_n(K)$ is a product of a finite number of constant matrices $Z_1 Z_2 \ldots Z_s$, where each Z_j has the property that for every $p = 1, 2, \ldots$ there is a pth root of Z_j in $GL_n(F)$. To prove this statement, since F is algebraically closed, we can assume that W_0 is a Jordan block:

$$W_0 = \begin{bmatrix} z_1 & 1 & 0 & \cdot & \cdot & \cdot & 0 \\ 0 & z_0 & 1 & & & & 0 \\ \cdot & \cdot & \cdot & & & & \cdot \\ \cdot & \cdot & & \cdot & & & \cdot \\ \cdot & \cdot & & & \cdot & & \cdot \\ \cdot & \cdot & & & & \cdot & 1 \\ 0 & 0 & \cdot & \cdot & \cdot & \cdot & z_0 \end{bmatrix}, z_0 \in F.$$

Then

$$W_0 = \begin{bmatrix} \alpha_1^{-1} & & & 0 \\ & \alpha_2^{-1} & & \\ 0 & & \ddots & \\ & & & \alpha_n^{-1} \end{bmatrix} \begin{bmatrix} \alpha_1 z_0 & \alpha_1 & 0 & \cdots & 0 \\ 0 & \alpha_2 z_0 & \alpha_2 & \cdots & 0 \\ \vdots & \vdots & & \ddots & \vdots \\ \vdots & \vdots & & & \alpha_{n-1} \\ 0 & 0 & \cdots & \cdots & \alpha_n z_0 \end{bmatrix},$$

where $\alpha_1, \cdots, \alpha_n$ are distinct non-zero elements in F, and it is easy to see (by passing to the Jordan form) that every factor in this representation of W_0 has p^{th} roots for all positive integers p.

STEP 4. *If $n \geq 3$, then $\mathcal{X}(W)$ is constant for every $W \in GL_m(\mathcal{R})$.*

For the proof we need the following lemma.

LEMMA 8.4. *Let $W(z)$ be an $n \times n$ rational matrix function with* $\det W \not\equiv 0$. *Then $W(z)$ can be written as a product (not necessarily minimal)*

$$W(z) = W_1(z) \ldots W_p(z)$$

where each factor $W_i(z)$ has degree 1.

PROOF: The lemma is easily verified when $n = 1$, and therefore also when $W(z)$ is a diagonal matrix function. Using the Smith-McMillan factorization for $W(z)$ (see, e.g., [GLR1]), we reduce the proof to the case when $W(z)$ is a polynomial with constant non-zero determinant. Such a polynomial can be written as a product of factors where each factor is either a constant matrix or has the form

$$I + \alpha z^p E_{ij} \qquad (i \neq j; p \text{ positive integer})$$

where $\alpha \in F \setminus \{0\}$ and E_{ij} is the $n \times n$ matrix with (i,j) entry 1 and all other entries zeros (see, e.g., [G]). Thus, the proof of the lemma is reduced further to the case when $n = 2$ and

$$W(z) = \begin{bmatrix} 1 & \alpha z^p \\ 0 & 1 \end{bmatrix}.$$

For such $W(z)$ the required factorization is given by

$$\begin{aligned} \begin{bmatrix} 1 & \alpha z^p \\ 0 & 1 \end{bmatrix} &= \begin{bmatrix} z^{p-1} & 0 \\ 0 & 1 \end{bmatrix} \begin{bmatrix} 1 & \alpha z \\ 0 & 1 \end{bmatrix} \begin{bmatrix} z^{-(p-1)} & 0 \\ 0 & 1 \end{bmatrix} \\ &= \begin{bmatrix} z & 0 \\ 0 & 1 \end{bmatrix} \cdots \begin{bmatrix} z & 0 \\ 0 & 1 \end{bmatrix} \begin{bmatrix} 1 & \alpha z \\ 0 & 1 \end{bmatrix} \begin{bmatrix} z^{-1} & 0 \\ 0 & 1 \end{bmatrix} \cdots \begin{bmatrix} z^{-1} & 0 \\ 0 & 1 \end{bmatrix}. \end{aligned}$$

(a total of $2p - 1$ factors). □

PROOF OF STEP 4: In view of Lemma 8.4 and the homomorphism property of $\mathcal{X}$ we only have to prove that $\mathcal{X}(W)$ is a constant for every $W \in GL_m(\mathcal{R})$ such that degree $(W) = 1$.

Let $W \in GL_m(\mathcal{R})$ be of degree 1. If W has no pole and no zero at infinity, then (up to a multiplication by a constant invertible matrix) $W(z) = I + (z - \lambda_0)^{-1}P$ for some $\lambda_0 \in F$ and some constant matrix P of rank 1. So by Steps 1 and 3 we obtain that $\chi(W)$ is a constant.

Assume now that W has a pole at infinity. Then (up to multiplication by a constant invertible matrix)

$$W(z) = I + (z - \lambda_0)P \tag{8.7}$$

where P is a constant matrix of rank 1, and $\lambda_0 \in F$. Indeed, since $\deg(W) = 1, W(z)$ cannot have any poles in F, so it is a polynomial. Furthermore, the pole of W at infinity is simple, so $W(z)$ must be of the form

$$W(z) = Q + zP,$$

where P is a rank 1 constant matrix. But $\det W(z) \not\equiv 0$; so $Q + \lambda_0 P$ is invertible for some $\lambda_0 \in F$. Rewrite $W(z)$ in the form

$$W(z) = Q + \lambda_0 P + (z - \lambda_0)P$$

and multiply by $(Q + \lambda_0 P)^{-1}$ to obtain the form (8.7). We can now argue as in the proof of Step 1 that $\chi(W)$ is a constant.

If W has a zero at infinity, then the preceding argument shows that $\chi(W^{-1})$ is a constant. Then $\chi(W)$ is a constant as well. □

The direct statement of Theorem 8.2 is completely proved for $n \geq 3$ by using Steps 1–4 above and the following proposition.

PROPOSITION 8.5. *Let $\chi : GL_n(\mathcal{R}) \to H$ be a group homomorphism (here $K = F(z)$) into a commutative group H. Then*

$$\det W \equiv 1 \Rightarrow \chi(W) = 1.$$

PROOF: If $n = 1$, the statement is trivial, so assume $n \geq 2$. We show the result first for an upper or lower triangular matrix polynomial $W(z)$ with 1's on the main diagonal. Any such $W(z)$ is a product of functions of the form $I + \alpha z^p E_{ij}$, where $\alpha \in F \setminus \{0\}, p$ is a nonnegative integer and $E_{ij} (i \neq j)$ is the matrix with 1 in the (i, j) position and zeros elsewhere. So it is sufficient to prove that

$$\chi(I + \alpha z^p E_{ij}) = 1.$$

For notational simplicity assume $n = 2$ and say, $i = 1, j = 2$. As

$$\begin{bmatrix} 1 & \alpha z^p \\ 0 & 1 \end{bmatrix} = \begin{bmatrix} \alpha & 0 \\ 0 & 1 \end{bmatrix} \begin{bmatrix} 1 & z^p \\ 0 & 1 \end{bmatrix} \begin{bmatrix} \alpha^{-1} & 0 \\ 0 & 1 \end{bmatrix},$$

it follows that

$$\chi\left(\begin{bmatrix} 1 & \alpha z^p \\ 0 & 1 \end{bmatrix}\right) = \chi\left(\begin{bmatrix} 1 & z^p \\ 0 & 1 \end{bmatrix}\right)$$

is independent of α. So we have

$$\chi\left(\begin{bmatrix} 1 & z^p \\ 0 & 1 \end{bmatrix}\right) = \chi\left(\begin{bmatrix} 1 & \frac{1}{2}z^p \\ 0 & 1 \end{bmatrix}\begin{bmatrix} 1 & \frac{1}{2}z^p \\ 0 & 1 \end{bmatrix}\right)$$
$$= \chi\left(\begin{bmatrix} 1 & \frac{1}{2}z^p \\ 0 & 1 \end{bmatrix}^2\right) = \left(\chi\left(\begin{bmatrix} 1 & z^p \\ 0 & 1 \end{bmatrix}\right)\right)^2.$$

(At this point we use the fact that the characteristic of F is different from 2.) We conclude that $\chi\left(\begin{bmatrix} 1 & z^p \\ 0 & 1 \end{bmatrix}\right) = 1$, as required.

Next, we show that $\chi(W) = 1$ for a constant diagonal W with $\det W = 1$. The proof of this statement reduces to the proof that

$$\chi(diag(1, \dots, 1, c, 1, \dots, 1, c^{-1}, 1, \dots, 1)) = 1,$$

where c appears in the i-th position and c^{-1} in the j-th position ($c \in F \setminus \{0\}$). Assuming again $n = 2$ for notational simplicity, we have

$$\chi\left(\begin{bmatrix} c & 0 \\ 0 & c^{-1} \end{bmatrix}\right) = \chi\left(\begin{bmatrix} c & 0 \\ 0 & 1 \end{bmatrix}\begin{bmatrix} 0 & 1 \\ 1 & 0 \end{bmatrix}\begin{bmatrix} c & 0 \\ 0 & 1 \end{bmatrix}^{-1}\begin{bmatrix} 0 & 1 \\ 1 & 0 \end{bmatrix}^{-1}\right)$$
$$= \chi\left(\begin{bmatrix} c & 0 \\ 0 & 1 \end{bmatrix}\right)\chi\left(\begin{bmatrix} 0 & 1 \\ 1 & 0 \end{bmatrix}\right)\left(\chi\left(\begin{bmatrix} c & 0 \\ 0 & 1 \end{bmatrix}\right)\right)^{-1}\left(\chi\left(\begin{bmatrix} 0 & 1 \\ 1 & 0 \end{bmatrix}\right)\right)^{-1} = 1.$$

We are now in position to prove that $\chi(W) = 1$ for every constant W with $det\, W = 1$. Every such W admits a factorization (see, e.g., [G G])

$$W = LE\, diag(\pm 1, 1, \dots, 1)U \tag{8.8}$$

where L is a lower triangular matrix, U is an upper triangular matrix, $det\, L = det\, U = 1$, and E is a permutation matrix (we need the factor $diag(-1, 1, \dots, 1)$ in (8.8) in case $det\, E = -1$). From what we have proved already we know that $\chi(L) = \chi(U) = 1$. On the other hand, the matrix

$$E\, diag(\pm 1, 1, \dots, 1)$$

can be written as a product of matrices of the form

$$I - Z_i,$$

where Z_i has 1 in the $(1,1)$, (i,i), $(i,1)$ entries Z_i has -1 in the $(1,i)$ entry and zeros elsewhere (here $i \neq 1$).

Thus, it is sufficient to prove that $\chi(I - Z_i) = 1$. Again, for simplicity of notation assume $n = 2$; then $I - Z_i = \begin{bmatrix} 0 & 1 \\ -1 & 0 \end{bmatrix}$. Let $\alpha \neq 0$, $\alpha \in F$. Then

$$\begin{bmatrix} 1 & \alpha \\ 0 & 1 \end{bmatrix}\begin{bmatrix} 0 & 1 \\ -1 & 0 \end{bmatrix} = \begin{bmatrix} 1 & 0 \\ \alpha^{-1} & 1 \end{bmatrix}\begin{bmatrix} -\alpha & 1 \\ 0 & -\alpha^{-1} \end{bmatrix}.$$

Using the already proved fact that $\chi(W) = 1$ for any upper or lower triangular constant matrix W, it follows that $\chi\left(\begin{bmatrix} 0 & 1 \\ -1 & 0 \end{bmatrix}\right) = 1$. We have prved that $\chi(W) = 1$ for every constant W with $det\, W = 1$.

Now every matrix polynomial $W(z)$ with constant determinant 1 is a product of constant matrices and upper and lower matrix polynomials with constant determinant 1 (see, e.g., [G]). Combining this fact with the already proved cases we conclude that $\chi(W) = 1$ for every matrix polynomial $W(z)$ with constant determinant 1. By using the Smith form of $W(z)$, the general case is reduced to the case when $W(z)$ is diagonal. This case in turn is easily reduced to the case when

$$W(z) = diag(1, \dots, c(z-\alpha), 1, \dots, c^{-1}(z-\alpha)^{-1}, 1, \dots, 1),$$

where $c \in F \setminus \{0\}, \alpha \in F$, the term $c^{-1}(z-\alpha)$ appears in the i-th position, and the term $c^{-1}(z-\alpha)^{-1}$ in the j-th position. Again assuming $n = 2$ for simplicity, we have

$$\begin{aligned} &\chi\left(\begin{bmatrix} c(z-\alpha) & 0 \\ 0 & c^{-1}(z-\alpha)^{-1} \end{bmatrix}\right) \\ =&\chi\left(\begin{bmatrix} c(z-\alpha) & 0 \\ 0 & 1 \end{bmatrix}\begin{bmatrix} 0 & 1 \\ 1 & 0 \end{bmatrix}\begin{bmatrix} c(z-\alpha) & 0 \\ 0 & 1 \end{bmatrix}^{-1}\begin{bmatrix} 0 & 1 \\ 1 & 0 \end{bmatrix}\right)^{-1} \\ =&\chi\left(\begin{bmatrix} c(z-\alpha) & 0 \\ 0 & 1 \end{bmatrix}\right)\chi\left(\begin{bmatrix} 0 & 1 \\ 1 & 0 \end{bmatrix}\right)\left(\chi\left(\begin{bmatrix} c(z-\alpha) & 0 \\ 0 & 1 \end{bmatrix}\right)\right)^{-1}\left(\chi\left(\begin{bmatrix} 0 & 1 \\ 1 & 0 \end{bmatrix}\right)\right)^{-1} = 1, \end{aligned}$$

and we are done. □

It remains to consider the case $n = 2$. We know already that there exist E, M, ψ as in Theorem 8.2 and there exists a group homomorphism $\eta : F(z)\setminus\{0\} \to F(z)\setminus\{0\}$ such that either

$$(\kappa(W))(z) = \eta(\det W) M W^E(\psi(z)) M^{-1} \tag{8.9}$$

for all $W \in GL_m(\mathcal{R})$, or

$$(\kappa(W))(z) = \eta(\det W) M (W^E(\psi(z)))^{T-1} M^{-1} \tag{8.10}$$

for all $W \in GL_m(\mathcal{R})$.

In the sequel we assume that (8.9) holds (the proof in the case (8.10) is analogous). Also, for the simplicity of notation we shall assume $M = I, \psi(z) = z$ and E is the identity.

STEP 5. *Assume $n = 2$ and the assumption made above. Then for any $a \in F$ either*

$$\eta(z-a) = L(z-a)^{-1} \tag{8.11}$$

for some constant L (depending on a) or

$$\eta(z-a) = L. \tag{8.12}$$

PROOF OF STEP 5: Consider

$$W(z) = \begin{bmatrix} z-a & 0 \\ 0 & 1 \end{bmatrix}.$$

Then $\deg W = 1$ and therefore $\deg(\kappa(W)) = 1$. But

$$\kappa(W) = \eta(z-a) \begin{bmatrix} z-a & 0 \\ 0 & 1 \end{bmatrix}.$$

The only way for $\kappa(W)$ to have degree 1 (in case $\eta(z-a)$ is not a constant) is when $\eta(z-a)$ is a constant multiple of $(z-a)^{-1}$. □

STEP 6. *Under the assumption of Step 5, either (8.11) holds for all $a \in F$ or (8.12) holds for all $a \in F$.*

PROOF OF STEP 6: Arguing by contradiction, assume that for some $a, b \in F$ we have

$$\eta(z-a) = L_1(z-a)^{-1}; \eta(z-b) = L_2,$$

where L_1 and L_2 are constants. For

$$W = \begin{bmatrix} \frac{z-a}{z-b} & 0 \\ 0 & 1 \end{bmatrix} = \begin{bmatrix} z-a & 0 \\ 0 & 1 \end{bmatrix} \begin{bmatrix} (z-b)^{-1} & 0 \\ 0 & 1 \end{bmatrix}$$

we have

$$\kappa(W) = L_1 L_2^{-1}(z-a)^{-1} \begin{bmatrix} \frac{z-a}{z-b} & 0 \\ 0 & 1 \end{bmatrix} = L_1 L_2^{-1} \begin{bmatrix} (z-b)^{-1} & 0 \\ 0 & (z-a)^{-1} \end{bmatrix}.$$

Now the degree of W is 1 but the degree of $\kappa(W)$ is 2, a contradiction with the degree 1 preservation property of κ. □

Steps 5 and 6 prove that in case $n = 2$, either the group automorphism $\kappa : GL_2(\mathcal{R}) \to GL_2(\mathcal{R})$ is given by (8.2) or by (8.3), or by one of the following formulas (where ξ, E, M, ψ have the same meaning as in Theorem 8.2)

$$(\kappa(W))(z) = \xi(\det W)(\det W^E(\psi(z)))^{-1} M W^E(\psi(z)) M^{-1} \tag{8.13}$$

for all $W \in GL_2(\mathcal{R})$;

$$(\kappa(W))(z) = \xi(\det W) \det W^E(\psi(z)) M (W^E(\psi(z))^{T^{-1}} M^{-1} \qquad \textit{forall } W \in GL_2(\mathcal{R}). \tag{8.14}$$

However, for a 2×2 matrix $A = \begin{bmatrix} a & b \\ c & d \end{bmatrix}$ we have

$$\begin{aligned} A^{-1} &= (\det A)^{-1} \begin{bmatrix} d & -b \\ -c & a \end{bmatrix} = (\det A)^{-1} \begin{bmatrix} 0 & -1 \\ 1 & 0 \end{bmatrix} \begin{bmatrix} a & b \\ c & d \end{bmatrix}^T \begin{bmatrix} 0 & 1 \\ -1 & 0 \end{bmatrix} \\ &= (\det A)^{-1} M_0 A^T M_0^{-1}, \end{aligned}$$

where $M_0 = \begin{bmatrix} 0 & -1 \\ 1 & 0 \end{bmatrix}$. Using this equality one easily sees that the formulas (8.13) and (8.14) are actually in the form (8.3) and (8.2), respectively.

This completes the proof of the direct statement of Theorem 8.2.

The converse statement of Theorem 8.2 is easily seen.

Theorem 8.2 is proved completely. □

The proof of Theorem 8.2 shows that the hypothesis on F being algebraically closed can be relaxed somewhat. Introduce the property $(*)_n$ of a field F:

$(*)_n$ *Every invertible matrix* $W \in GL_n(F)$ *can be written as a product of a finite number of invertible matrices* $Z_1, \ldots, Z_s$*, where each* Z_j *has the property that there is a* p^{th} *root of* Z_j *in* $GL_n(F)$ *for infinitely many positive integers* p. Only the properties $(*)_1$ and $(*)_n$ of F, besides being of characteristic $\neq 2$, were used in the proof of Theorem 8.2. (The concept of local degree used in Step 1 did not require F being algebraically closed in an essential way; one could repeat the proof of Step 1 using the local degree defined over the algebraic closure of F). As shown in the proof of Step 3, every algebraically closed field F has the properties $(*)_n$ for all n. However, there exist fields which are not algebraically closed with the properties $(*)_n$. For example, the field of real numbers is such; indeed, as it is easily seen from the real Jordan form, every invertible real $n \times n$ matrix has real p^{th} roots for every odd positive integer p.

9. CONTINUOUS AND ANALYTIC AUTOMORPHISMS OF $GL_m(\mathcal{R})$.

In this section the ground field is $\mathbb{C}$. We have obtained in the previous section a description of all degree preserving automorphisms of $GL_m(\mathcal{R})$. However, a degree preserving automorphism need not be continuous. In this section we impose natural continuity and analyticity properties on the degree preserving automorphism, and describe all continuous and analytic degree preserving automorphisms.

Everywhere in this section it will be assumed that $\kappa : GL_n(\mathcal{R}) \to GL_n(\mathcal{R})$, where $\mathcal{R} = \mathbb{C}(z)$, is a degree one preserving group automorphism: if $\deg(W) = 1, W \in GL_n(\mathcal{R})$, then also $\deg(\kappa(W)) = 1$. We know from Theorem 8.2 that any such automorphism κ has the form $\kappa = \kappa(\xi, E, M, \varphi)$ where

(1) $\xi : \mathcal{R} \setminus \{0\} \to \mathbb{C} \setminus \{0\}$ is a group homomorphism

(2) $x \to x^E$ is a field isomorphism on $\mathbb{C}$

(3) M is a constant invertible matrix over $\mathbb{C}$

and

(4) φ is a linear fractional (Möbius) transformation of the extended complex plane $\mathbb{C} \cup \{\infty\}$,

and where $\kappa = \kappa(\xi, E, M, \varphi)$ is given by either

(DPAa) $$(\kappa(W))(z) = \xi(\det W) M W^E(\varphi(z)) M^{-1}, W \in GL_n(\mathcal{R})$$

or

$$\text{(DPAb)} \qquad (\kappa(W))(z) = \xi(\det W)M(W^E(\varphi(z)))^{T-1}M^{-1}, W \in GL_n(\mathcal{R})$$

We start with analyzing the consequences of continuity of κ on constant matrices of determinant 1. From the form of $k(\xi, E, M, \varphi)$ we see that $\deg(k(W)) = \deg(W)$ for every $W \in GL_n(\mathcal{R})$. In particular, $k(W)$ is constant for every constant invertible matrix W and $\det k(W) = 1$ if W is a constant matrix with $\det W = 1$.

THEOREM 9.1. *Let $\kappa = \kappa(\xi, E, M, \varphi)$ be a degree one preserving group automorphism of $GL_n(\mathcal{R})$ as above, and suppose that the restriction $\hat{\kappa}$ of κ to the set of constant matrices of determinant 1 is continuous (as a map of $\mathbb{C}^{n\times n}$ into $\mathbb{C}^{n\times n}$). Then E has one of the two forms*

$$\lambda^E = \lambda \quad or \quad \lambda^E = \overline{\lambda}, \qquad for\ \lambda \in \mathbb{C}.$$

Hence κ has one of the four forms

(i) $(\kappa(W))(z) = \xi(\det W)MW(\varphi(z))M^{-1}$

(ii) $(\kappa(W))(z) = \xi(\det W)M\overline{W(\varphi(z))}M^{-1}$

(iii) $(\kappa(W))(z) = \xi(\det W)MW(\varphi(z))^{-1T}M^{-1}$

(iv) $(\kappa(W))(z) = \xi(\det W)M\overline{W(\varphi(z))^{-1T}}M^{-1}$

for all $W \in GL_n(\mathcal{R})$.

For the proof use the well-known fact that the only continuous field automorphisms of $\mathbb{C}$ are the identity and complex conjugation, and use Theorem 8.2.

From now on it will be assumed that the restriction of κ to the set of constant matrices of determinant 1 is continuous, so κ has one of the forms (i)–(iv) in Theorem 9.1.

We now focus on the group homomorphism $\xi : \mathcal{R}\setminus\{0\} \to \mathbb{C}\setminus\{0\}$. We will require certain analyticity properties of ξ. Consider the set D_1 of scalar rational functions (with coefficients in $\mathbb{C}$) of degree 1. Such a function has a unique representation in the form

$$f(z) = \frac{\alpha z + \beta}{\gamma z + \delta}$$

with $\det \begin{bmatrix} \alpha & \beta \\ \gamma & \delta \end{bmatrix} = 1$, and therefore can be identified with the special linear group $SL_2(\mathbb{C})$. The group $SL_2(\mathbb{C})$, being a classical Lie group, has a well-known analytic structure. We pull back this structure to D_1, using the correspondence

$$f(z) = \frac{\alpha z + \beta}{\gamma z + \delta} \in D_1 \Longleftrightarrow \begin{bmatrix} \alpha & \beta \\ \gamma & \delta \end{bmatrix} \in SL_2(\mathbb{C}),$$

and so D_1 becomes an analytic manifold.

THEOREM 9.2. *Let $\xi : \mathcal{R}\backslash\{0\} \to \mathbb{C}\backslash\{0\}$ be a group homomorphism. Assume that ξ is a continuous function when restricted to the set D_0 of constant $\neq 0$ rational functions, and assume that ξ is analytic when restricted to the analytic manifold D_1. Then ξ is trivial, i.e. $\xi(f(z)) = 1$ for every $f(z) \in \mathcal{R} \setminus \{0\}$.*

PROOF: The group homomorphism property of ξ implies that $|\xi(x)| = 1$ for every $x \in \mathbb{C}$ which is a root of 1. Since the roots of 1 are dense on the unit circle, the continuity of ξ implies that $|\xi(x)| = 1$ for every $x \in \mathbb{C}$, $|z| = 1$. Let $s > 2$ be such that

$$\frac{1}{s} \leq |\xi(x)| \leq s$$

for every $x \in \mathbb{C}$ with $\frac{1}{2} \leq |x| \leq 2$ (the existence of such an s is guaranteed again by the continuity of ξ). Define the number r by $r = \frac{\ln s}{\ln 2}$ (observe that $r \geq 1$). Now for every $x \in \mathbb{C}$, $|x| \geq 1$ let $q = q(x)$ be a nonnegative integer such that

$$2^q \leq |x| \leq 2^{q+1}.$$

Then

$$\ln |x| \geq q \ln 2,$$

so $r \ln |x| \geq rq \ln 2 = q \ln s$; and $s^q \leq |x|^r$.

We have

$$|\xi(x)| = \left|\xi\left(\frac{x}{2^q}\right)\xi(2)\ldots\xi(2)\right| \geq \frac{1}{s^{q+1}} \geq \frac{1}{s}|x|^{-r}. \tag{9.1}$$

This inequality will be used later on.

We now return to the restriction of ξ on D_1. Let $h(\alpha) = \xi(z - \alpha)$, $\alpha \in \mathbb{C}$. The function $h(\alpha)$ is an entire function that omits zero, so it must be of the form $h(\alpha) = \exp(g(\alpha))$, where $g(\alpha)$ is an entire function. The group homomorphism property of ξ implies

$$\xi\left((z - \alpha)(z - \beta)^{-1}\right) = \exp\left[g(\alpha) - g(\beta)\right], \qquad \alpha, \beta \in \mathbb{C},\ \alpha \neq \beta.$$

Consider now the family of functions

$$f_q(z) = z(qz + 1)^{-1},$$

where $q \in \mathbb{C}$ is a parameter (the case $q = 0$ is not excluded). We have for $q \neq 0$:

$$\begin{aligned}
\xi\left(f_q(z)\right) &= \xi\left(z\left(z + \frac{1}{q}\right)^{-1} \cdot \frac{1}{q}\right) \\
&= \xi\left(z\left(z + \frac{1}{q}\right)^{-1}\right)\xi\left(\frac{1}{q}\right) \\
&= \exp\left[g(0) - g\left(-\frac{1}{q}\right)\right] \cdot \xi\left(\frac{1}{q}\right).
\end{aligned}$$

But this analytic function of q has a limit when $q \to 0$:

$$\lim_{q \to 0} \left[\exp\left[g(0) - g\left(-\frac{1}{q}\right)\right] \cdot \xi\left(\frac{1}{q}\right)\right] = \xi\left(f_0(z)\right) = \exp\left[g(0)\right],$$

so it must be bounded in a neighborhood of 0. We obtain that the function

$$\exp\left[g(0) - g(-x)\right]\xi(x), \qquad x \in \mathbb{C} \setminus \{0\}$$

is bounded in a neighborhood of infinity. The inequality (9.1) implies that $\exp\left[g(0) - g(-x)\right]$ as a function of x has at most polynomial growth at infinity and therefore is a constant. Thus

$$g(x) = C$$

for all $x \in \mathbb{C}$.

Return now to the function $f_q(z)$. As we have observed already $\xi\left(f_q(z)\right)$ has a limit when $q \to 0$. Consequently, $\xi(x)$ has a limit when $x \to \infty$. In particular, $\xi(x)$ is bounded for $|x| \geq 1$:

$$|\xi(x)| \leq M, \qquad for\ |x| \geq 1$$

(the constant $M \geq 1$ is independent of x). But

$$|\xi(x)| = |\xi(x^n)|^{\frac{1}{n}} \leq M^{\frac{1}{n}}$$

for $|x| \geq 1$ and $n = 1, 2, \ldots$, so it must be that

$$|\xi(x)| = 1, \qquad for\ |x| \geq 1.$$

As $\xi(x^{-1}) = (\xi(x))^{-1}$, it follows that

$$|\xi(x)| = 1, \qquad for\ all\ x \in \mathbb{C} \setminus \{0\}.$$

But looking again at the family of functions $f_q(z)$ and recalling that

$$\xi\left(f_q(z)\right) = (const.)\xi\left(\frac{1}{q}\right),$$

we realize that ξ is an analytic function on $\mathbb{C} \setminus \{0\}$. Since the only way an analytic function can have a constant modulus is when the function is a constant, we conclude that $\xi(x) = 1$ for every $x \in \mathbb{C} \setminus \{0\}$.

We have proved that $\xi(x) = 1$ for every $x \in \mathbb{C} \setminus \{0\}$ and that $\xi(z - \alpha) = N$ for all $\alpha \in \mathbb{C}$, where the constant $N \in \mathbb{C} \setminus \{0\}$ is independent of α. But

$$\xi\left(z(qz+1)^{-1}\right) = \begin{cases} 1, & q \neq 0 \\ N, & q = 0. \end{cases}$$

The continuity of ξ now ensures that $N = 1$. Finally, using the group homomorphism property of ξ we conclude that $\xi(z) = 1$ for every $f(z) \in \mathcal{R}$.□

Putting Theorems 9.1 and 9.2 together, we arrive at the following result:

THEOREM 9.3. *Let $\kappa : GL_n(\mathcal{R}) \to GL_n(\mathcal{R})$, where $\mathcal{R} = \mathbb{C}(z)$, be a degree one preserving group automorphism. Furthermore, assume that κ, when restricted to the set of constant invertible matrices, is continuous (the degree one preservation property implies that $\kappa(M)$ is a constant for every constant nonsingular M). By Theorem 9.1, κ must have one of the forms (i)–(iv) in that theorem. Assume, in addition, that ξ is analytic when restricted to the analytic manifold of degree 1 scalar rational functions. Then κ has one of the forms Ia, Ib, IIa or IIb, as described in Section 2.1.*

REFERENCES

[AG] D. Alpay and I. Gohberg, Unitary rational matrix functions, in Topics in Interpolation Theory of Rational Matrix-Valued Functions (ed. by I. Gohberg), Birkhauser OT33, Basel, 1988, pp. 175–222.

[BGR] J. A. Ball, I. Gohberg and L. Rodman, Interpolation of Rational Matrix Functions, Birkhäuser, 1990.

[BGK] H. Bart, I. Gohberg and M. A. Kaashoek, Minimal Factorization of Matrix and Operator Functions, Birkhauser OT1, Basel, 1979.

[BBV] J. W. Brewer, J. W. Bunce and F. S. van Vleck, Linear Systems over Commutative, Rings, Lecture Notes in Pure and Applied Math. Vol 104, Marcel Dekker, New York, 1986.

[D] J. Dieudonné, La Géométrie des Groupes Classiques, Springer–Verlag, Berlin, 1962.

[DE] Yu. M. Dyukarev and V. E. Katsnelśon, Multiplicative and additive classes of Stieltjes analytic matrix-valued functions, and interpolation problems associated with, them I, Amer. Math. Soc. Transl. (2) 131 (1986), 55–70.

[EP] A. V. Efimov and V. P. Potapov, J–expanding matrix–valued functions and their role in the analytic theory of electric circuits, Russian Math. Surveys 28 (1973), 69–140.

[G] F. R. Gantmacher, The Theory of Matrices Vols. I and II, Chelsea, New York, 1959.

[GG] I. Gohberg and S. Goldberg, Finite dimensional Wiener-Hopf equations and factorizations of matrices, Linear Algebra and its Appl. 48 (1982), 219–236.

[GLR] I. Gohberg, P. Lancaster and L. Rodman, Invariant Subspaces of Matrices with Applications, Wiley, New York, 1986.

[KFA] R. E. Kalman, P. L. Falb and M. A. Arbib, Topics in Mathematical System Theory, McGraw Hill, New York, 1969.

[OVW] C. M. Özveren, G. C. Verghese and A. S. Willsky, Asymptotic orders of reachability in perturbed linear systems, IEEE Trans. Auto. Control AC–33 (1988), 915–923.

[R] A. C. M. Ran, Minimal factorizations of selfadjoint rational matrix functions, Integral Equations and Operator Theory 5 (1982), 850–869.

[RR1] A. C. M. Ran and L. Rodman, Stable real invariant semidefinite subspaces and stable factorizations of symmetric rational matrix functions, Linear and Multilinear Algebra 22 (1987), 25–55.

[RR2] A. C. M. Ran and L. Rodman, Stability of neutral invariant subspaces in indefinite inner products and stable symmetric factorizations, Integral Equations and Operator Theory 6 (1983), 536–571.

[RR3] A. C. M. Ran and L. Rodman, Stable invariant Lagrangian subspaces: Factorization of symmetric rational matrix functions and other applications, Lin. Alg. and Appl., to appear.

Daniel Alpay
Department of Theoretical Mathematics
Weizmann Institute of Science
Rehovot, ISRAEL

Joseph A. Ball
Department of Mathematics
Virginia Tech
Blacksburg, VA 24061 USA

Israel Gohberg
School of Mathematical Sciences
Raymond and Beverly Scakler
Faculty of Exact Sciences
Tel Aviv University
Ramat Aviv 69978, ISRAEL

Leiba Rodman
Department of Mathematics
College of William and Mary
Williamsburg, VA 23187–8795 USA

Operator Theory
Advances and Applications, Vol. 47

LOSSLESS INVERSE SCATTERING AND REPRODUCING KERNELS FOR UPPER TRIANGULAR OPERATORS

Daniel Alpay, Patrick Dewilde and Harry Dym

Dedicated to the memory of M.G. Krein,
a mathematical giant.

In this paper the topics mentioned in the title are studied in the algebra of upper triangular bounded linear operators acting on the space $\ell^2_{\mathcal{N}}$ of "square summable" sequences $f = (\ldots, f_{-1}, f_0, f_1, \ldots)$ with components f_i in a complex separable Hilbert space $\mathcal{N}$. Enroute analogues of point evaluation (where in this case the points are operators and the functions are operator valued functions of operators) and simple Blaschke products are developed in this more general context. These tools are then used to establish a theory of structured reproducing kernel Hilbert spaces in the class of upper triangular Hilbert Schmidt operators on $\ell^2_{\mathcal{N}}$. (An application of these tools and spaces to solve a Nevanlinna Pick interpolation problem wherein both the interpolation points and the values assigned at those points are block diagonal operators, will be considered in a future publication.)

CONTENTS

The authors wish to acknowledge with thanks the support of the Department of Electrical Engineering, of the Commission of the EEC under the ESPRIT NANA project and of FOM under DEL.77.160, all at the Delft University of Technology, where a large part of this research was carried out. H. Dym would also like to thank Renee and Jay Weiss for endowing the chair which supports his research at the Weizmann Institute.

1. INTRODUCTION

The classical lossless inverse scattering (LIS) problem of network theory is to find all possible representations of a given Schur function $s(z)$ (i.e., a function which is analytic and contractive in the open unit disc $\mathbb{D}$) in terms of an appropriately restricted class of linear fractional transformations. These linear fractional transformations correspond to lossless, causal, time-invariant two port networks and from this point of view, $s(z)$ may be interpreted as the input transfer function of such a network with a suitable load. More precisely, the sought for representation is of the form

$$s(z) = -\{-A(z)s_L(z) + B(z)\}\{-C(z)s_L(z) + D(z)\}^{-1} , \tag{1.1}$$

where "the load" $s_L(z)$ is again a Schur function and

$$\Theta(z) = \begin{bmatrix} A(z) & B(z) \\ C(z) & D(z) \end{bmatrix} \tag{1.2}$$

is a 2×2 J inner function with respect to the signature matrix

$$J = \begin{bmatrix} 1 & 0 \\ 0 & -1 \end{bmatrix} .$$

This means that Θ is meromorphic in $\mathbb{D}$ and

$$\Theta(z)^* J \Theta(z) \leq J \tag{1.3}$$

for every point $z \in \mathbb{D}$ at which Θ is analytic with equality at almost every point on the boundary $|z| = 1$.

A more general formulation starts with an admissible matrix valued function $X(z) = [a(z) \quad b(z)]$ which is one with entries $a(z)$ and $b(z)$ which are analytic and bounded in $\mathbb{D}$ and in addition are subject to the constraint that, for every n, the $n \times n$ matrix with ij entry equal to

$$\frac{X(z_i)JX(z_j)^*}{1 - z_i z_j^*} , \qquad i, j = 1, \ldots, n ,$$

is positive semidefinite for every choice of points $z_1, \ldots, z_n$ in $\mathbb{D}$. In these terms, the objective is to find all J inner matrix functions $\Theta(z)$ such that

$$[c(z) \quad d(z)] = X(z)\Theta(z)$$

is again admissible. The admissibility constraint guarantees the existence of a pair of Schur functions $s(z)$ and $s_L(z)$ such that $b(z) = a(z)s(z)$ and $d(z) = c(z)s_L(z)$ and hence, leads again to the representation (1.1) at every point $z \in \mathbb{D}$ at which $a(z)$ is invertible and $\Theta(z)$ is analytic. (The minus signs in (1.1) can be eliminated by replacing s by $-s$ and s_L by $-s_L$; the present formulation is kept because it is closer to the physical model.)

Many problems of interpolation, linear least squares estimation and constrained approximation can be fitted into this general framework. Details for both scalar and matrix versions may be found in [DVK], [DD1], [DD2], [DD3] for the case of rational Θ; see [D] for a quick introduction. The more general problem wherein $a(z)$ is a nonsquare matrix is treated in [ADD] for rational Θ which are analytic in the closed disc. The case of general Θ for square invertible $a(z)$ is solved in [AD1] by reproducing kernel methods.

In [DD1]-[DD3] the LIS problem is first solved recursively by a sequence of successive extractions based on the Schur algorithm. These extractions are carried out at arbitrary points in the disc. Deprettere [Dep] observed that the simple Schur algorithm corresponding to extraction at the point zero, could be adapted to invert finite positive definite matrices. In a later refinement with Dewilde [DeDe] it was shown that this adaptation, wherein the multiplication of analytic functions by z is replaced by the shifting of the diagonals of finite upper triangular matrices could be used to generate the maximum entropy extensions introduced by Dym and Gohberg [DG]. This work partially motivated the analysis initiated in [ADe] and subsequently the present rather more general study.

Another motivation rests on the observation that if the LIS problem is studied in the "time domain", then the multiplication

$$y(z) = u(z)h(z)$$

of, say, the analytic functions $u(z)$ and $h(z)$ is replaced by the convolution of their Fourier coefficients:

$$y_k = \sum_{j=k}^{\infty} u_j h_{j-k} \ , \qquad k = 0, 1, \dots \ .$$

The latter can be pictured as the product of an infinite row vector with an infinite upper triangular Toeplitz matrix:

$$[y_0 \ y_1 \ y_2 \cdots] = [u_0 \ u_1 \ u_2 \cdots] \begin{bmatrix} h_0 & h_1 & h_2 & \dots \\ 0 & h_0 & h_1 & \dots \\ 0 & 0 & h_0 & \dots \\ \vdots & \vdots & \vdots & \end{bmatrix}$$

and leads naturally to the question as to what can be said when the Toeplitz assumption: $h_{jk} = h_{j-k}$, is dropped and the entries in the upper triangular matrix are no longer presumed to be constant along diagonals.

The objective of this paper is to develop the theory of the LIS problem in the more general setting of upper triangular operators. An important step is the development of a suitable replacement for the classical Blaschke factor and for elementary J inner factors. These tools are then used to solve an approximation problem for positive operators and to develop a general theory of reproducing kernel Hilbert spaces in the more general setting. In order to explain this properly, it is necessary to explain the setting.

Let $\mathcal{X}$ denote the set of bounded linear operators on the space

$$\ell^2_{\mathcal{N}} = \bigoplus_{i=-\infty}^{\infty} \mathcal{N}_i$$

of "square summable" sequences $f = (\ldots, f_{-1}, f_0, f_1, \ldots)$ with components f_i in a complex separable Hilbert space $\mathcal{N}_i = \mathcal{N}$: $f \in \ell^2_{\mathcal{N}}$ if and only if

$$\|f\|^2_{\ell^2_{\mathcal{N}}} = \sum_{i=-\infty}^{\infty} \|f_i\|^2_{\mathcal{N}} < \infty .$$

Let Z be the shift operator:

$$(Zf)_i = f_{i+1} , \qquad i = \ldots, -1, 0, 1, \ldots ,$$

and let

$$\pi : u \in \mathcal{N} \longrightarrow f \in \ell^2_{\mathcal{N}} \text{ where } \begin{cases} f_0 = u \\ f_i = 0 , & i \neq 0 \end{cases}$$

with adjoint

$$\pi^* : f \in \ell^2_{\mathcal{N}} \longrightarrow f_0 \in \mathcal{N} ,$$

so that

$$\pi^* Z^i f = f_i .$$

It is readily seen that Z is unitary on $\ell^2_{\mathcal{N}}$: $ZZ^* = Z^*Z = I$, and that

$$\pi^* Z^j \pi = \begin{cases} I \text{ (on } \mathcal{N}) & \text{if } j = 0 \\ 0 \text{ (on } \mathcal{N}) & \text{if } j \neq 0 . \end{cases} \tag{1.4}$$

The latter implies in particular that

$$(Z^s\pi)^*(Z^t\pi) = 0 \text{ if } s \neq t$$

and hence that

$$P_{ab} = \sum_{s=a}^{b} Z^s \pi\pi^* Z^{s*}$$

is a selfadjoint projection:

$$P_{ab} = P_{ab}^* = (P_{ab})^2 .$$

Moreover, P_{ab} converges strongly to the identity

$$I = \sum_{s=-\infty}^{\infty} Z^s \pi\pi^* Z^{*s} = \sum_{s=-\infty}^{\infty} Z^{*s} \pi\pi^* Z^s \tag{1.5}$$

in $\ell_{\mathcal{N}}^2$ as $a \downarrow -\infty$ and $b \uparrow \infty$.

Next, we define the space of upper triangular operators

$$\mathcal{U} = \{A \in \mathcal{X} : \ \pi^* Z^i A Z^{*j} \pi = 0 \ \text{ for } \ i > j\} ,$$

the space of lower triangular operators

$$\mathcal{L} = \{A \in \mathcal{X} : \ \pi^* Z^i A Z^{*j} \pi = 0 \ \text{ for } \ i < j\} ,$$

and the space of diagonal operators

$$\mathcal{D} = \mathcal{U} \cap \mathcal{L} .$$

The terminology upper triangular is consistent with picturing $A \in \mathcal{X}$ as a doubly infinite matrix of operators

$$\begin{bmatrix} & \vdots & \vdots & \vdots & \vdots & \\ \cdots & A_{-1,-1} & A_{-1,0} & A_{-1,1} & A_{-1,1} & \cdots \\ \cdots & A_{0,-1} & \boxed{A_{00}} & A_{01} & A_{02} & \cdots \\ & A_{1,-1} & A_{10} & A_{11} & A_{12} & \cdots \\ & \vdots & \vdots & \vdots & \vdots & \end{bmatrix}$$

with

$$A_{ij} = \pi^* Z^i A Z^{*j} \pi ,$$

where the 00 element is distinguished by a surrounding square.

Correspondingly, $f \in \ell_{\mathcal{N}}^2$ should be pictured as a doubly infinite column vector

$$f = \begin{bmatrix} \vdots \\ f_{-1} \\ \boxed{f_0} \\ f_1 \\ \vdots \end{bmatrix} .$$

The formula

$$(Af)_i = \sum_{j=-\infty}^{\infty} A_{ij} f_j \ , \qquad i = \dots, -1, 0, 1, \dots \ ,$$

then fits the usual rules for matrix-vector multiplication.

It follows immediately from (1.4) that $Z \in \mathcal{U}$. In fact if $A = Z$, then

$$A_{ij} = \begin{cases} I & \text{if } j = i+1 \\ 0 & \text{if } j \neq i+1 \end{cases}$$

corresponding to the picture

$$Z = \begin{bmatrix} \ddots & \ddots & & & 0 \\ & 0 & I & & \\ & & \boxed{0} & I & \\ 0 & & & 0 & \ddots \\ & & & & \ddots \end{bmatrix}$$

In the present setting a *linear lossless system* is characterized by a scattering operator S which is unitary and belongs to $\mathcal{U}$, whereas a *two port linear lossless system* is characterized by a 2×2 matrix of operators

$$\Sigma = \begin{bmatrix} \Sigma_{11} & \Sigma_{12} \\ \Sigma_{21} & \Sigma_{22} \end{bmatrix}$$

which is unitary on $\ell^2_{\mathcal{N}} \times \ell^2_{\mathcal{N}}$:

$$\Sigma\Sigma^* = \Sigma^*\Sigma = I \ ,$$

with entries Σ_{ij} in $\mathcal{U}$. In scattering theory, Σ relates a set of input waves $[a^1 \quad b^2] \in \ell^2_{\mathcal{N}} \times \ell^2_{\mathcal{N}}$ to a set of output waves $[b^1 \quad a^2] \in \ell^2_{\mathcal{N}} \times \ell^2_{\mathcal{N}}$ by the rule

$$[a^1 \quad b^2]\Sigma = [b^1 \quad a^2]$$

(where the indicated products follow the rules of matrix multiplication) which in turn corresponds to the schematic diagram depicted in Figure 1.

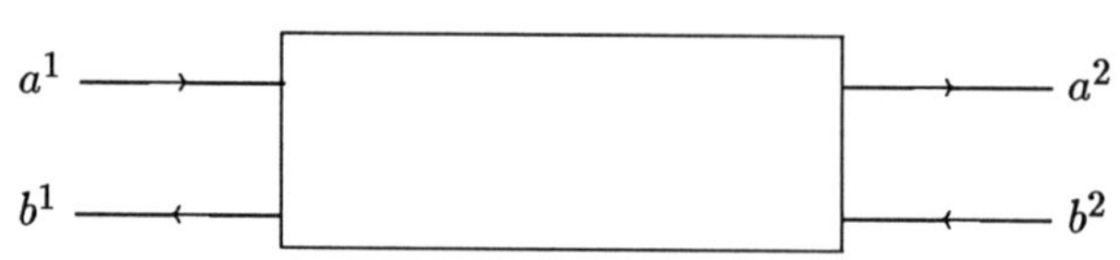

Figure 1.

Σ is a causal norm preserving map of $\ell^2_{\mathcal{N}} \times \ell^2_{\mathcal{N}}$ onto itself. The causality stems from the convention of incorporating the inputs as row vectors applied to the right of Σ. The point is that any operator $S \in \mathcal{U}$ is causal in the sense that if the i-th coefficient a_i of the row vector a vanishes for $i < k$, then the response $(aS)_i = \sum_{j=-\infty}^{\infty} a_j S_{ji} = 0$ for $i < k$ when $S \in \mathcal{U}$.

In this paper we shall work with matrices of operators as inputs and outputs to the lossless system characterized by Σ:

$$[A_1 \quad B_2]\Sigma = [B_1 \quad A_2] ,$$

where the components A_1, B_2 of the input and hence also (as will become clear from the subsequent analysis) the components B_1 and A_2 of the output all belong to $\mathcal{U}$. The components of Σ have physically meaningful interpretations in terms of input and output scattering (or reflection) operators S_i and S_o and input and output transmission operators T_i and T_o:

$$\Sigma = \begin{bmatrix} S_i & T_i \\ T_o & S_o \end{bmatrix} .$$

More precisely, S_i represents the input scattering operator when the input B_2 is perfectly absorbing, i.e., when $B_2 = 0$ and S_o represents the output scattering operator when the input A_1 is perfectly absorbing: $A_1 = 0$.

The chain scattering operator Θ corresponding to Σ is defined by the formula

$$[A_1 \quad B_1]\Theta = [A_2 \quad B_2] .$$

It turns out that Θ exists with components $\Theta_{ij} \in \mathcal{X}$ if and only if $T_o^{-1} \in \mathcal{X}$ and moreover, in this case,

$$\begin{aligned} \Theta &= \begin{bmatrix} I & -S_i \\ 0 & I \end{bmatrix} \begin{bmatrix} T_i & 0 \\ 0 & T_o^{-1} \end{bmatrix} \begin{bmatrix} I & 0 \\ S_o & I \end{bmatrix} \\ &= \begin{bmatrix} T_i - S_i T_o^{-1} S_o & -S_o T_o^{-1} \\ T_o^{-1} S_o & T_o^{-1} \end{bmatrix} \\ &= \begin{bmatrix} T_o^{-1} & -S_i T_o^{-1} \\ -S_i^* T_i^{-*} & T_o^{-1} \end{bmatrix} \end{aligned}$$

and is J unitary for

$$J = \begin{bmatrix} I & 0 \\ 0 & -I \end{bmatrix} :$$

$$\Theta^* J \Theta = \Theta J \Theta^* = J .$$

In this paper we shall confine ourselves to the more restrictive class of chain scattering operators Θ with all its block entries Θ_{ij} in $\mathcal{U}$ and $\Theta_{22}^{-1} \in \mathcal{U}$. We shall distinguish this class by the symbol $\mathcal{A}$, i.e., we shall say that a 2×2 matrix of operators Θ with components Θ_{ij}, $i, j = 1, 2$, belong to $\mathcal{A}$ if and only if

(1) $\Theta_{ij} \in \mathcal{U}$ for $i, j = 1, 2$

(2) $\Theta_{22}^{-1} \in \mathcal{U}$

(3) $\Theta^* J \Theta = \Theta J \Theta^* = J$.

If $\Theta \in \mathcal{A}$, then

$$S_i^* S_i + T_o^* T_o = I$$

from which it follows that $I - S_i$ is invertible in $\mathcal{U}$ and hence that

$$\Phi_i = (I + S_i)(I - S_i)^{-1} ,$$

$$F_i = T_o(I - S_i)^{-1} \quad \text{and}$$

$$R_i = (I - S_i)^{-1} T_i$$

all belong to $\mathcal{U}$. Moreover,

$$\frac{\Phi_i + \Phi_i^*}{2} = F_i^* F_i = R_i R_i^*$$

and Θ can be expressed in the immittance form

$$\Theta = \frac{1}{2} \begin{bmatrix} (I + \Phi_i^*) R_i^{-*} & (I - \Phi_i) F_i^{-1} \\ (I - \Phi_i^*) R_i^{-*} & (I + \Phi_i) F_i^{-1} \end{bmatrix} .$$

Notice that F_i is minimal phase in the sense that both F_i and F_i^{-1} belong to $\mathcal{U}$, whereas R_i is maximal phase in the sense that both R_i and R_i^{-*} belong to $\mathcal{U}$. The immittance form of Θ is useful in the study and approximation of spectral factorization and related approximation problems much as in the classical matrix case which was treated in [DD2], and its adaptation to matrices in [DeDe].

The body of the paper begins in Section 2 with a collection of preliminary facts about the classes $\mathcal{L}$, $\mathcal{U}$ and $\mathcal{D}$ of lower triangular, upper triangular and diagonal operators respectively. In Section 3 the operator valued function $\hat{F}(W)$ is introduced (and explored) for $F \in \mathcal{U}$ and suitably restricted $W \in \mathcal{D}$. Roughly speaking $\hat{F}(W)$ plays the same role as $f(\omega)$ for a conventional function of a complex variable ω and the diagonal operators play the role of scalars. Perhaps the main result of this section is Theorem 3.4 which sharpens this analogy. An analogue of the classical elementary

Blaschke factor for $\mathbb{D}$ is introduced in Section 4 and then used in Section 5 to solve a one step LIS problem in the present setting. The latter also contains some preliminaries on linear fractional transformations. Section 6 treats an approximation problem and Section 7 introduces reproducing kernel Hilbert spaces in the setting of Hilbert Schmidt operators. The subsequent analysis depends heavily upon the two basic formulas (7.1) (which is roughly speaking the counterpart of Cauchy's formula in conventional Hardy space theory) and (7.2) (which is the counterpart of an important evaluation in classical interpolation theory; see e.g. Lemma 2.1 of [D]). Section 8 is devoted to the theory of operator ranges which provides an elegant framework for studying reproducing kernel spaces and for defining admissibility.

A word on usage: We shall say that an operator A on a complex Hilbert space $\mathcal{H}$ with inner product $\langle\ ,\ \rangle_{\mathcal{H}}$ is positive definite if $\langle Ax, x\rangle_{\mathcal{H}} > 0$ for every nonzero $x \in \mathcal{H}$ and positive semidefinite if $\langle Ax, x\rangle_{\mathcal{H}} \geq 0$ for every $x \in \mathcal{H}$.

2. PRELIMINARIES

In this section we shall establish a number of elementary properties of the spaces $\mathcal{L}$, $\mathcal{D}$ and $\mathcal{U}$, which will prove useful in the sequel.

LEMMA 2.1.

(1) $Z\mathcal{U} \subset \mathcal{U}$ *and* $Z^*\mathcal{L} \subset \mathcal{L}$.

(2) $\mathcal{U}Z \subset \mathcal{U}$ *and* $\mathcal{L}Z^* \subset \mathcal{L}$.

(3) $\mathcal{U} \cdot \mathcal{U} \subset \mathcal{U}$, $\mathcal{L} \cdot \mathcal{L} \subset \mathcal{L}$ *and* $\mathcal{D} \cdot \mathcal{D} \subset \mathcal{D}$.

(4) $\mathcal{L}^* = \mathcal{U}$.

PROOF. Let A and B belong to $\mathcal{U}$. Then

$$\pi^* Z^i A Z^{*s} \pi = 0 \quad \text{for} \quad s < i$$

and

$$\pi^* Z^s B Z^{*j} \pi = 0 \quad \text{for} \quad s > j.$$

Therefore, using (1.5),

$$\pi^* Z^i A B Z^{*j} \pi = \sum_{s=-\infty}^{\infty} \pi^* Z^i A Z^{*s} \pi \pi^* Z^s B Z^{*j} \pi = 0$$

for $i > j$. This proves the first part of (3) and hence also the first halves of (1) and (2) since $Z \in \mathcal{U}$.

The second parts of (1)–(3) are immediate from the corresponding first halves and (4), which in turn is selfevident from the definitions of $\mathcal{U}$ and $\mathcal{L}$.

∎

LEMMA 2.2.

(1) $\mathcal{D} = \{A \in \mathcal{X}: \pi^* Z^i A Z^{*j} \pi = 0 \text{ for } i \neq j\} = \mathcal{D}^*$.

(2) $Z^*\mathcal{D}Z = \mathcal{D}$.

(3) $Z\mathcal{D}Z^* = \mathcal{D}$.

(4) *If $D \in \mathcal{D}$ and D is invertible in $\mathcal{X}$, then $D^{-1} \in \mathcal{D}$.*

(5) *If $D \in \mathcal{D}$ and $D \geq 0$, then $D^{\frac{1}{2}} \in \mathcal{D}$.*

PROOF. Assertion (1) is immediate from the definitions of $\mathcal{L}$, $\mathcal{U}$ and $\mathcal{D}$. The inclusions

$$Z^*\mathcal{D}Z \subset \mathcal{D} \quad \text{and} \quad Z\mathcal{D}Z^* \subset \mathcal{D}$$

drop out easily from (1) and together yield (2) and (3).

Next, let $B = D^{-1}$. Then, since $D \in \mathcal{D}$,

$$(DB)_{tt} = D_{tt}B_{tt} = I \quad \text{and} \quad (BD)_{tt} = B_{tt}D_{tt} = I\ ,$$

and therefore the operator $C \in \mathcal{D}$ with $C_{tt} = B_{tt}$ inverts D. Thus, by the uniqueness of the inverse, $B = C$ and hence $D^{-1} \in \mathcal{D}$. This proves (4).

Finally, if $D \geq 0$, then each of its block components D_{tt} is positive semidefinite on $\mathcal{N}$ and hence admits a unique positive semidefinite square root $E_{tt} = D^{\frac{1}{2}}$. The bound

$$\begin{aligned} \|E_{tt}f\|_{\mathcal{N}}^2 &= \langle D_{tt}f, f\rangle_{\mathcal{N}} \\ &\leq \|D_{tt}\|\ \|f\|_{\mathcal{N}}^2 \end{aligned}$$

insures that the diagonal operator E with block components E_{tt} is bounded and hence belongs to $\mathcal{D}$. This establishes (5) since E is clearly a positive semidefinite square root of D and there is only one such.

∎

Next we introduce the transformations

$$A^{(j)} = Z^{*j}AZ^j$$

for $A \in \mathcal{X}$ and every integer $j = \ldots, -1, 0, 1, \ldots$. Clearly

$$(AB)^{(j)} = A^{(j)}B^{(j)} \tag{2.1}$$

and

$$A^{(j+k)} = \{A^{(j)}\}^{(k)}. \tag{2.2}$$

LEMMA 2.3. *For any integer j:*

(1) $A \in \mathcal{U}$ *if and only if* $A^{(j)} \in \mathcal{U}$.

(2) $A \in \mathcal{L}$ *if and only if* $A^{(j)} \in \mathcal{L}$.

(3) $A \in \mathcal{D}$ *if and only if* $A^{(j)} \in \mathcal{D}$.

PROOF. This is an easy consequence of the definitions of $\mathcal{U}$, $\mathcal{L}$ and $\mathcal{D}$. ∎

LEMMA 2.4.

(1) $\mathcal{L} \cap Z\mathcal{U} = 0$.

(2) $(Z\mathcal{U})^* = Z^*\mathcal{L}$.

(3) $Z^*\mathcal{L} \cap \mathcal{U} = 0$.

PROOF. Let $A \in \mathcal{L} \cap Z\mathcal{U}$. Then, since $Z^*A \in \mathcal{U}$,

$$\pi^* Z^i Z^* A Z^{*j} \pi = \pi^* Z^{i-1} A Z^{*j} \pi = 0 \quad \text{for} \quad i > j,$$

which implies in turn that

$$\pi^* Z^i A Z^{*j} \pi = 0 \quad \text{for} \quad i \geq j.$$

Thus, since $A \in \mathcal{L}$ also

$$\pi^* Z^i A Z^{*j} \pi = 0 \quad \text{for all} \quad i, j.$$

This proves (1).

Suppose next that $B \in Z^*\mathcal{L}$. Then $(ZB)^* = B^*Z^*$ belongs to $\mathcal{U}$, as does

$$Z^*(B^*Z^*)Z = Z^*B^*,$$

by Lemma 2.3. Thus $B^* \in Z\mathcal{U}$, i.e.,

$$(Z^*\mathcal{L})^* \subset Z\mathcal{U}.$$

Since the opposite inclusion goes through in much the same way, (2) may be considered proved.

Finally, (3) is immediate from (1) and (2).

∎

LEMMA 2.5. *Let* $A \in \mathcal{X}$. *Then* $\|A_{ij}\| \leq \|A\|$.

PROOF. Let $u \in \mathcal{N}$. Then

$$\begin{aligned} \|A_{ij}u\|_{\mathcal{N}} &= \|\pi^* Z^i A Z^{*j} \pi u\|_{\mathcal{N}} \\ &\leq \|Z^i A Z^{*j} \pi u\|_{\ell^2_{\mathcal{N}}} \\ &= \|A Z^{*j} \pi u\|_{\ell^2_{\mathcal{N}}} \\ &\leq \|A\| \|Z^{*j} \pi u\|_{\ell^2_{\mathcal{N}}} \\ &= \|A\| \|\pi u\|_{\ell^2_{\mathcal{N}}} \\ &= \|A\| \|u\|_{\mathcal{N}}. \end{aligned}$$

∎

LEMMA 2.6. *If* $B \in \mathcal{D}$, *then*

$$\|B\| = \sup_i \|B_{ii}\|.$$

PROOF. Let

$$\beta = \sup_i \|B_{ii}\|.$$

Then $\beta < \infty$ by Lemma 2.5 and clearly

$$\begin{aligned} \|Bf\|^2_{\ell^2_{\mathcal{N}}} &= \sum_{i=-\infty}^{\infty} \|B_{ii} f_i\|^2_{\mathcal{N}} \\ &\leq \beta^2 \sum_{i=-\infty}^{\infty} \|f_i\|^2_{\mathcal{N}} \\ &= \beta^2 \|f\|^2_{\ell^2_{\mathcal{N}}}. \end{aligned}$$

This proves that

$$\|B\| \leq \beta.$$

Now, if B is the zero operator, then the asserted equality is plain. Suppose therefore that B is nonzero. Then for any $\varepsilon > 0$ there exists an integer i such that

$$\|B_{ii}\| \geq \beta - \varepsilon.$$

For this choice of i, there exists a $u \in \mathcal{N}$ with $\|u\|_{\mathcal{N}} = 1$ such that

$$\|B_{ii} u\|_{\mathcal{N}} \geq \|B_{ii}\| - \varepsilon.$$

Thus if $f \in \ell^2_{\mathcal{N}}$ has its i'th component equal to u and all its other components equal to zero, it follows that

$$\|Bf\|_{\ell^2_{\mathcal{N}}} = \|B_{ii}u\|_{\mathcal{N}} \geq \beta - 2\varepsilon.$$

Therefore, since $\|f\|_{\ell^2_{\mathcal{N}}} = 1$, this proves that

$$\|B\| \geq \beta - 2\varepsilon,$$

and hence since ε is an arbitrary positive number, the proof is complete.

∎

For $A \in \mathcal{X}$, let $A_{[j]}$ denote the diagonal operator with block entries

$$(A_{[j]})_{ii} = A_{i-j,i}. \tag{2.3}$$

LEMMA 2.7. *If* $A \in \mathcal{X}$, *then*

$$\|A_{[j]}\| = \sup_i \|A_{i-j,i}\| \leq \|A\|.$$

PROOF. This is an immediate corollary of Lemmas 2.6 and 2.5.

∎

LEMMA 2.8. $\mathcal{U} = \mathcal{D} \dot{+} Z\mathcal{U}$ *and* $\mathcal{L} = \mathcal{D} \dot{+} Z^*\mathcal{L}$.

PROOF. Let $A \in \mathcal{U}$. Then clearly

$$A = A_{[0]} + (A - A_{[0]}),$$

where $A_{[0]}$ is the diagonal component of A which is defined in (2.3). $A_{[0]} \in \mathcal{D}$ (i.e., it is bounded) by Lemma 2.7 and

$$\begin{aligned}\pi^* Z^i\{Z^*(A - A_{[0]})\}Z^{*j}\pi &= \pi^* Z^{i-1}(A - A_{[0]})Z^{*j}\pi \\ &= \begin{cases} \pi^* Z^{i-1} A Z^{*j}\pi & \text{for} \quad i > j+1 \\ 0 & \text{for} \quad i = j+1, \end{cases}\end{aligned}$$

by the definition of $A_{[0]}$. Since $A \in \mathcal{U}$,

$$\pi^* Z^{i-1} A Z^{*j}\pi = 0$$

for $i > j + 1$. Thus

$$A - A_{[0]} \in Z\mathcal{U}.$$

Since also $\mathcal{D} \cap Z\mathcal{U} = 0$ by Lemma 2.4, the indicated sum decomposition of $A \in \mathcal{U}$ is direct. This completes the proof of the first assertion. The second follows from the first by taking adjoints.

∎

LEMMA 2.9. *Let $A \in \mathcal{U}$. Then there exists a unique set of diagonal operators $D_j \in \mathcal{D}$, $j = 0, 1, \ldots$ such that*

$$A - \sum_{j=0}^{n-1} Z^j D_j \in Z^n \mathcal{U}\,, \qquad n = 1, \ldots \tag{2.4}$$

In fact

$$D_j = A_{[j]}, \qquad j = 0, 1, \ldots,$$

where the $A_{[j]}$ are defined by (2.3).

PROOF. By a double application of Lemma 2.8,

$$A - D_0 = ZB_1$$

for some choice of $D_0 \in \mathcal{D}$ and $B_1 \in \mathcal{U}$, and

$$B_1 = D_1 + ZB_2$$

for some choice of $D_1 \in \mathcal{D}$ and $B_2 \in \mathcal{U}$. Thus

$$A - D_0 - ZD_1 = Z^2 B_2.$$

Formula (2.4) drops out by iterating this procedure.

Next, it follows from (2.4), that

$$A_{s-j,s} = \pi^* Z^{s-j} A Z^{*s} \pi = \pi^* Z^s D_j Z^{*s} \pi = (D_j)_{ss}$$

and hence that $D_j = A_{[j]}$, as claimed.

∎

LEMMA 2.10. *If A is a positive semidefinite operator in $\mathcal{X}$ such that its main diagonal vanishes (i.e., $\pi^* Z^t A Z^t \pi = 0$ for $t = 0, \pm 1, \ldots$), then $A = 0$.*

PROOF. If $A \geq 0$ on $\ell^2_{\mathcal{N}}$ and $A_{tt} = 0$ for $t = 0, \pm 1, \ldots$, then for any $\epsilon > 0$, the operator matrix

$$\begin{bmatrix} \epsilon I & A_{st} \\ A_{ts} & 0 \end{bmatrix} \geq \begin{bmatrix} 0 & A_{st} \\ A_{ts} & 0 \end{bmatrix} = \begin{bmatrix} A_{ss} & A_{st} \\ A_{ts} & A_{tt} \end{bmatrix} \geq 0\,.$$

Therefore, the Schur complement,

$$0 - A_{ts}\epsilon^{-1}A_{st} \geq 0$$

and hence, since $A_{st} = A_{ts}^*$, we must have $A_{st} = 0$ for every choice of s and t.

∎

LEMMA 2.11. *$\mathcal{U}$, $\mathcal{L}$ and $\mathcal{D}$ are closed subspaces of $\mathcal{X}$ with respect to pointwise convergence: if $A_1, A_2, \ldots$ belong to $\mathcal{U}$ [resp. $\mathcal{L}$, $\mathcal{D}$] and*

$$\|A_n f - Bf\| \to 0 \quad \text{as} \quad n \uparrow \infty$$

for some linear operator B and every $f \in \ell^2_{\mathcal{N}}$, then B belongs to $\mathcal{U}$ [resp. $\mathcal{L}$, $\mathcal{D}$].

PROOF. It follows from the given assumption that

$$\sup_n \|A_n f\| \leq M_f < \infty$$

for every $f \in \ell^2_{\mathcal{N}}$. Therefore, by the uniform boundedness principle,

$$\sup_n \|A_n\| \leq M < \infty ,$$

and hence $B \in \mathcal{X}$. Suppose next that $A_1, A_2, \ldots \in \mathcal{U}$. Then, for any choice of $u \in \mathcal{N}$ and $i > j$,

$$\begin{aligned} \|\pi^* Z^i B Z^{*j} \pi u\|_{\mathcal{N}} &= \|\pi^* Z^i (B - A_n) Z^{*j} \pi u\|_{\mathcal{N}} \\ &\leq \|Z^i (B - A_n) Z^{*j} \pi u\|_{\ell^2_{\mathcal{N}}} \\ &= \|(B - A_n) Z^{*j} \pi u\|_{\ell^2_{\mathcal{N}}} . \end{aligned}$$

Therefore, since the exhibited upper bound tends to zero as $n \uparrow \infty$, it follows that $B \in \mathcal{U}$ also.

The same argument works in $\mathcal{L}$ and in $\mathcal{D}$ also.

∎

LEMMA 2.12. *Each of the spaces $\mathcal{L}$, $\mathcal{D}$ and $\mathcal{U}$ is a Banach subalgebra of (the Banach algebra) $\mathcal{X}$ with identity.*

PROOF. Lemma 2.11 guarantees that $\mathcal{L}$, $\mathcal{D}$ and $\mathcal{U}$ are each closed in operator norm: if $\|A_n - B\| \to 0$ as $n \uparrow \infty$ and the A_n all belong to one of these three spaces, then so does B. The rest is immediate from (3) of Lemma 2.1.

∎

3. A PAIR OF TRANSFORMS

Recall that in Section 2 we showed that for every $F \in \mathcal{U}$, there is a unique set of operators $F_{[j]} \in \mathcal{D}$, $j = 0, 1, \ldots$, such that

$$F - \sum_{j=0}^{n-1} Z^j F_{[j]} \in Z^n \mathcal{U}.$$

Correspondingly we now introduce the operator transform

$$\hat{F}(W) = \sum_{j=0}^{\infty} W^{[j]} F_{[j]} \tag{3.1}$$

based on the sequence of operators

$$W^{[0]} = I \quad \text{and} \quad W^{[j+1]} = W(W^{[j]})^{(1)} \quad \text{for} \quad j = 0, 1, \ldots$$

for any $W \in \mathcal{X}$ for which the indicated sum converges.

Let

$$r(W) = \max\{|\lambda|: \lambda \in \sigma(W)\}$$

denote the spectral radius of W. It is well known that

$$r(W) = \lim_{n\uparrow\infty} \|W^n\|^{1/n},$$

where the limit exists for every $W \in \mathcal{X}$.

LEMMA 3.1.*If $W \in \mathcal{X}$, then*

$$r(WZ^*) = \lim_{n\uparrow\infty} \|W^{[n]}\|^{1/n} = r(W^*Z).$$

PROOF. Clearly

$$(WZ^*)^n = W^{[n]} Z^{*n}$$

and hence, by the definition of the spectral radius,

$$\begin{aligned} r(WZ^*) &= \lim_{n\uparrow\infty} \|W^{[n]} Z^{*n}\|^{1/n} \\ &= \lim_{n\uparrow\infty} \|W^{[n]}\|^{1/n}, \end{aligned}$$

since Z is unitary. This establishes the first formula. The second comes from the first and the observation that

$$\|(WZ^*)^n\| = \|(ZW^*)^n\| = \|(W^*Z)^n\|,$$

where the second equality in the preceding line depends upon the fact that Z is unitary. ∎

Lemma 3.1 guarantees that the notation

$$\ell_W = \lim_{n\uparrow\infty} \|W^{[n]}\|^{1/n}$$

is meaningful for every $W \in \mathcal{X}$, i.e., that the indicated limit exists. In general

$$\ell_W \leq \|W\|$$

and

$$\ell_W = r(W) \quad \text{if} \quad W = W^{(1)},$$

i.e., if W is Toeplitz. The condition $\ell_W < 1$ guarantees the convergence of $\hat{F}(W)$ for $F \in \mathcal{U}$ because, as follows from the elementary inequality

$$\|W^{[j]}F_{[j]}\| \leq \|W^{[j]}\|\|F\|,$$

$$\limsup_{j\uparrow\infty} \|W^{[j]}F_{[j]}\|^{1/j} \leq \ell_W.$$

In the sequel we shall be particularly interested in $\hat{F}(W)$ for $W \in \mathcal{D}$.

THEOREM 3.1. *Let $F \in \mathcal{U}$. Then*

$$\|\hat{F}(W)\| \leq \sup_i \left\{ \sum_{j=0}^{\infty} \|W^{[j]}\|\|F_{i-j,i}\| : i = 0, \pm 1, \ldots \right\} < \infty$$

for every choice of $W \in \mathcal{D}$ with $\ell_W < 1$. In particular,

$$\sup_{W\in\mathcal{D}} \{\|\hat{F}(W)\| : \|W\| < 1\} \leq \sup_i \left\{ \sum_{j=0}^{\infty} \|F_{i-j,i}\| : i = 0, 1, \ldots \right\}$$

with equality if $\mathcal{N} = \mathbb{C}$.

PROOF. By Lemma 2.6,

$$\|\hat{F}(W)\| = \sup_i \{\|\hat{F}(W)_{ii}\| : i = 0, \pm 1, \ldots\}$$

for any choice of $W \in \mathcal{D}$ with $\ell_W < 1$. Therefore, since $W^{[j]}$ and $F_{[j]}$ belong to $\mathcal{D}$,

$$\begin{aligned}
\|\hat{F}(W)_{ii}\| &= \left\| \sum_{j=0}^{\infty} (W^{[j]})_{ii} (F_{[j]})_{ii} \right\| \\
&\leq \sum_{j=0}^{\infty} \|W^{[j]}\|\|F_{i-j,i}\|,
\end{aligned}$$

which establishes the first of the asserted inequalities. The second is immediate from the first.

Finally, if $\mathcal{N} = \mathbb{C}$ we can choose $W_{kk} = \rho e^{i\alpha_k}$, with $0 < \rho < 1$ in such a way that

$$|\hat{F}(W)_{ii}| = |F_{ii}| + \rho|F_{i-1,i}| + \rho^2|F_{i-2,i}| + \cdots .$$

Thus

$$\sup_{W\in\mathcal{D}} \{\|\hat{F}(W)\|: \|W\| < 1\} \geq \sum_{j=0}^{\infty} \rho^j |F_{i-j,i}|$$

for $0 < \rho < 1$ which implies in turn that

$$\sup_{W\in\mathcal{D}} \{\|\hat{F}(W)\|: \|W\| < 1\} \geq \sum_{j=0}^{\infty} |F_{i-j,i}|$$

for every integer i. This completes the proof. ∎

LEMMA 3.2. *If* $F \in \mathcal{U}$, *then*

$$(ZF)^{\wedge}(W) = W\hat{F}(W^{(1)}) \tag{3.2}$$

and

$$(FZ)^{\wedge}(W) = W\{\hat{F}(W)\}^{(1)} \tag{3.3}$$

for every choice of $W \in \mathcal{D}$ *with* $\ell_W < 1$.

PROOF. The diagonal operators $F_{[j]}$ are characterized by the relationship

$$F - \sum_{j=0}^{n-1} Z^j F_{[j]} \in Z^n\mathcal{U}.$$

But this in turn implies that

$$ZF - \sum_{j=0}^{n-1} Z^{j+1} F_{[j]} \in Z^{n+1}\mathcal{U}$$

and hence that

$$(ZF)_{[0]} = 0 \quad \text{and} \quad (ZF)_{[j]} = F_{[j-1]} \quad \text{for} \quad j \geq 1.$$

Therefore

$$(ZF)^{\wedge}(W) = \sum_{j=1}^{\infty} W^{[j]} F_{[j-1]}$$
$$= \sum_{j=0}^{\infty} W^{[j+1]} F_{[j]}$$
$$= W \sum_{j=0}^{\infty} (W^{[j]})^{(1)} F_{[j]},$$

which is the same as the right hand side of (3.2).

Next, since

$$F - \sum_{j=0}^{n-1} Z^j F_{[j]} = Z^n B$$

with $B \in \mathcal{U}$,

$$FZ - \sum_{j=0}^{n-1} Z^j F_{[j]} Z = Z^n B Z = Z^{n+1} B^{(1)} \in Z^{n+1} \mathcal{U}$$

and

$$F_{[j]} Z = Z (F_{[j]})^{(1)},$$

it follows much as before that

$$(FZ)_{[0]} = 0 \quad \text{and} \quad (FZ)_{[j]} = (F_{[j-1]})^{(1)} \quad \text{for} \quad j \geq 1.$$

Therefore

$$(FZ)^{\wedge}(W) = \sum_{j=0}^{\infty} W^{[j+1]} (F_{[j]})^{(1)}$$
$$= W \sum_{j=0}^{\infty} (W^{[j]})^{(1)} (F_{[j]})^{(1)},$$

which agrees with the right hand side of (3.3).

∎

LEMMA 3.3. *If V and W belong to $\mathcal{D}$ and $\ell_V < 1$, then*

$$(FW)^{\wedge}(V) = \hat{F}(V) W, \tag{3.4}$$

and

$$(WF)^{\wedge}(V) = \sum_{j=0}^{\infty} V^{[j]} W^{(j)} F_{[j]} \tag{3.5}$$

for every $F \in \mathcal{U}$.

PROOF. The first formula is an easy calculation since

$$(FW)_{[j]} = F_{[j]}W,$$

whereas the second rests on the identification

$$(WF)_{[j]} = W^{(j)}F_{[j]}.$$

∎

In general (3.5) cannot be simplified unless extra constraints are imposed on W. Two particular cases of interest are summarized in the next two lemmas.

LEMMA 3.4. *If* $V \in \mathcal{D}$ *with* $\ell_V < 1$ *and* $F \in \mathcal{U}$, *then*

$$(VF)^{\wedge}(V) = V\hat{F}(V^{(1)}). \tag{3.6}$$

PROOF. By (3.5),

$$\begin{aligned}(VF)^{\wedge}(V) &= \sum_{j=0}^{\infty} V^{[j+1]}F_{[j]} \\ &= V\sum_{j=0}^{\infty}\{V^{(1)}\}^{[j]}F_{[j]}.\end{aligned}$$

∎

LEMMA 3.5. *If* V *and* $W \in \mathcal{D}$ *and* $\ell_V < 1$ *and* $W^{-1} \in \mathcal{D}$, *then*

$$(WF)^{\wedge}(V) = W\hat{F}(W^{-1}VW^{(1)}) \tag{3.7}$$

for every $F \in \mathcal{U}$.

PROOF. The proof is immediate from (3.5) and the observation that if W is invertible in $\mathcal{D}$, then

$$V^{[j]}W^{(j)} = W\{W^{-1}VW^{(1)}\}^{[j]}.$$

∎

THEOREM 3.2. *If* $F \in \mathcal{U}$, *then* $\hat{F}(V) = 0$ *for every* $V \in \mathcal{D}$ *with* $\ell_V < 1$ *if and only if* $F = 0$.

PROOF. Suppose that $\hat{F}(V) = 0$ for every $V \in \mathcal{D}$ with $\ell_V < 1$. Then

$$F_{[0]} = \hat{F}(0) = 0.$$

Next, fix an integer k and choose $V \in \mathcal{D}$ such that

$$V_{ii} = \begin{cases} I & \text{if} \quad i = k \\ 0 & \text{if} \quad i \neq k. \end{cases}$$

Then $V^{[j]} = 0$ for $j \geq 2$ and hence

$$\hat{F}(V) = VF_{[1]} = (F_{[1]})_{kk} = F_{k-1,k} = 0.$$

This proves that $F_{[1]} = 0$, since k is arbitrary.

The next step is to again fix an integer k, but this time to choose $V \in \mathcal{D}$ such that

$$V_{ii} = \begin{cases} I & \text{if } i = k \text{ or } i = k-1 \\ 0 & \text{otherwise.} \end{cases}$$

Then

$$(V^{[2]})_{ii} = \begin{cases} I & \text{if} \quad i = k \\ 0 & \text{if} \quad i \neq k \end{cases}$$

whereas $V^{[j]} = 0$ for $j > 2$ and hence, for this choice of V,

$$\hat{F}(V) = V^{[2]}F_{[2]} = (F_{[2]})_{kk} = F_{k-2,k} = 0.$$

Thus, since k is arbitrary, $F_{[2]} = 0$. Moreover, the same argument can clearly be extended to prove that $F_{[j]} = 0$ for all j and hence that $F = 0$. This completes the proof in one direction. The other is selfevident.

∎

LEMMA 3.6. *Let $W \in \mathcal{D}$. Then, for every positive integer n,*

$$Z^n - W^{[n]} = (Z - W)P_n(Z),$$

where $P_n(Z)$ is a polynomial in Z with coefficients in $\mathcal{D}$ such that

$$\|P_n(Z)\| \leq 1 + \|W\| + \cdots + \|W\|^{n-1}.$$

PROOF. The statement is clearly true for $n = 1$. Suppose that it is true for $n = k$. Then

$$\begin{aligned} Z^{k+1} - W^{[k+1]} &= (Z^k - W^{[k]})Z + W^{[k]}Z - W^{[k+1]} \\ &= (Z - W)P_k(Z)Z + (Z - W)(W^{[k]})^{(1)} \\ &= (Z - W)\{P_k(Z)Z + (W^{[k]})^{(1)}\}. \end{aligned}$$

This establishes the first assertion by induction and serves further to exhibit the recursion

$$P_{k+1}(Z) = P_k(Z)Z + (W^{[k]})^{(1)}.$$

The rest is plain since $P_1(Z) = I$.

∎

THEOREM 3.3. *If $F \in \mathcal{U}$, $D \in \mathcal{D}$ and $W \in \mathcal{D}$ with $\ell_W < 1$, then the operator $(Z-W)^{-1}(F-D)$ belongs to $\mathcal{U}$ if and only if $D = \hat{F}(W)$.*

PROOF. We first show that $(Z-W)^{-1}(F-D) \in \mathcal{U}$ for at most one choice of $D \in \mathcal{D}$. To this end suppose that both $(Z-W)^{-1}(F-D_1)$ and $(Z-W)^{-1}(F-D_2)$ belong to $\mathcal{U}$ for D_1 and D_2 in $\mathcal{D}$. Then $(Z-W)^{-1}(D_2-D_1) \in \mathcal{U}$.

On the other hand, since

$$\begin{aligned}(Z-W)^{-1} &= \{(I-WZ^*)Z\}^{-1}\\ &= Z^*\sum_{n=0}^{\infty}(WZ^*)^n \in Z^*\mathcal{L},\end{aligned}$$

we see that

$$(Z-W)^{-1}(D_2-D_1) \in Z^*\mathcal{L}\cap\mathcal{U} = \{0\}.$$

Therefore $D_2 = D_1$.

The next step is to show that there exists at least one $D \in \mathcal{D}$ such that $(Z-W)^{-1}(F-D) \in \mathcal{U}$: $D = \hat{F}(W)$. It suffices to show that

$$\pi^* Z^i (Z-W)^{-1}\{F-\hat{F}(W)\}Z^{*j}\pi u = 0$$

for every choice of $i > j$ and $u \in \mathcal{N}$. To this end it is convenient to let

$$F_n = \sum_{j=0}^{n} Z^j F_{[j]}$$

and to also introduce the shorthand notation

$$\varphi_{ij}(G) = \pi^* Z^i (Z-W)^{-1} G Z^{*j}\pi u$$

for $G \in \mathcal{X}$. Then

$$\varphi_{ij}(F-\hat{F}(W)) = \varphi_{ij}(F_n - \hat{F}_n(W)) + \varphi_{ij}(F-F_n) + \varphi_{ij}(\hat{F}_n(W) - \hat{F}(W)).$$

Now, for $i > j$, the first of the three terms on the right is clearly equal to zero by Lemma 3.6, whereas the norm of the third is subject to the bound

$$\|\varphi_{ij}(\hat{F}_n(W) - \hat{F}(W))\|_{\mathcal{N}} \leq \|(Z - W)^{-1}\| \sum_{t=n+1}^{\infty} \|W^{[t]}\|\|F\|\|u\|_{\mathcal{N}},$$

which clearly tends to zero as $n \uparrow \infty$ since

$$\sum_{t=0}^{\infty} \|W^{[t]}\| < \infty \quad \text{when} \quad \ell_W < 1.$$

Finally, since

$$\|\varphi_{ij}(F - F_n)\|_{\mathcal{N}} \leq \|(Z - W)^{-1}\|\|(F - F_n)Z^{*j}\pi u\|_{\ell^2_{\mathcal{N}}},$$

it remains to show that the upper bound in the last inequality tends to zero as $n \uparrow \infty$. But this follows from the identifications

$$\begin{aligned} \|(F - F_n)Z^{*j}\pi u\|^2_{\ell^2_{\mathcal{N}}} &= \sum_{t=-\infty}^{j} \|\pi^* Z^t (F - F_n) Z^{*j}\pi u\|^2_{\mathcal{N}} \\ &= \sum_{t=-\infty}^{j} \|(F - F_n)_{tj} u\|^2_{\mathcal{N}} \\ &= \sum_{t=-\infty}^{j-(n+1)} \|F_{tj} u\|^2_{\mathcal{N}} , \end{aligned}$$

since the full sum is finite. ∎

COROLLARY. *If $F \in \mathcal{U}$ and $W \in \mathcal{D}$ with $\ell_W < 1$, then $\hat{F}(W) = 0$ if and only if $F \in (Z - W)\mathcal{U}$.*

LEMMA 3.7. *If F and $G \in \mathcal{U}$ and $W \in \mathcal{D}$ with $\ell_W < 1$, then*

$$(FG)^{\wedge}(W) = \{\hat{F}(W)G\}^{\wedge}(W).$$

PROOF. By Theorem 3.3,

$$\begin{aligned} (\{F - \hat{F}(W)\}G)^{\wedge}(W) &= \{(Z - W)F_1 G\}^{\wedge}(W) \\ &= \{(Z - W)E\}^{\wedge}(W) \end{aligned}$$

for some choice of $F_1 \in \mathcal{U}$ and $E = F_1 G \in \mathcal{U}$. But this is equal to zero by the preceding corollary.

∎

COROLLARY 2. *If $F \in \mathcal{U}$ and F is invertible in $\mathcal{U}$, then $\hat{F}(W) \neq 0$ for every $W \in \mathcal{D}$ with $\ell_W < 1$.*

PROOF. By Lemma 3.7,

$$\{\hat{F}(W)F^{-1}\}^{\wedge}(W) = (FF^{-1})^{\wedge}(W) = I\ .$$

∎

For the sake of completeness we introduce the auxiliary transform

$$F^{\Delta}(W) = \sum_{j=0}^{\infty} \{F_{[j]}\}^{(-j)}\{W^{[j]}\}^{(1-j)}$$

for $F \in \mathcal{U}$. It also clearly converges in norm if $\ell_W < 1$, because

$$\begin{aligned} \|\{F_{[j]}W^{[j]}\}^{(-j)}\| &= \|F_{[j]}W^{[j]}\| \\ &\leq \|F\|\|W^{[j]}\|. \end{aligned}$$

This transform plays the same role for division on the right by $Z - W$ that $F^{\wedge}(W)$ does for division on the left:

THEOREM 3.4. *If $F \in \mathcal{U}$, $D \in \mathcal{D}$ and $W \in \mathcal{D}$ with $\ell_W < 1$, then the operator $(F - D)(Z - W)^{-1}$ belongs to $\mathcal{U}$ if and only if $D = F^{\Delta}(W)$.*

PROOF. The proof is much the same as the proof of Theorem 3.3 except that now a key role is played by the identity

$$Z^{n+1} - (V^{[n+1]})^{(-n)} = \left\{\sum_{j=0}^{n} Z^{n-j}(V^{[j]})^{(-j)}\right\}(Z - V).$$

The details are left to the reader.

∎

4. AN ELEMENTARY BLASCHKE FACTOR

4.1. Motivation.

The simplest elementary Blaschke factor is of the form

$$U = (I - WW^*)^{-\frac{1}{2}}(Z - W)(I - W^*Z)^{-1}(I - W^*W)^{\frac{1}{2}}\ ,$$

where $W \in \mathcal{D}$ and $\|W\| < 1$. Indeed, as will follow from the discussion in the next section, U is just equal to the linear fractional transformation $T_\Theta[S]$ given by the formula (5.5), when $\Theta = H(-W)$, as defined in (5.2), and $S = Z$. Our present objective is to reexpress U in a form which exhibits the "left zero" explicitly. The first step is to rewrite the "numerator" of U as

$$\begin{aligned}(I - WW^*)^{-\frac{1}{2}}(Z - W) &= Z\{(I - WW^*)^{-\frac{1}{2}}\}^{(1)} - (I - WW^*)^{-\frac{1}{2}}W \\ &= (Z - V)\{(I - WW^*)^{-\frac{1}{2}}\}^{(1)}\end{aligned}$$

with

$$V = (I - WW^*)^{-\frac{1}{2}}W\{(I - WW^*)^{\frac{1}{2}}\}^{(1)} \,.$$

This exhibits $Z - V$ as a left divisor of U. In order to express U in terms of V it is necessary to express W in terms of V. The last formula yields

$$\begin{aligned}V\{(I - WW^*)^{-1}\}^{(1)}V^* &= (I - WW^*)^{-\frac{1}{2}}WW^*(I - WW^*)^{-\frac{1}{2}} \\ &= (I - WW^*)^{-1} - I\end{aligned}$$

which can be solved recursively for $(I - WW^*)^{-1}$ to obtain

$$(I - WW^*)^{-1} = \sum_{j=0}^{\infty} V^{[j]}V^{[j]*} \,.$$

The sum converges because $\ell_V < 1$, and hence, upon setting

$$G_V = \left\{\sum_{j=0}^{\infty} V^{[j]}V^{[j]*}\right\}^{\frac{1}{2}} ,$$

the last formula enables us to express first

$$W = G_V^{-1}VG_V^{(1)}$$

and then U totally in terms of V. To carry this out to the end, it is useful to set

$$M_V = G_V^{(1)}W^*G_V^{-1} \qquad (= \{G_V^2\}^{(1)}V^*G_V^{-2})$$

and to note that

$$G_V^{(1)}(I - W^*Z)^{-1} = (I - M_VZ)^{-1}G_V^{(1)} \,.$$

We shall, however, proceed in a different way starting directly from V in the next subsection.

4.2. A new start.

To begin with, it is convenient to introduce the operators

$$B_V = \sum_{j=0}^{\infty} V^{[j]} V^{[j]*} \tag{4.1}$$

and

$$L_V = B_V^{(1)} - B_V^{(1)} V^* B_V^{-1} V B_V^{(1)} \tag{4.2}$$

for $V \in \mathcal{X}$ with $\ell_V < 1$. The indicated inverse exists because B_V is a positive definite operator with

$$B_V \geq I. \tag{4.3}$$

LEMMA 4.1. *If* $V \in \mathcal{X}$ *and* $\ell_V < 1$, *then*

$$I + V B_V^{(1)} V^* = B_V. \tag{4.4}$$

PROOF. By definition,

$$\begin{aligned} V B_V^{(1)} V^* &= V \left\{ \sum_{j=0}^{\infty} V^{[j]} V^{[j]*} \right\}^{(1)} V^* \\ &= \sum_{j=0}^{\infty} V^{[j+1]} V^{[j+1]*} \\ &= B_V - I. \end{aligned}$$

∎

LEMMA 4.2. *If* $V \in \mathcal{X}$ *and* $\ell_V < 1$, *then*

(1) $V L_V = B_V^{-1} V B_V^{(1)}$, (4.5)

(2) $L_V V^* = B_V^{(1)} V^* B_V^{-1}$, (4.6)

(3) $V L_V V^* = I - B_V^{-1}$, (4.7)

(4) $(I + V^*V - Z^* V L_V V^* Z) L_V = I$, (4.8)

(5) L_V *is invertible and*

$$L_V^{-1} = V^*V + Z^* B_V^{-1} Z, \tag{4.9}$$

(6) L_V *is positive definite.*

PROOF. By Lemma 4.1,

$$VL_V = VB_V^{(1)} - (B_V - I)B_V^{-1}VB_V^{(1)},$$

which yields (1); (2) and (3) follow easily from (1) and a dose of Lemma 4.1.

Next, by (3) and (1), the left hand side of (4) is readily seen to be equal to

$$\begin{aligned}\{I + V^*V - Z^*(I - B_V^{-1})Z\}L_V &= V^*VL_V + \{B_V^{(1)}\}^{-1}L_V \\ &= V^*B_V^{-1}VB_V^{(1)} + I - V^*B_V^{-1}VB_V^{(1)} \\ &= I,\end{aligned}$$

as needed to establish (4). This proves that L_V is left invertible. The right invertibility follows by taking adjoints, since both of the factors on the left hand side of (4.8) are selfadjoint. Moreover, by (4) and (3),

$$\begin{aligned}L_V^{-1} &= I + V^*V - Z^*(I - B_V^{-1})Z \\ &= V^*V + Z^*B_V^{-1}Z,\end{aligned}$$

which yields (4.9). Finally to obtain (6) it suffices to prove that L_V^{-1} is positive definite, but that is immediate from (4.9).

∎

LEMMA 4.3. *If $V \in \mathcal{X}$ and $\ell_V < 1$, then*

$$r(L_V V^* Z) = \ell_V.$$

PROOF. By (4.5),

$$(VL_V)^{[n]} = B_V^{-1}V^{[n]}B_V^{(n)} \tag{4.10}$$

and therefore

$$\|(VL_V)^{[n]}\| \leq \|B_V^{-1}\|\|V^{[n]}\|\|B_V\|.$$

Thus, by Lemma 3.1 and the last inequality,

$$\begin{aligned}r(L_V V^* Z) &= \lim_{n\uparrow\infty} \|(VL_V)^{[n]}\|^{1/n} \\ &\leq \lim_{n\uparrow\infty} \|V^{[n]}\|^{1/n} \\ &= \ell_V.\end{aligned}$$

On the other hand, (4.10) implies that

$$V^{[n]} = B_V (VL_V)^{[n]} (B_V^{-1})^{(n)}$$

and hence that

$$\|V^{[n]}\| \leq \|B_V\| \|(VL_V)^{[n]}\| \|B_V^{-1}\|,$$

which in turn implies that

$$\ell_V \leq \lim_{n \uparrow \infty} \|(VL_V)^{[n]}\|^{1/n} = r(L_V V^* Z).$$

∎

LEMMA 4.4. *If $V \in \mathcal{D}$ and $\ell_V < 1$, then $I - L_V V^* Z$ is an invertible operator in $\mathcal{U}$.*

PROOF. If $V \in \mathcal{D}$, then $L_V V^* \in \mathcal{D}$ and

$$\sum_{j=0}^{n} (L_V V^* Z)^j \in \mathcal{U}$$

for $n = 0, 1, \ldots$. Moreover the sum converges in $\mathcal{X}$ to $(I - L_V V^* Z)^{-1}$ since

$$r(L_V V^* Z) = \ell_V < 1,$$

by Lemma 4.3. Finally $(I - L_V V^* Z)^{-1} \in \mathcal{U}$ since each of the approximating sums does and $\mathcal{U}$ is a Banach algebra.

∎

For future use it is convenient to define the operator

$$\rho_V = I - ZV^*. \tag{4.11}$$

LEMMA 4.5. *If $V \in \mathcal{L}$ and $\ell_V < 1$, then ρ_V is invertible in $\mathcal{U}$.*

PROOF. The partial sums

$$\sum_{j=0}^{n} (ZV^*)^j$$

belong to $\mathcal{U}$ for $n = 0, 1, \ldots$. They converge to $(I - ZV^*)^{-1}$ in $\mathcal{X}$ and hence also in $\mathcal{U}$ as $n \uparrow \infty$ since

$$r(ZV^*) = r(VZ^*) = \ell_V < 1,$$

by Lemma 3.1.

∎

THEOREM 4.1. *If* $V \in \mathcal{D}$ *and* $\ell_V < 1$, *then*

$$U_V = (Z - V)(I - L_V V^* Z)^{-1}(L_V)^{1/2} \tag{4.12}$$

belongs to $\mathcal{U}$ *and is unitary.*

PROOF. By Lemma 4.4, $U_V \in \mathcal{U}$. Moreover, with the help of (4.8), it is readily checked that

$$(Z - V)^*(Z - V) = (I - Z^* V L_V)(L_V)^{-1}(I - L_V V^* Z)$$

and hence that U_V is isometric:

$$U_V^* U_V = I.$$

To complete the proof it suffices to show that U_V is invertible in $\mathcal{X}$. But this is immediate from Lemma 4.5 and the observation that

$$Z - V = (I - VZ^*)Z = \rho_V^* Z.$$

∎

If $V \in \mathcal{D}$ and $\ell_V < 1$, then

$$B_V = (\rho_V^{-1})^{\wedge}(V). \tag{4.13}$$

If $\mathcal{D}$ is a commutative algebra, then, by Lemma 4.1, formula (4.2) simplifies to

$$\begin{aligned} L_V &= B_V^{(1)}\{I - B_V^{-1} V B_V^{(1)} V^*\} \\ &= B_V^{(1)}\{I - B_V^{-1}(B_V - I)\} \\ &= B_V^{(1)} B_V^{-1} \end{aligned}$$

for $V \in \mathcal{D}$ with $\ell_V < 1$. If $\mathcal{D}$ is commutative and $V \in \mathcal{D}$ is also Toeplitz, i.e., $V = V^{(1)}$, then $L_V = I$. The same conclusion holds under the weaker assumptions that $V \in \mathcal{X}$, $VV^* = V^*V$, $V = V^{(1)}$ and $\ell_V < 1$, whether or not $\mathcal{D}$ is commutative. In this latter case

$$B_V = \sum_{j=0}^{\infty} V^j V^{*j} = \sum_{j=0}^{\infty} (VV^*)^j = (I - VV^*)^{-1}.$$

The indicated inverse exists because under these assumptions,

$$\|V\| = r(V) = \ell_V < 1.$$

THEOREM 4.2. *If* $F \in \mathcal{U}$ *and* $V \in \mathcal{D}$ *with* $\ell_V < 1$, *then* $\hat{F}(V) = 0$ *if and only if* $F \in U_V\mathcal{U}$.

PROOF. If $F \in U_V\mathcal{U}$, then, by Lemma 4.4, $F \in (Z - V)\mathcal{U}$. Therefore $\hat{F}(V) = 0$ by Theorem 3.3. Conversely, if $\hat{F}(V) = 0$, then by another application of Theorem 3.3,

$$\begin{aligned} F &= (Z - V)E \\ &= U_V L_V^{-1/2}(I - L_V V^* Z)E \end{aligned}$$

for some choice of $E \in \mathcal{U}$. The rest is plain.

∎

5. LIS AND LINEAR FRACTIONAL TRANSFORMATIONS

We shall say that the operator $X = [A \quad B]$ is admissible if

(1) A and B belong to $\mathcal{U}$, and

(2) $$\sum_{i,j=1}^{n} \langle \left\{ \hat{X}(V_i) J \hat{X}(V_j)^* \rho_{V_j}^{-1} \right\}^{\wedge} (V_i) f_i, f_j \rangle_{\ell^2_{\mathcal{N}}} \geq 0 \tag{5.1}$$

for every choice of $f_j \in \ell^2_{\mathcal{N}}$ and $V_j \in \mathcal{D}$ with $\ell_{V_j} < 1$, for $j = 1, \ldots, n$, where

$$J = \begin{bmatrix} I & 0 \\ 0 & -I \end{bmatrix}.$$

This definition will be rephrased in terms of operator ranges in Section 8.

From now on we let

$$\mathcal{S} = \{F \in \mathcal{U} : \ \|F\| \leq 1\}$$

denote the analogue of the Schur class in the present setting.

THEOREM 5.1. *If* A *and* B *belong to* $\mathcal{U}$ *and* $B = AS$ *for some* $S \in \mathcal{S}$, *then* $X = [A \quad B]$ *is admissible.*

The proof is deferred to Section 8. It appears just after Theorem 8.5.

The basic lossless inverse scattering (LIS) problem for a given admissible $X = [A \quad B]$ is to find all lossless chain scattering operators $\Theta \in \mathcal{A}$ such that

$$[A \quad B]\Theta = [A' \quad B']$$

is again admissible. In this paper we shall restrict ourselves to those Θ which arise as finite products of the "elementary forms"

$$H(S)\begin{bmatrix} U_W & 0 \\ 0 & I \end{bmatrix}$$

for suitably chosen $S \in \mathcal{D}$ with $\|S\| < 1$ and $W \in \mathcal{D}$ with $\ell_W < 1$, where

$$H(S) = \begin{bmatrix} I & S \\ S^* & I \end{bmatrix} \begin{bmatrix} (I - SS^*)^{-\frac{1}{2}} & 0 \\ 0 & (I - S^*S)^{-\frac{1}{2}} \end{bmatrix} . \tag{5.2}$$

LEMMA 5.1. *If $S \in \mathcal{D}$ and $\|S\| < 1$, then*

(1) *the block entries of $H(S)$ belong to $\mathcal{D}$, and*

(2) $H(S)^{-1} = H(-S)$.

PROOF. The first assertion is immediate from parts (4) and (5) of Lemma 2.2. The second is a straightforward calculation.

∎

THEOREM 5.2. *Let $X = [A \;\; B]$ be admissible, let $V \in \mathcal{D}$ with $\ell_V < 1$ and suppose further that the operator*

$$\hat{X}(V)J\hat{X}(V)^* = \hat{A}(V)\hat{A}(V)^* - \hat{B}(V)\hat{B}(V)^*$$

is positive definite on $\ell^2_{\mathcal{N}}$ and invertible in $\mathcal{X}$. Then:

(1) *$\hat{A}(V)$ is invertible in $\mathcal{D}$.*

(2) $S_V = \hat{A}(V)^{-1}\hat{B}(V) \in \mathcal{D}$.

(3) $\|S_V\| < 1$.

(4) *The operator*

$$W = \{\hat{A}(V)(I - S_V S_V^*)^{\frac{1}{2}}\}^{-1} V \{\hat{A}(V)(I - S_V S_V^*)^{\frac{1}{2}}\}^{(1)} \tag{5.3}$$

is well defined and belongs to $\mathcal{D}$.

(5) $\ell_W = \ell_V$.

(6) *The block row operator*

$$XH(-S_V)\begin{bmatrix} U_W & 0 \\ 0 & I \end{bmatrix} = U_V[C \;\; D]$$

with C and D in $\mathcal{U}$.

(7) $[C \quad D]$ *is admissible.*

PROOF. Since $\hat{X}(V)J\hat{X}(V)^* > 0$ and invertible in $\mathcal{X}$, it admits a unique positive definite square root M which is also invertible in $\mathcal{X}$. Therefore, there exists an $\epsilon > 0$ such that

$$\|Mf\|^2_{\ell^2_{\mathcal{N}}} \geq \epsilon\|f\|^2_{\ell^2_{\mathcal{N}}}$$

for every $f \in \ell^2_{\mathcal{N}}$. Thus

$$\hat{X}(V)J\hat{X}(V)^* = \hat{A}(V)\hat{A}(V)^* - \hat{B}(V)\hat{B}(V)^* \geq \epsilon I .$$

But this in turn implies that $\hat{A}(V)$ is invertible in $\mathcal{X}$ and that $\|S_V\| < 1$. The fact that $\hat{A}(V)^{-1} \in \mathcal{D}$ and hence that also $S_V \in \mathcal{D}$ follows from Lemma 2.2. This completes the proof of (1)-(3) and so too (4).

Now, let

$$E = \hat{A}(V)(I - S_V S_V^*)^{\frac{1}{2}}$$

so that

$$W = E^{-1}VE^{(1)}$$

and

$$W^{[n]} = E^{-1}V^{[n]}E^{(n)} .$$

Then it follows from the bounds

$$\|W^{[n]}\| \leq \|E^{-1}\| \; \|V^{[n]}\| \; \|E\|$$

and

$$\|V^{[n]}\| \leq \|E\| \; \|W^{[n]}\| \; \|E^{-1}\| ,$$

just as in the proof of Lemma 4.3, that

$$\ell_W \leq \ell_V \leq \ell_W .$$

This establishes (5) and so guarantees that U_W is well defined.

Next, it is readily checked that

$$[A \quad B]H(-S_V) = [A_1 \quad B_1] ,$$

where

$$A_1 = (A - BS_V^*)(I - S_V S_V^*)^{-\frac{1}{2}} ,$$
$$B_1 = (B - AS_V)(I - S_V^* S_V)^{-\frac{1}{2}} ,$$

and hence, by Lemma 3.3, that

$$\hat{A}_1(V) = \hat{A}(V)(I - S_V S_V^*)^{\frac{1}{2}}$$

and

$$\hat{B}_1(V) = \{\hat{B}(V) - \hat{A}(V)S_V\}\{I - S_V^* S_V\}^{-\frac{1}{2}} = 0 .$$

Moreover, if W is given by (5.3), then, by (3.3) and (3.4),

$$\{A_1(Z - W)\}^{\wedge}(V) = V\{\hat{A}_1(V)\}^{(1)} - \hat{A}_1(V)W = 0 .$$

Therefore, by Theorem 4.2, both $A_1(Z - W)$ and B_1 belong to $U_V\mathcal{U}$, which yields (6).

(7) will be established in Theorem 8.7.

∎

Clearly the extraction procedure discussed in Theorem 5.2 can be iterated if, in the notation of that theorem, $C(V')C(V')^* - D(V')D(V')^*$ is positive definite and invertible in $\mathcal{X}$ for some $V' \in \mathcal{D}$ with $\ell_{V'} < 1$.

We complete this section with some general remarks on linear fractional transformations. For classical analogues see e.g., Chapter 3 of [D] and the references cited there.

LEMMA 5.2. *If*

$$\Theta = \begin{bmatrix} A & B \\ C & D \end{bmatrix} \tag{5.4}$$

is a matrix of operators A, B, C and D all of which belong to $\mathcal{X}$ and if

$$\Theta^* J \Theta = \Theta J \Theta^* = J ,$$

then:

(1) *A and D are invertible in $\mathcal{X}$.*

(2) $\|A^{-1}B\| < 1$ *and* $\|D^{-1}C\| < 1$.

(3) $\|CA^{-1}\| < 1$ *and* $\|BD^{-1}\| < 1$.

(4) $\|A^{-1}\| \leq 1$ *and* $\|D^{-1}\| \leq 1$.

(5) $\|A - BD^{-1}C\| \leq 1$.

PROOF. The identity $\Theta^* J\Theta = J$ implies that

$$A^*A = I + C^*C \quad \text{and} \quad D^*D = I + B^*B \ ,$$

and hence that A and D both have zero kernel and closed range.

The supplementary formulas

$$I + BB^* = AA^* \quad \text{and} \quad I + CC^* = DD^* \ ,$$

which follow from $\Theta J\Theta^* = J$, imply that the same holds true for A^* and D^*. Thus A and D are both one to one maps of $\ell^2_{\mathcal{N}}$ onto itself and hence (1) follows from the open mapping theorem (see e.g., Theorem 5.10 of [Ru]); (2), (3) and (4) follow easily from the preceding formulas.

Finally, since D is invertible, we can express

$$\Theta = \begin{bmatrix} I & \sigma \\ 0 & I \end{bmatrix} \begin{bmatrix} \Delta & 0 \\ 0 & I \end{bmatrix} \begin{bmatrix} I & 0 \\ C & D \end{bmatrix}$$

with

$$\Delta = A - BD^{-1}C \quad \text{and} \quad \sigma = BD^{-1} \ .$$

Thus, upon substituting this formula for Θ into the identity $\Theta J\Theta^* = J$ and multiplying the resulting expression through by the inverses of the outer triangular factors, we obtain

$$\begin{bmatrix} \Delta & 0 \\ 0 & I \end{bmatrix} \begin{bmatrix} I & 0 \\ C & -D \end{bmatrix} \begin{bmatrix} I & C^* \\ 0 & D^* \end{bmatrix} \begin{bmatrix} \Delta^* & 0 \\ 0 & I \end{bmatrix} = \begin{bmatrix} I & \sigma \\ 0 & -I \end{bmatrix} \begin{bmatrix} I & 0 \\ -\sigma^* & I \end{bmatrix} .$$

The 11 entry of this last identity yields

$$\Delta\Delta^* = I - \sigma\sigma^* \leq I \ ,$$

as needed. ∎

We next introduce the linear fractional transformation

$$T_\Theta[S] = (AS + B)(CS + D)^{-1} \tag{5.5}$$

for Θ as in Lemma 5.2 and those $S \in \mathcal{X}$ for which the indicated inverse exists.

LEMMA 5.3. *If Θ is a lossless chain scattering operator in $\mathcal{A}$ and if $S \in \mathcal{S}$, then $CS + D$ is invertible in $\mathcal{U}$ and $T_\Theta[S] \in \mathcal{S}$. $T_\Theta[S]$ is isometric [resp. unitary] if and only if S is isometric [resp. unitary].*

PROOF. By Lemma 5.2, D^{-1} exists and $\|D^{-1}C\| < 1$. Thus

$$CS + D = D(I + D^{-1}CS)$$

is invertible in $\mathcal{X}$. Since $\Theta \in \mathcal{A}$, C, D and D^{-1} all belong to $\mathcal{U}$ and hence $CS + D$ is in fact invertible in $\mathcal{U}$. Thus $T_\Theta[S]$ clearly belongs to $\mathcal{U}$. The fact that $T_\Theta[S] \in \mathcal{S}$ and that it is isometric if and only if S is, is an easy consequence of the identities

$$\begin{aligned}(AS+B)^*(AS+B) - (CS+D)^*(CS+D) &= [S^* \quad I]\Theta^* J\Theta \begin{bmatrix} S \\ I \end{bmatrix} \\ &= [S^* \quad I]J \begin{bmatrix} S \\ I \end{bmatrix} .\end{aligned}$$

If S is also unitary, then, by (2) of Lemma 5.2,

$$AS + B = A(I + A^{-1}BS^*)S$$

is invertible. Thus $T_\Theta[S]$ is invertible and so too, unitary. Conversely, if $T_\Theta[S]$ is unitary, then the very same argument applied to the linear fractional transformation based on $\Theta^{-1} = J\Theta^* J$ implies that $S = T_{\Theta^{-1}}[T_\Theta[S]]$ is unitary also.

∎

Just as in the classical case, the LIS problem can be reexpressed in terms of linear fractional transformations. Part of the connecting link is provided by the next lemma.

LEMMA 5.4. *If* $S = T_\Theta[S_L]$ *for some choice of* $S_L \in \mathcal{S}$ *and* $\Theta \in \mathcal{A}$, *then* $[I \quad -S]\Theta$ *is admissible.*

PROOF. If Θ is expressed in the block form (5.4), then clearly the entries in

$$[I \quad -S]\Theta = [A - SC \quad B - SD]$$

belong to $\mathcal{U}$, since $S \in \mathcal{S}$ by Lemma 5.3. Moreover, $A - SC$ is invertible in $\mathcal{X}$ by Lemma 5.2 and, as follows easily from the given formula for S,

$$S_L = -(A - SC)^{-1}(B - SD) .$$

Thus

$$[I \quad -S]\Theta = (A - SC)[I \quad -S_L]$$

is admissible by Theorem 5.1, since $S_L \in \mathcal{S}$.

∎

LEMMA 5.5. *If Θ is a lossless chain scattering operator in $\mathcal{A}$ which admits a block representation of the form (5.4), then it can also be expressed in the form*

$$\Theta = \frac{1}{2}\begin{bmatrix} (I+\Phi^*)R^{-*} & (I-\Phi)F^{-1} \\ (I-\Phi^*)R^{-*} & (I+\Phi)F^{-1} \end{bmatrix}, \tag{5.6}$$

where $F^{\pm 1}$, R, R^{-*}, and Φ all belong to $\mathcal{U}$,

$$\frac{\Phi+\Phi^*}{2} = F^*F = RR^* \tag{5.7}$$

is a positive invertible operator and

$$V = FR^{-*} \tag{5.8}$$

belongs to $\mathcal{U}$ and is unitary.

PROOF. In view of Lemma 5.2,

$$B + D = (I + BD^{-1})D \quad \text{and} \quad A + C = (I + CA^{-1})A$$

are both invertible in $\mathcal{X}$. Let

$$F = (B+D)^{-1} \qquad \text{and} \qquad R = (A+C)^{-*}\,. \tag{5.9}$$

Then, since $\Theta \in \mathcal{A}$, it is readily checked that $F^{\pm 1} \in \mathcal{U}$ and $R^{-*} \in \mathcal{U}$ and further, with a little more effort, that

$$(I + BD^{-1})^{-1}(A - BD^{-1}C) = (A+C)^{-*}\,,$$

which exhibits R as an element of $\mathcal{U}$ also. Next let

$$\Phi = (D-B)(D+B)^{-1}\,. \tag{5.10}$$

Then, in view of the preceding analysis, Φ also belongs to $\mathcal{U}$. The rest is a straightforward calculation which is largely based on the identities $\Theta J \Theta^* = \Theta^* J \Theta = J$.

∎

6. AN APPROXIMATION PROBLEM

In this section we shall show how to adapt the approximation problem treated in [DD2] and [DeDe] to the present setting.

We begin with a given positive definite invertible operator G in $\mathcal{X}$ which admits both an additive decomposition of the form

$$G = \frac{1}{2}\{\Phi + \Phi^*\}$$

with $\Phi \in \mathcal{U}$ and a multiplicative decomposition of the form

$$G = F^* F$$

with $F^{\pm 1} \in \mathcal{U}$. Then, by Theorem 5.1,

$$X = [\Phi + I \quad \Phi - I]$$

is admissible.

Suppose next that Θ is a lossless chain scattering operator in $\mathcal{A}$ which solves the LIS problem for a given unitary $U \in \mathcal{U}$:

$$[\Phi + I \quad \Phi - I]\Theta = U[E \quad H] , \tag{6.1}$$

where $[E \quad H]$ is admissible. We shall say that Θ is minimal with respect to U if

$$U^*(\Theta_{11} + \Theta_{21}) \in \mathcal{L} \text{ and } U^*(\Theta_{12} + \Theta_{22}) \in \mathcal{L} .$$

We shall say that Θ is normalized if $\hat{\Theta}_{11}(0) + \hat{\Theta}_{21}(0) = 0$. As noted in Section 5, every lossless chain scattering operator $\Theta \in \mathcal{A}$ can be expressed in the form

$$\Theta = \frac{1}{2}\begin{bmatrix} (I + \Phi_i^*)R_i^{-*} & (I - \Phi_i)F_i^{-1} \\ (I - \Phi_i^*)R_i^{-*} & (I + \Phi_i)F_i^{-1} \end{bmatrix} \tag{6.2}$$

with F_i, R_i and Φ_i as in (5.9) and (5.10). Thus in view of Lemma 5.5, $F_i^{\pm 1}$, R_i^{-*}, R_i, and Φ_i all belong to $\mathcal{U}$,

$$\frac{\Phi_i + \Phi_i^*}{2} = F_i^* F_i = R_i R_i^* \tag{6.3}$$

is positive definite and invertible in $\mathcal{X}$ and

$$V_i = F_i R_i^{-*} \tag{6.4}$$

belongs to $\mathcal{U}$ and is unitary:

$$V_i^* V_i = V_i V_i^* = I .$$

THEOREM 6.1. *Let Θ be a lossless chain scattering operator in $\mathcal{A}$ which solves the LIS problem (6.1) for a given unitary $U \in \mathcal{U}$ and suppose that E is invertible in $\mathcal{U}$ and that $\|E^{-1}H\| < 1$. Then Θ is minimal with respect to U. If $\|\hat{V}_i(0)\| < 1$, then $\Theta H(-\hat{V}_i(0))$ is both minimal with respect to U and normalized.*

PROOF. A direct calculation based on (6.2) yields

$$[\Phi + I \quad \Phi - I]\Theta = [(\Phi + \Phi_i^*)R_i^{-*} \quad (\Phi - \Phi_i)F_i^{-1}] \, . \tag{6.5}$$

Therefore, by (6.1), (6.3) and (6.4),

$$\begin{aligned} UE &= (\Phi + \Phi_i^*)R_i^{-*} \\ &= (\Phi - \Phi_i)F_i^{-1}F_iR_i^{-*} + (\Phi_i + \Phi_i^*)R_i^{-*} \\ &= UHV_i + 2R_i \\ &= UEE^{-1}HV_i + 2R_i \, . \end{aligned}$$

Thus

$$UE = 2R_i(I - E^{-1}HV_i)^{-1}$$

which in turn implies that

$$R_i^{-1}U = 2(I - E^{-1}HV_i)^{-1}E^{-1}$$

belongs to $\mathcal{U}$ since $\|E^{-1}H\| < 1$ and $E^{-1} \in \mathcal{U}$ by assumption. Therefore

$$U^*R_i^{-*} \in \mathcal{L}$$

as does

$$\begin{aligned} U^*F_i^{-1} &= U^*R_i^{-*}R_i^*F_i^{-1} \\ &= U^*R_i^{-*}V_i^* \end{aligned}$$

and so Θ is minimal with respect to U.

Next, it is readily checked that $\Theta H(K)$ is also minimal for every choice of $K \in \mathcal{D}$ with $\|K\| < 1$ and that

$$\begin{aligned} &\{\Theta H(K)\}_{11}^{\wedge}(0) + \{\Theta H(K)\}_{21}^{\wedge}(0) \\ &= \{(R_i^{-*})^{\wedge}(0) + (F_i^{-1})^{\wedge}(0)K^*\}(I - KK^*)^{-\frac{1}{2}} \, . \end{aligned}$$

But now as $F_i^{\pm 1} \in \mathcal{U}$ and $F_iF_i^{-1} = I$, it follows from Lemma 3.7 that

$$(F_i^{-1})^{\wedge}(0) = \hat{F}_i(0)^{-1}$$

and hence, since

$$\hat{F}_i(0)(R_i{}^{-*})^\wedge(0) = (F_iR_i{}^{-*})^\wedge(0) = \hat{V}_i(0)$$

by the very same lemma, Θ is normalized if $K = -\hat{V}_i(0)^*$.

∎

Next, for unitary $U \in \mathcal{U}$ as above, we introduce the space

$$\mathcal{U}_U = \{G \in \mathcal{U} : \; U^*G \in Z^*\mathcal{L}\} \ ,$$

which will play the role here which the space $H^2 \ominus UH^2$ plays in the more conventional Hardy space setting. We also introduce the symbol P_0 to denote the projection of $E \in \mathcal{X}$ onto its main diagonal:

$$P_0E = \hat{E}(0)$$

and further define

$$\{M, N\}_G = P_0(N^*GM)$$

for M, N and G in $\mathcal{X}$.

LEMMA 6.1. *Let $\Theta \in \mathcal{A}$ be a minimal (but not necessarily normalized) solution of the LIS problem (6.1) for a given unitary $U \in \mathcal{U}$ and let*

$$G = \frac{\Phi + \Phi^*}{2} \quad \text{and} \quad G_i = \frac{\Phi_i + \Phi_i^*}{2}$$

be defined in terms of the entries in X and Θ, respectively. Then

$$\{M, N\}_G = \{M, N\}_{G_i}$$

for every choice of M and N in $\mathcal{U}_U$.

PROOF. By definition,

$$\begin{aligned}\{M, N\}_G - \{M, N\}_{G_i} &= P_0\{N^*(G - G_i)M\} \\ &= \frac{1}{2}P_0\{N^*(\Phi - \Phi_i)M + N^*(\Phi^* - \Phi_i^*)M\} \ .\end{aligned}$$

But now, upon comparing the right hand sides of (6.1) and (6.5), it follows that

$$(\Phi - \Phi_i)F_i^{-1} = UH$$

and hence that

$$P_0\{N^*(\Phi - \Phi_i)M\} = P_0\{N^*UHF_iM\} = 0 \ ,$$

since $N^*U \in ZU$ and $HF_iM \in \mathcal{U}$. Similarly,

$$P_0\{N^*(\Phi^* - \Phi_i^*)M\} = 0 \; ,$$

since the term inside the curly brackets belongs to $Z^*\mathcal{L}$.

∎

LEMMA 6.2. *Let $\Theta \in \mathcal{A}$ be a normalized minimal solution of the LIS problem (6.1) for a given unitary $U \in \mathcal{U}$. Then $F_i^{-1}M \in \mathcal{U}_U$ for every choice of $M \in \mathcal{D}$.*

PROOF. Since $V_i = F_i R_i^{-*}$ is unitary,

$$\begin{aligned} U^*F_i^{-1} &= U^*R_i^{-*}R_i^*F_i^{-1} \\ &= U^*R_i^{-*}V_i^* \; . \end{aligned}$$

Moreover, $U_i^*R_i^{-*} \in \mathcal{L}$ since Θ is minimal, whereas $V_i^* \in Z^*\mathcal{L}$ since

$$\hat{V}(0) = \hat{F}_i(0)(R_i^{-*})^\wedge(0) = 0$$

because Θ was presumed to be normalized. Thus

$$U^*F_i^{-1}M = U^*R_i^{-*}V_i^*M$$

clearly belongs to $Z^*\mathcal{L}$ for every choice of $M \in \mathcal{D}$ as claimed.

∎

LEMMA 6.3. *Let $\Theta \in \mathcal{A}$ be a normalized minimal solution of the LIS problem (6.1) for a given unitary $U \in \mathcal{U}$. Then*

$$\{M, F_i^{-1}\hat{F}_i(0)^{-*}\}_G = \hat{M}(0)$$

for every $M \in \mathcal{U}_U$.

PROOF. By Lemma 6.2, $F_i^{-1}\hat{F}_i(0)^{-*} \in \mathcal{U}_U$, and therefore, by Lemma 6.1,

$$\begin{aligned} \{M, F_i^{-1}\hat{F}_i(0)^{-*}\}_G &= \{M, F_i^{-1}\hat{F}_i(0)^{-*}\}_{G_i} \\ &= P_0\{\hat{F}_i(0)^{-1}F_i^{-*}F_i^*F_iM\} \\ &= P_0\{\hat{F}_i(0)^{-1}F_iM\} \\ &= \hat{F}_i(0)^{-1}(F_iM)^\wedge(0) \\ &= \hat{M}(0) \; . \end{aligned}$$

∎

THEOREM 6.2. *Let $\Theta \in \mathcal{A}$ be a normalized minimal solution of the LIS problem (6.1) for a given unitary $U \in \mathcal{U}$. Then*

$$\{I - FM, I - FM\}_I \geq I - \hat{F}(0)\hat{F}_i(0)^{-1}\hat{F}_i(0)^{-*}\hat{F}(0)^*$$

for every choice of $M \in \mathcal{U}_U$. Equality prevails if and only if

$$M = F_i^{-1}\hat{F}_i(0)^{-*}\hat{F}(0)^* .$$

PROOF. Let

$$N = F_i^{-1}\hat{F}_i(0)^{-*}\hat{F}(0)^* .$$

Then, by Lemma 6.2, $N \in \mathcal{U}_U$ and hence, by Lemma 6.1,

$$\begin{aligned}\{F^{-1} - N, M\}_G &= \{F^{-1}, M\}_G - \{N, M\}_G \\ &= \{I, FM\}_I - \{N, M\}_{G_i} \\ &= (FM)^\wedge(0)^* - \{\hat{F}_i(0)^{-*}\hat{F}(0)^*, F_iM\}_I \\ &= (FM)^\wedge(0)^* - (F_iM)^\wedge(0)^*\hat{F}_i(0)^{-*}\hat{F}(0)^* \\ &= 0 .\end{aligned}$$

Thus,

$$\begin{aligned}\{I - FM, I - FM\}_I &= \{F^{-1} - M, F^{-1} - M\}_G \\ &= \{F^{-1} - N, F^{-1} - N\}_G + \{F^{-1} - N, N - M\}_G \\ &\quad + \{N - M, F^{-1} - N\}_G + \{N - M, N - M\}_G \\ &= \{F^{-1} - N, F^{-1} - N\}_G + \{N - M, N - M\}_G ,\end{aligned}$$

since the crossterms vanish by the preceding evaluation. Therefore,

$$\{I - FM, I - FM\}_I \geq \{F^{-1} - N, F^{-1} - N\}_G$$

with equality if and only if

$$\{N - M, N - M\}_G = 0 .$$

By another application of the evaluation in the first half of the proof,

$$\begin{aligned}\{F^{-1} - N, F^{-1} - N\}_G &= \{F^{-1} - N, F^{-1}\}_G \\ &= \{I - FN, I\}_G \\ &= I - \hat{F}(0)\hat{N}(0)\end{aligned}$$

which, by the definition of N, is the same as the stated lower bound.

Finally, equality prevails in the stated inequality if and only if

$$\{N - M, N - M\}_G = 0$$

or equivalently, in view of Lemma 2.10, if and only if $N = M$, as claimed.

∎

7. REPRODUCING KERNEL SUBSPACES OF HILBERT SCHMIDT OPERATORS

In this section we prepare the grounds for a more abstract study of the LIS problem. We begin in subsection 7.1 with a number of definitions and results on Hilbert Schmidt operators and also establish a pair of identities (formulas (7.1) and (7.2)) which will play a central role in the sequel. The next two subsections deal with reproducing kernel Hilbert spaces adapted to the present framework.

7.1. Preliminaries.

An operator $A \in \mathcal{X}$ is said to be a Hilbert-Schmidt operator if

$$\sum_{i=1}^{\infty} \|Ax_i\|^2 < \infty$$

for some orthonormal basis $x_1, x_2, \ldots$ of $\ell^2_{\mathcal{N}}$. It turns out that this sum is independent of the choice of the basis, and that this class of operators, which we shall designate by $\mathcal{X}_2$, is a Hilbert space with respect to the inner product

$$\langle A, B\rangle_{HS} = \sum_{i=1}^{\infty} \langle Ax_i, Bx_i\rangle_{\ell^2_{\mathcal{N}}}.$$

The latter sum is also independent of the choice of orthonormal basis and since B^*A is trace class is often written as

$$\langle A, B\rangle_{HS} = \operatorname{trace} B^*A.$$

We now summarize a number of well known facts for the convenience of the reader. All these and much more can be found in the monograph [GK] of Gohberg and Krein.

Let $\mathcal{X}_2$ denote the set of operators in $\mathcal{X}$ which are Hilbert-Schmidt. It is well known that:

(1) $\mathcal{X}_2$ is a two sided ideal in the algebra of bounded operators: If $A \in \mathcal{X}_2$ and $B \in \mathcal{X}$, then both AB and BA belong to $\mathcal{X}_2$.

(2) $\mathcal{X}_2$ is closed in $\mathcal{X}$: If $\{F_n\} \in \mathcal{X}_2$ and $F_n \to F$ in $\mathcal{X}$ as $n \uparrow \infty$, then $F \in \mathcal{X}_2$.

(3) $\mathcal{X}_2$ is closed under adjoints: If $F \in \mathcal{X}_2$, then $F^* \in \mathcal{X}_2$, where F^* denotes the adjoint of F as an operator on $\ell^2_{\mathcal{N}}$.

(4) If $F \in \mathcal{X}_2$, then

$$\|F\|^2_{HS} = \langle F, F\rangle_{HS} = \|F^*\|^2_{HS}.$$

(5) If E and $F \in \mathcal{X}_2$ and $G \in \mathcal{X}$, then

$$\langle E, FG\rangle_{HS} = \langle EG^*, F\rangle_{HS}$$

and

$$\langle E, GF\rangle_{HS} = \langle G^*E, F\rangle_{HS}.$$

LEMMA 7.1. *If $F \in \mathcal{X}_2$, then F_{st} is a Hilbert Schmidt operator on $\mathcal{N}$ and*

$$\|F\|^2_{HS} = \sum_{s,t=-\infty}^{\infty} \|F_{st}\|^2_{HS}.$$

PROOF. Since $F_{st} = \pi^* Z^s F Z^{*t}\pi$, F_{st} is of Hilbert Schmidt class when F is. Let $\varphi_0, \varphi_1, \ldots$ be an orthonormal basis for $\mathcal{N}$. Then

$$\{Z^{*t}\pi\varphi_j,\ -\infty < t < \infty,\ 0 \le j < \infty\}$$

is an orthonormal basis for $\ell^2_{\mathcal{N}}$. Thus

$$\begin{aligned}
\|F\|^2_{HS} &= \sum_{t=-\infty}^{\infty}\sum_{j=0}^{\infty}\langle FZ^{*t}\pi\varphi_j, FZ^{*t}\pi\varphi_j\rangle_{\ell^2_{\mathcal{N}}} \\
&= \sum_{t=-\infty}^{\infty}\sum_{j=0}^{\infty}\sum_{s=-\infty}^{\infty} \langle \pi^* Z^s F Z^{*t}\pi\varphi_j, \pi^* Z^s F Z^{*t}\pi\varphi_j\rangle_{\mathcal{N}} \\
&= \sum_{t=-\infty}^{\infty}\sum_{j=0}^{\infty}\sum_{s=-\infty}^{\infty} \langle F_{st}\varphi_j, F_{st}\varphi_j\rangle_{\mathcal{N}} \\
&= \sum_{s,t=-\infty}^{\infty} \|F_{st}\|^2_{HS}.
\end{aligned}$$

∎

We now let

$$\mathcal{U}_2 = \mathcal{U} \cap \mathcal{X}_2, \quad \mathcal{L}_2 = \mathcal{L} \cap \mathcal{X}_2 \quad \text{and} \quad \mathcal{D}_2 = \mathcal{L}_2 \cap \mathcal{U}_2.$$

THEOREM 7.1. *The Hilbert space $\mathcal{X}_2$ admits the orthogonal decomposition*

$$\mathcal{X}_2 = Z^*\mathcal{L}_2 \oplus \mathcal{D}_2 \oplus Z\mathcal{U}_2.$$

PROOF. Let us first check the indicated orthogonality.

If $F \in \mathcal{L}_2$ and $G \in \mathcal{U}_2$, then

$$\begin{aligned}\langle G, Z^*F\rangle_{HS} &= \text{trace}\{F^*ZG\} \\ &= \sum_{i=-\infty}^{\infty} \text{trace}(F^*ZG)_{ii}.\end{aligned}$$

However,

$$\begin{aligned}(F^*ZG)_{ii} &= \sum_{j=-\infty}^{\infty} (F^*)_{ij}(ZG)_{ji} \\ &= \sum_{j=-\infty}^{\infty} (F^*)_{ij}G_{j+1,i} = 0,\end{aligned}$$

since $G_{j+1,i} = 0$ for $j+1 > i$, i.e., for $j \geq i$, whereas,

$$(F^*)_{ij} = 0 \quad \text{for} \quad i > j$$

because both F^* and G belong to $\mathcal{U}$. This proves that $Z^*\mathcal{L}_2$ is orthogonal to $\mathcal{U}_2$. Much the same argument serves to show that $\mathcal{D}_2$ is orthogonal to $Z\mathcal{L}_2$ and hence that the three subspaces $Z^*\mathcal{L}_2$, $\mathcal{D}_2$ and $Z\mathcal{U}_2$ are mutually orthogonal.

It remains to show that every $F \in \mathcal{X}_2$ can be split into three components, but this is clear upon splitting the doubly infinite matrix of operators $[F_{ij}]$ into its strictly lower triangular, diagonal and strictly upper triangular parts, each of which again corresponds to a Hilbert-Schmidt operator because of Lemma 7.1.

∎

LEMMA 7.2.

(1) $\mathcal{U} \cdot \mathcal{U}_2 \subset \mathcal{U}_2$ *and* $\mathcal{U}_2 \cdot \mathcal{U} \subset \mathcal{U}_2$.

(2) $\mathcal{L} \cdot \mathcal{L}_2 \subset \mathcal{L}_2$ *and* $\mathcal{L}_2 \cdot \mathcal{L} \subset \mathcal{L}_2$.

(3) $\mathcal{D} \cdot \mathcal{D}_2 \subset \mathcal{D}_2$ *and* $\mathcal{D}_2 \cdot \mathcal{D} \subset \mathcal{D}_2$.

PROOF. This is selfevident from (3) of Lemma 2.1 and the fact that $\mathcal{X}_2$ is a two sided ideal in $\mathcal{X}$.

∎

LEMMA 7.3. *If $\mathcal{M}$ is a closed subspace of $\mathcal{X}_2$ and if $A \in \mathcal{M}$ and* trace AB^* $= 0$ *for every $B \in \mathcal{M}$, then $A = 0$.*

PROOF. This is immediate from the fact that $\mathcal{M}$ is a Hilbert space and

$$\langle A, B\rangle_{HS} = \operatorname{trace} B^*A = \operatorname{trace} AB^*.$$

∎

LEMMA 7.4. *If $F \in \mathcal{U}_2$ and if $V \in \mathcal{D}$ with $\ell_V < 1$, then both $\hat{F}(V)$ and $F^{\Delta}(V)$ belong to $\mathcal{D}_2$.*

PROOF. Clearly

$$\begin{aligned}\|\hat{F}(V)\|_{HS} &= \left\|\sum_{j=0}^{\infty} V^{[j]}F_{[j]}\right\|_{HS}\\ &\le \sum_{j=0}^{\infty} \|V^{[j]}\| \|F_{[j]}\|_{HS}\\ &\le \sum_{j=0}^{\infty} \|V^{[j]}\| \; \|F\|_{HS},\end{aligned}$$

thanks to Lemma 7.1. This establishes the first assertion. The proof for $F^{\Delta}(V)$ goes through in much the same way since

$$\begin{aligned}\|\{F_{[j]}\}^{(-j)}\{V^{[j]}\}^{(1-j)}\|_{HS} &= \|F_{[j]}\{V^{[j]}\}^{(1)}\|_{HS}\\ &\le \|V^{[j]}\| \; \|F_{[j]}\|_{HS}.\end{aligned}$$

∎

In view of Lemma 4.5, $\rho_V^{-1}G \in \mathcal{U}_2$ for any choice of $V \in \mathcal{L}$ with $\ell_V < 1$ and $G \in \mathcal{U}_2$. We shall restrict attention to $V \in \mathcal{D}$ and shall establish the following important evaluation:

THEOREM 7.2. *If $V \in \mathcal{D}$ with $\ell_V < 1$ and if $F \in \mathcal{U}_2$ and $G \in \mathcal{D}_2$, then $\rho_V^{-1}G \in \mathcal{U}_2$ and*

$$\langle F, \rho_V^{-1}G\rangle_{HS} = \operatorname{trace}\{G^*\hat{F}(V)\}. \tag{7.1}$$

Moreover, the set of all such $\rho_V^{-1}G$ is dense in $\mathcal{U}_2$.

PROOF. The orthogonal projection of $Z^{*n}F$ onto $\mathcal{D}_2$ is equal to $F_{[n]}$. Thus

$$\begin{aligned}\langle F, \rho_V^{-1}G\rangle_{HS} &= \sum_{n=0}^{\infty}\langle F, Z^n V^{[n]*}G\rangle_{HS}\\ &= \sum_{n=0}^{\infty}\langle F_{[n]}, V^{[n]*}G\rangle_{HS}\\ &= \sum_{n=0}^{\infty}\operatorname{trace}\{G^* V^{[n]} F_{[n]}\}\end{aligned}$$

which is readily seen to yield (7.1).

Next, if F is orthogonal to $\rho_V^{-1}G$ for all the permitted choices of V and G, then

$$\operatorname{trace}\{G^*\hat{F}(V)\} = \langle \hat{F}(V), G\rangle_{HS} = 0,$$

which in turn implies that

$$\hat{F}(V) = 0$$

for every choice of $V \in \mathcal{D}$, with $\ell_V < 1$ since $\mathcal{D}_2$ is a Hilbert space. Therefore, by Theorem 3.2, $F = 0$. The proof is complete. ∎

Let $\underline{\underline{p}}$ denote the orthogonal projection of $\mathcal{X}_2$ onto $\mathcal{U}_2$.

THEOREM 7.3. *If $F \in \mathcal{U}$, $G \in \mathcal{D}_2$ and $V \in \mathcal{D}$ with $\ell_V < 1$, then*

$$\underline{\underline{p}} F^* \rho_V^{-1} G = \hat{F}(V)^* \rho_V^{-1} G. \tag{7.2}$$

PROOF. Since $\hat{F}(V)^*\rho_V^{-1}G \in \mathcal{U}_2$, it suffices to show that

$$\underline{\underline{p}}\{(F - \hat{F}(V))^* \rho_V^{-1} G\} = 0.$$

But, since

$$F - \hat{F}(V) = (Z - V)E$$

for some $E \in \mathcal{U}$ by Theorem 3.3, the operator

$$\begin{aligned}(F - \hat{F}(V))^* \rho_V^{-1} G &= E^*(Z^* - V^*)\rho_V^{-1} G\\ &= E^* Z^* G\end{aligned}$$

clearly belongs to $Z^*\mathcal{L}_2$. Therefore the projection indicated at the beginning of the proof vanishes, as needed. ∎

We have already observed that if $V \in \mathcal{D}$ and $\ell_V < 1$, then the elementary Blaschke type product U_V which is defined by formula (4.12) belongs to $\mathcal{U}$. Therefore $U_V\mathcal{U}_2 \subset \mathcal{U}_2$. In fact, since U_V is unitary, $U_V\mathcal{U}_2$ is a closed subspace at $\mathcal{U}_2$. Its orthogonal complement: $\mathcal{U}_2 \ominus U_V\mathcal{U}_2$, will turn out to be of special interest.

LEMMA 7.5. *If* $V \in \mathcal{D}$ *with* $\ell_V < 1$, *then*

$$(U_V)^{\wedge}(V) = 0 \tag{7.3}$$

and

$$\langle U_V F, \rho_V^{-1} G\rangle_{HS} = 0 \tag{7.4}$$

for every choice of $F \in \mathcal{U}_2$ *and* $G \in \mathcal{D}_2$.

PROOF. The evaluation (7.3) is immediate from Theorem 3.3 and the observation that $(Z - V)^{-1}U_V \in \mathcal{U}$. The rest drops out from Theorem 7.2.

∎

LEMMA 7.6. *If* $V \in \mathcal{D}$ *with* $\ell_V < 1$ *and* U_V *is given by* (4.12), *then*

$$\mathcal{U}_2 \ominus U_V\mathcal{U}_2 = \{\rho_V^{-1}G : G \in \mathcal{D}_2\}$$

. PROOF. The fact that $\rho_V^{-1}G \in \mathcal{U}_2 \ominus U_V\mathcal{U}_2$ for every $G \in \mathcal{D}_2$ is immediate from Theorem 7.2 and Lemma 7.5.

Suppose next that $F \in \mathcal{U}_2 \ominus U_V\mathcal{U}_2$. Then $U_V^*F \in Z^*\mathcal{L}_2$ from which in turn it readily follows that $(Z^* - V^*)F \in Z^*\mathcal{L}_2$. Therefore,

$$(I - ZV^*)F = \rho_V F = G$$

for some choice of $G \in \mathcal{L}_2$. However, since $F \in \mathcal{U}_2$, G must also belong to $\mathcal{U}_2$. Therefore

$$F = \rho_V^{-1}G$$

for some choice of G in $\mathcal{U}_2 \cap \mathcal{L}_2 = \mathcal{D}_2$.

∎

COROLLARY 1. $\{\rho_V^{-1}G : G \in \mathcal{D}_2\}$ *is a closed subspace of* $\mathcal{U}_2$.

PROOF. $\mathcal{U}_2 \ominus U_V\mathcal{U}_2$ is a closed subspace of $\mathcal{U}_2$.

∎

One can also prove Corollary 1 directly, since $\mathcal{D}_2$ is a closed subspace of $\mathcal{U}_2$ and multiplication by ρ_V is a bounded invertible operator from $\mathcal{U}_2$ into itself.

COROLLARY 2. *If $V \in \mathcal{D}$ with $\ell_V < 1$ and if $F \in \mathcal{U}_2 \ominus U_V \mathcal{U}_2$, then*

$$F = \rho_V^{-1} B_V^{-1} \hat{F}(V), \tag{7.5}$$

where B_V is given by (4.1).

PROOF. If $F \in \mathcal{U}_2 \ominus U_V \mathcal{U}_2$, then, by the lemma,

$$F = \rho_V^{-1} G$$

for some choice of $G \in \mathcal{D}_2$. Therefore, by Lemma 3.3,

$$\begin{aligned} \hat{F}(V) &= (\rho_V^{-1})^{\wedge}(V) G \\ &= \left\{ \sum_{j=0}^{\infty} V^{[j]} V^{[j]*} \right\} G \\ &= B_V G. \end{aligned}$$

The rest is selfevident. ∎

7.2. Reproducing Kernel Hilbert Spaces in $\mathcal{U}_2$.

Our next objective is to identify $\mathcal{U}_2 \ominus U_V \mathcal{U}_2$ as a reproducing kernel Hilbert space. We begin with the relevant definitions for the present setting.

Let $\mathcal{M}$ be a closed subspace of $\mathcal{U}_2$. We shall say that $\mathcal{M}$ is a reproducing kernel Hilbert space if for every choice of $V \in \mathcal{D}$ with $\ell_V < 1$ there exists an operator $K_V \in \mathcal{U}$ such that

(1) $K_V D \in \mathcal{M}$

and

(2) $\langle F, K_V D \rangle_{HS} = \operatorname{trace} D^* \hat{F}(V)$

for every $D \in \mathcal{D}_2$ and $F \in \mathcal{M}$.

In view of Theorem 7.2 the full space $\mathcal{U}_2$ is itself readily seen to be a reproducing kernel Hilbert space with reproducing kernel ρ_V^{-1}. This is completely analogous to the classical H^2 space theory; see e.g., the discussion in Section 2 of [D]. This analogy is reinforced by Lemma 7.6 and Theorem 7.5, both of which have well known counterparts in the Hardy space setting.

THEOREM 7.4. *Let $\mathcal{M}$ be a reproducing kernel Hilbert space which is included isometrically in $\mathcal{U}_2$. Then:*

(1) *$\mathcal{M}$ has exactly one reproducing kernel K_V.*

(2)
$$\hat{K}_V(W) = \hat{K}_W(V)^* \tag{7.6}$$
for every choice of V and W in $\mathcal{D}$ with $\ell_V < 1$ and $\ell_W < 1$.

(3)
$$\text{trace}\left\{\sum_{i,j=1}^{n} D_i^* \hat{K}_{V_j}(V_i) D_j\right\} \geq 0$$
for every choice of $V_j \in \mathcal{D}$ with $\ell_{V_j} < 1$ and $D_j \in \mathcal{D}_2$, $j = 1, \ldots, n$.

PROOF. Suppose that M_V is also a reproducing kernel for $\mathcal{M}$. Then

$$\begin{aligned}\text{trace}\, E^* \hat{K}_V(W) D &= \text{trace}\, E^* (K_V D)^{\wedge}(W) \\ &= \langle K_V D, M_W E\rangle_{HS} \\ &= \{\langle M_W E, K_V D\rangle_{HS}\}^* \\ &= \{\text{trace}\, D^* \hat{M}_W(V) E\}^* \\ &= \text{trace}\, E^* \hat{M}_W(V)^* D\end{aligned}$$

for every choice of E and D in $\mathcal{D}_2$. Therefore, by Lemma 7.3,

$$\hat{K}_V(W) = \hat{M}_W(V)^*,$$

which supplies both the asserted uniqueness and formula (7.6).

Finally, (3) is immediate since the indicated trace is equal to $\| \sum_{j=1}^{n} K_{V_j} D_j \|_{HS}^2$.

∎

THEOREM 7.5. *If Σ is inner, i.e., if $\Sigma \in \mathcal{U}$ and is unitary, then $\mathcal{U}_2 \ominus \Sigma\mathcal{U}_2$ is a reproducing kernel Hilbert space with reproducing kernel*

$$K_W = \{I - \Sigma\hat{\Sigma}(W)^*\}\rho_W^{-1}$$

for every $W \in \mathcal{D}$ with $\ell_W < 1$.

PROOF. The operator $K_W D$ clearly belongs to $\mathcal{U}_2$ for every choice of $D \in \mathcal{D}_2$. Moreover, for any $F \in \mathcal{U}_2$,

$$\begin{aligned}\langle \Sigma F, K_W D\rangle_{HS} &= \langle \Sigma F, \rho_W^{-1} D\rangle_{HS} - \langle F, \hat{\Sigma}(W)^* \rho_W^{-1} D\rangle_{HS} \\ &= \text{trace}\{D^*(\Sigma F)^{\wedge}(W) - D^*\big(\hat{\Sigma}(W) F\big)^{\wedge}(W)\} \\ &= 0,\end{aligned}$$

by Lemma 3.7. Thus $K_W D \in \mathcal{U}_2 \ominus \Sigma\mathcal{U}_2$.

Suppose next that $F \in \mathcal{U}_2 \ominus \Sigma\mathcal{U}_2$. Then, since

$$\Sigma\hat{\Sigma}(W)^* \rho_W^{-1} D \in \Sigma\mathcal{U}_2$$

for $D \in \mathcal{D}_2$,

$$\langle F, K_W D\rangle_{HS} = \langle F, \rho_W^{-1} D\rangle_{HS} = \operatorname{trace} D^* \hat{F}(W),$$

by Theorem 7.2, as needed to complete the proof.

∎

COROLLARY. *If $V \in \mathcal{D}$ with $\ell_V < 1$ and U_V is defined by (4.12), then $\mathcal{U}_2 \ominus U_V\mathcal{U}_2$ is a reproducing kernel Hilbert space with reproducing kernel*

$$K_W = \{I - U_V\hat{U}_V(W)^*\}\rho_W^{-1}$$

for every $W \in \mathcal{D}$ with $\ell_W < 1$.

7.3. Reproducing Kernel Hilbert Subspaces in $\mathcal{U}_2^{2\times 1}$.

Let $\mathcal{U}_2^{2\times 1}$ [resp., $\mathcal{L}_2^{2\times 1}$, $\mathcal{D}_2^{2\times 1}$] denote the set of block columns $\underline{\underline{F}} = \begin{bmatrix} F_1 \\ F_2 \end{bmatrix}$ with $F_i \in \mathcal{U}_2$ [resp., $\mathcal{L}_2$, $\mathcal{D}_2$]. $\mathcal{U}_2^{2\times 1}$ is a Hilbert space with inner product

$$\begin{aligned} \langle \underline{\underline{F}}, \underline{\underline{G}}\rangle_{HS} &= \langle F_1, G_1\rangle_{HS} + \langle F_2, G_2\rangle_{HS} \\ &= \operatorname{trace}\{\underline{\underline{G}}^* \underline{\underline{F}}\}. \end{aligned}$$

We shall also need the (indefinite) J inner product based on

$$J = \begin{bmatrix} I & 0 \\ 0 & -I \end{bmatrix}$$

which is defined by the rule:

$$\langle J\underline{\underline{F}}, \underline{\underline{G}}\rangle_{HS} = \langle F_1, G_1\rangle_{HS} - \langle F_2, G_2\rangle_{HS}.$$

Next, for every lossless chain scattering operator Θ in $\mathcal{A}$, we define

$$\mathcal{H}(\Theta) = \mathcal{U}_2^{2\times 1} \boxminus \Theta\mathcal{U}_2^{2\times 1},$$

where $\boxminus$ indicates that the orthogonal complement is taken with respect to the J inner product.

The definition of a reproducing kernel Hilbert space which was furnished at the beginning of Subsection 7.2 is easily adapted to the present setting. These spaces are analogues of a class of reproducing kernel Hilbert spaces which were introduced by de Branges [dB1] in a more classical setting.

THEOREM 7.6. *If Θ is a lossless chain scattering operator in $\mathcal{A}$, then $\mathcal{H}(\Theta)$ is a reproducing kernel Hilbert space with respect to the J inner product. Its reproducing kernel (with some abuse of notation)*†

$$K_W = \{J - \Theta J\hat{\Theta}(W)^*\}\rho_W^{-1}$$

for $W \in \mathcal{D}$ with $\ell_W < 1$.

PROOF. Let $\underline{\underline{D}} \in \mathcal{D}_2^{2\times 1}$. Then $K_W\underline{\underline{D}} \in \mathcal{U}_2^{2\times 1}$ and for any $\underline{\underline{F}} \in \mathcal{U}_2^{2\times 1}$,

$$\begin{aligned}\langle J\Theta\underline{\underline{F}}, K_W\underline{\underline{D}}\rangle_{HS} &= \langle \Theta\underline{\underline{F}}, \rho_W^{-1}\underline{\underline{D}}\rangle_{HS} - \langle \underline{\underline{F}}, \hat{\Theta}(W)^*\rho_W^{-1}\underline{\underline{D}}\rangle_{HS} \\ &= \operatorname{trace}\{\underline{\underline{D}}^*(\Theta\underline{\underline{F}})^\wedge(W) - \underline{\underline{D}}^*(\hat{\Theta}(W)\underline{\underline{F}})^\wedge(W)\} \\ &= 0,\end{aligned}$$

since

$$(\Theta\underline{\underline{F}})^\wedge(W) = (\hat{\Theta}(W)\underline{\underline{F}})^\wedge(W)$$

just as before. Therefore $K_W\underline{\underline{D}} \in \mathcal{H}(\Theta)$.

Suppose next that $\underline{\underline{F}} \in \mathcal{H}(\Theta)$. Then, since $\Theta J\hat{\Theta}(W)^*\rho_W^{-1}\underline{\underline{D}} \in \Theta\mathcal{U}_2^{2\times 1}$ for $\underline{\underline{D}} \in \mathcal{D}_2^{2\times 1}$,

$$\begin{aligned}\langle J\underline{\underline{F}}, \{J - \Theta J\hat{\Theta}(W)^*\}\rho_W^{-1}\underline{\underline{D}}\rangle_{HS} &= \langle J\underline{\underline{F}}, J\rho_W^{-1}\underline{\underline{D}}\rangle_{HS} \\ &= \operatorname{trace} D^*\hat{\underline{\underline{F}}}(W).\end{aligned}$$

∎

THEOREM 7.7. *Suppose that F, S and V are elements in $\mathcal{D}$ such that F is invertible with $F^{-1} \in \mathcal{D}$, $\|S\| < 1$ and $\ell_V < 1$ and let Θ be the lossless chain scattering operator which is given by the formula*

$$\Theta = H(S)\begin{bmatrix} U_W & 0 \\ 0 & I \end{bmatrix},$$

where

$$W = \{F(I - SS^*)^{1/2}\}^{-1}V\{F(I - SS^*)^{1/2}\}^{(1)}.$$

† $\hat{\Theta}(W)$ is short for $[\hat{\Theta}_{ij}(W)]$, $i, j = 1, 2$, and ρ_W^{-1} is short for the 2×2 block diagonal matrix each of whose diagonal entries is equal to ρ_W^{-1}.

Then

$$\mathcal{H}(\Theta) = \left\{ \begin{bmatrix} I \\ S^* \end{bmatrix} F^* \rho_V^{-1} D : V \in \mathcal{D} \text{ with } \ell_V < 1 \text{ and } D \in \mathcal{D}_2 \right\}.$$

PROOF. Let

$$\underline{\underline{E}} = \begin{bmatrix} I \\ S^* \end{bmatrix} F^*.$$

Then, by Theorem 5.2,

$$\underline{\underline{E}}^* J\Theta = U_V[G\ H]$$

for some admissible pair $[G \quad H]$ (where in fact $H = 0$). Therefore, for any $\underline{\underline{C}} \in \mathcal{U}_2^{2\times 1}$ and $D \in \mathcal{D}_2$

$$\begin{aligned}
\langle J\Theta\underline{\underline{C}}, \underline{\underline{E}}\rho_V^{-1}D\rangle_{HS} &= \langle U_V[G\ H]\underline{\underline{C}}, \rho_V^{-1}D\rangle_{HS} \\
&= \operatorname{trace} D^* \{U_V[G\ H]\underline{\underline{C}}\}^\wedge(V) \\
&= \operatorname{trace} D^* \{\hat{U}_V(V)[G\ H]\underline{\underline{C}}\}^\wedge(V) \\
&= 0
\end{aligned}$$

by successive applications of Lemma 3.7, first to pass from line 2 to line 3 and then to verify that $\hat{U}_V(V) = 0$. Thus

$$\underline{\underline{E}}\rho_V^{-1}D \in \mathcal{U}_2^{2\times 1} \boxminus \Theta\mathcal{U}_2^{2\times 1} = \mathcal{H}(\Theta)$$

for every choice of $D \in \mathcal{D}_2$. This proves inclusion in one direction.

Now, to get the reverse inclusion, suppose that $\underline{\underline{C}} \in \mathcal{H}(\Theta)$. Then

$$\langle J\underline{\underline{C}}, \Theta\underline{\underline{G}}\rangle_{HS} = 0$$

for every $\underline{\underline{G}} \in \mathcal{U}_2^{2\times 1}$. Therefore the components of

$$\Theta^* J\underline{\underline{C}} = \begin{bmatrix} U_W^* & 0 \\ 0 & I \end{bmatrix} H(S)^* J\underline{\underline{C}}$$

belong to $Z^*\mathcal{L}_2$. In other words if

$$\begin{bmatrix} X \\ Y \end{bmatrix} = H(S)^* J\underline{\underline{C}}\,,$$

then

$$U_W^* X \in Z^*\mathcal{L}_2 \quad \text{and} \quad Y \in Z^*\mathcal{L}_2.$$

Since Y also clearly belongs to $\mathcal{U}_2$ we see that $Y = 0$, whereas, by Lemma 7.6,

$$X = \rho_W^{-1} D$$

for some choice of $D \in \mathcal{D}_2$. Thus

$$H(S)^* J\underline{\underline{C}} = \begin{bmatrix} X \\ Y \end{bmatrix} = \begin{bmatrix} \rho_W^{-1} D \\ 0 \end{bmatrix},$$

which in turn implies that

$$\begin{aligned} \underline{\underline{C}} &= JH(-S)^* \begin{bmatrix} \rho_W^{-1} D \\ 0 \end{bmatrix} \\ &= \begin{bmatrix} I \\ S^* \end{bmatrix} (I - SS^*)^{-1/2} \rho_W^{-1} D. \end{aligned}$$

The supplementary formula

$$(I - SS^*)^{-1/2} \rho_W^{-1} = F^* \rho_V^{-1} \{(I - SS^*)^{1/2} F^*\}^{-1} ,$$

which is obtained by a lengthy but straightforward calculation serves to exhibit $\underline{\underline{C}}$ in the requisite form:

$$\underline{\underline{C}} = \begin{bmatrix} I \\ S^* \end{bmatrix} F^* \rho_V^{-1} D'$$

with

$$D' = \{(I - SS^*)^{1/2} F^*\}^{-1} D$$

in $\mathcal{D}_2$, as needed.

∎

8. OPERATOR RANGES

The reproducing kernel Hilbert spaces introduced in Section 7 can be expressed in an elegant way in terms of operator ranges as we shall explain in this section. For additional information, the papers of Schwartz [Sc] and Fillmore and Williams [FW] are suggested.

The notation $\operatorname{ran} A$ for the range of an operator A and $\ker A$ for its kernel (or nullspace) will prove useful.

THEOREM 8.1. *Let Γ be a bounded positive operator on a complex separable Hilbert space $\mathcal{H}$ with norm $\| \ \|_{\mathcal{H}}$ and let P denote the orthogonal projection of $\mathcal{H}$ onto the kernel of Γ. Then:*

(1) $\operatorname{ran} \Gamma^{1/2}$ *is a Hilbert space with respect to the norm*

$$\|\Gamma^{1/2} u\|_\Gamma = \|(I - P)u\|_{\mathcal{H}}.$$

(2) $\operatorname{ran}\Gamma$ *is dense in* $\operatorname{ran}\Gamma^{1/2}$ *in the norm* $\| \ \|_\Gamma$.

(3) $\|\Gamma u\|_\Gamma = \|\Gamma^{1/2}u\|_{\mathcal{H}}$ *for every* $u \in H$.

(4) If $\Gamma = \Gamma_1 + \Gamma_2$ *is a decomposition of* Γ *into the sum of two positive operators, then*

$$\operatorname{ran}\Gamma^{1/2} = \operatorname{ran}\Gamma_1^{1/2} + \operatorname{ran}\Gamma_2^{1/2} .$$

PROOF. Since $\Gamma \geq 0$, it is both well known and readily checked that $\ker\Gamma^{1/2} = \ker\Gamma$. Thus if $\Gamma^{1/2}u_1 = \Gamma^{1/2}u_2$, then $u_1 - u_2 \in \ker\Gamma$ and hence

$$P(u_1 - u_2) = u_1 - u_2.$$

This shows that

$$(I-P)u_1 = (I-P)u_2$$

and hence that the number $\| \ \|_\Gamma$ is well defined: if $\Gamma^{1/2}u_1 = \Gamma^{1/2}u_2$, then

$$\|(I-P)u_1\|_{\mathcal{H}} = \|(I-P)u_2\|_{\mathcal{H}}.$$

Moreover, if

$$\|\Gamma^{1/2}u\|_\Gamma = 0,$$

then

$$(I-P)u = 0,$$

which implies that u is in $\ker\Gamma = \ker\Gamma^{1/2}$ and hence that $\Gamma^{1/2}u = 0$. Thus $\| \ \|_\Gamma$ does indeed define a norm on $\operatorname{ran}\Gamma^{1/2}$.

Now suppose that $\{\Gamma^{1/2}u_n\}$ is a Cauchy sequence in $\operatorname{ran}\Gamma^{1/2}$ with respect to $\| \ \|_\Gamma$. Then $\{(I-P)u_n\}$ is a Cauchy sequence in the underlying Hilbert space $\mathcal{H}$. Therefore, since $\mathcal{H}$ is complete, $(I-P)u_n$ tends to a limit $h \in \mathcal{H}$ and $(I-P)^2u_n \to (I-P)h$ in $\mathcal{H}$ as $n \uparrow \infty$. Thus, as $(I-P)^2 = I-P$, it follows that

$$(I-P)h = h$$

and

$$\begin{aligned}\|\Gamma^{1/2}(u_n - h)\|_\Gamma &= \|(I-P)(u_n-h)\|_{\mathcal{H}} \\ &= \|(I-P)u_n - h\|_{\mathcal{H}} \to 0\end{aligned}$$

as $n \uparrow \infty$. Therefore $\Gamma^{1/2}u_n \to \Gamma^{1/2}h$ in $\| \ \|_\Gamma$ as $n \uparrow \infty$, i.e., $\operatorname{ran}\Gamma^{1/2}$ is complete. It is also closed because if $\Gamma^{1/2}u_n \to v$ in $\| \ \|_\Gamma$ as $n \uparrow \infty$, then $\Gamma^{1/2}u_n$ is a Cauchy sequence and hence by the preceding analysis $\|\Gamma^{1/2}u_n - \Gamma^{1/2}h\|_\Gamma \to 0$ for some $h \in \mathcal{H}$. Therefore $v = \Gamma^{1/2}h$ belongs to $\operatorname{ran}\Gamma^{1/2}$, i.e., $\operatorname{ran}\Gamma^{1/2}$ is closed. This completes the proof of (1).

We turn next to the proof of (2). Clearly $\operatorname{ran}\Gamma \subset \operatorname{ran}\Gamma^{1/2}$. Suppose that there is a $u \in \mathcal{H}$ such that

$$\langle \Gamma v, \Gamma^{1/2}u\rangle_\Gamma = 0$$

for every $v \in \mathcal{H}$, where $\langle\ ,\ \rangle_\Gamma$ denotes the inner product based on $\|\ \|_\Gamma$. But this means that

$$\langle \Gamma^{1/2}v, (I-P)u\rangle_{\mathcal{H}} = 0$$

for every $v \in \mathcal{H}$ and hence, since Pu belongs to $\ker\Gamma = \ker\Gamma^{1/2}$, that

$$\langle v, \Gamma^{1/2}u\rangle_{\mathcal{H}} = 0$$

for every $v \in \mathcal{H}$. Thus $\Gamma^{1/2}u = 0$, as needed to establish (2).

The proof of (3) is easy:

$$\|\Gamma u\|_\Gamma = \|(I-P)\Gamma^{1/2}u\|_{\mathcal{H}} = \|\Gamma^{1/2}u\|_{\mathcal{H}}$$

because

$$P\Gamma^{1/2}u = 0,$$

since $\operatorname{ran}\Gamma^{1/2}$ is orthogonal to $\ker\Gamma^{1/2} = \ker\Gamma$.

Suppose next that

$$\Gamma = \Gamma_1 + \Gamma_2$$

with $\Gamma_i \geq 0$. Then clearly

$$\Gamma^{1/2}\Gamma^{1/2} \geq \Gamma_1^{1/2}\Gamma_1^{1/2}$$

and hence, by a theorem of Douglas [Do], there exists a contraction C such that

$$\Gamma_1^{1/2} = \Gamma^{1/2}C.$$

Therefore $\operatorname{ran}\Gamma_1^{1/2}$ is included in $\operatorname{ran}\Gamma^{1/2}$, as is $\operatorname{ran}\Gamma_2^{1/2}$ by the very same argument, and hence

$$\operatorname{ran}\Gamma_1^{1/2} + \operatorname{ran}\Gamma_2^{1/2} \subset \operatorname{ran}\Gamma^{1/2}.$$

Our next objective is to show that equality prevails. Let $h \in \mathcal{H}$. Then, by (2), there exists a sequence $v_1, v_2, \ldots$ in $\mathcal{H}$ such that

$$\begin{aligned}\|\Gamma^{1/2}h - \Gamma v_n\|_\Gamma &= \|(I-P)(h - \Gamma^{1/2}v_n)\|_{\mathcal{H}}\\ &= \|(I-P)h - \Gamma^{1/2}v_n\|_{\mathcal{H}}\end{aligned}$$

tends to zero as $n \uparrow \infty$. Therefore $\{\Gamma^{1/2}v_n\}$ is a Cauchy sequence in $\mathcal{H}$, as are $\{\Gamma_1^{1/2}v_n\}$ and $\{\Gamma_2^{1/2}v_n\}$, thanks to the identity

$$\|\Gamma^{1/2}(v_n - v_m)\|^2_{\mathcal{H}} = \|\Gamma_1^{1/2}(v_n - v_m)\|^2_{\mathcal{H}} + \|\Gamma_2^{1/2}(v_n - v_m)\|^2_{\mathcal{H}}.$$

Let

$$a = \lim_{n\uparrow\infty} \Gamma_1^{1/2}v_n \quad \text{and} \quad b = \lim_{n\uparrow\infty} \Gamma_2^{1/2}v_n.$$

Then, upon letting P_i denote the orthogonal projection onto $\ker\ \Gamma_i$, we see that

$$P_1 a = 0, \qquad P_2 b = 0$$

and

$$\begin{aligned}\Gamma v_n &= \Gamma_1 v_n + \Gamma_2 v_n \\ &= \Gamma_1^{1/2}\Gamma_1^{1/2}v_n + \Gamma_2^{1/2}\Gamma_2^{1/2}v_n \\ &\to \Gamma_1^{1/2}a + \Gamma_2^{1/2}b\end{aligned}$$

in $\mathcal{H}$ as $n \uparrow \infty$. Thus

$$\begin{aligned}\Gamma^{1/2}h &= \Gamma^{1/2}(I-P)h \\ &= \lim_{n\uparrow\infty} \Gamma v_n \qquad (\text{in } \mathcal{H}) \\ &= \Gamma_1^{1/2}a + \Gamma_2^{1/2}b\ ,\end{aligned}$$

which serves to complete the proof.

∎

Item (4) in Theorem 8.1 was first proved by T. Crimmins; see p.260 of [FW].

We remark that the supplementary identity (which is expressed in terms of the notation used in the proof of (4) of the last theorem)

$$\begin{aligned}\|\Gamma^{1/2}h\|^2_{\Gamma} &= \|(I-P)h\|^2_{\mathcal{H}} \\ &= \lim_{n\uparrow\infty} \langle \Gamma v_n, v_n\rangle_{\mathcal{H}} \\ &= \lim_{n\uparrow\infty} \|\Gamma_1^{1/2}v_n\|^2_{\mathcal{H}} + \lim_{n\uparrow\infty} \|\Gamma_2^{1/2}v_n\|^2_{\mathcal{H}} \\ &= \|a\|^2_{\mathcal{H}} + \|b\|^2_{\mathcal{H}} \\ &= \|(I-P_1)a\|^2_{\mathcal{H}} + \|(I-P_2)b\|^2_{\mathcal{H}} \\ &= \|\Gamma_1^{1/2}a\|^2_{\Gamma_1} + \|\Gamma_2^{1/2}a\|^2_{\Gamma_2}.\end{aligned}$$

is the main ingredient in the proof that ran $\Gamma_1^{\frac{1}{2}}$ and ran $\Gamma_2^{\frac{1}{2}}$ are complementary subspaces of ran $\Gamma^{\frac{1}{2}}$ in the sense of de Branges and Rovnyak [dBR1], [dBR2]; for further refinements see also [dB2].

Let Θ be a lossless chain scattering operator in $\mathcal{A}$ and let M_Θ denote the operator of multiplication by Θ restricted to $\mathcal{U}_2^{2\times 1}$. Then it is readily checked that

$$M_\Theta^* = \underline{\underline{p}}\Theta^* \qquad (= \underline{\underline{p}}M_{\Theta^*}) \; , \tag{8.1}$$

where $\underline{\underline{p}}$ denotes the orthogonal projection of $\mathcal{X}_2^{2\times 1}$ onto $\mathcal{U}_2^{2\times 1}$ (as well as $\mathcal{X}_2$ onto $\mathcal{U}_2$, where the meaning will always be clear from the context).

LEMMA 8.1. *If Θ is a lossless chain scattering operator in $\mathcal{A}$, and if $V \in \mathcal{D}$ with $\ell_V < 1$ and $\underline{\underline{D}} \in \mathcal{D}_2^{2\times 1}$, then*

(1)
$$M_\Theta^* \rho_V^{-1} \underline{\underline{D}} = \hat{\Theta}(V)^* \rho_V^{-1} \underline{\underline{D}} \tag{8.2}$$

and

(2)
$$(J - M_\Theta J M_\Theta^*)\rho_V^{-1} \underline{\underline{D}} = K_V \underline{\underline{D}}, \tag{8.3}$$

where

$$K_V = \{J - \Theta J \hat{\Theta}(V)^*\}\rho_V^{-1} \tag{8.4}$$

is the reproducing kernel for $\mathcal{H}(\Theta)$ (which was introduced in Theorem 7.6).

PROOF. This is an easy consequence of Theorem 7.3.

∎

THEOREM 8.2. *If Θ is a lossless chain scattering operator in $\mathcal{A}$, then the operator*

$$\Gamma_\Theta = J - M_\Theta J M_\Theta^* \tag{8.5}$$

is positive semidefinite on $\mathcal{U}_2^{2\times 1}$.

PROOF. Let

$$P = \begin{bmatrix} I & 0 \\ 0 & 0 \end{bmatrix} \quad \textit{and} \quad Q = \begin{bmatrix} 0 & 0 \\ 0 & I \end{bmatrix} \; .$$

Then, just as in the classical case,

$$\Sigma = (P\Theta - Q)(P - Q\Theta)^{-1}$$

is unitary with

$$\Sigma_{22} = \Theta_{22}^{-1}$$

and

$$\Theta = (P + \Sigma Q)^{-1}(\Sigma P + Q) \; .$$

Let $\alpha = P + \Sigma Q$ and $\beta = \Sigma P + Q$. Then, since the block entries of

$$\alpha^{-1} = \begin{bmatrix} I & -\Sigma_{12}\Sigma_{22}^{-1} \\ 0 & \Sigma_{22}^{-1} \end{bmatrix}$$

all belong to $\mathcal{U}$,

$$M_{\alpha^{-1}} = M_\alpha^{-1} .$$

Thus

$$M_\Theta = M_\alpha^{-1} M_\beta$$

and

$$\begin{aligned} \Gamma_\Theta &= M_\alpha^{-1}\{M_\alpha J M_\alpha^* - M_\beta J M_\beta^*\}(M_\alpha^{-1})^* \\ &= M_\alpha^{-1}\{(P + M_\Sigma Q)(P - Q)(P + QM_\Sigma^*) \\ &\quad - (M_\Sigma P + Q)(P - Q)(PM_\Sigma^* + Q)\}(M_\alpha^{-1})^* \\ &= M_\alpha^{-1}\{I - M_\Sigma M_\Sigma^*\}(M_\alpha^{-1})^* . \end{aligned}$$

But now for any $\underline{\underline{F}} \in \mathcal{U}_2^{2\times 1}$,

$$\begin{aligned} \langle (I - M_\Sigma M_\Sigma^*)\underline{\underline{F}}, \underline{\underline{F}}\rangle_{HS} &= \|\underline{\underline{F}}\|_{HS}^2 - \|M_\Sigma^* \underline{\underline{F}}\|_{HS}^2 \\ &= \|\underline{\underline{F}}\|_{HS}^2 - \|\underline{\underline{p}}\Sigma^* \underline{\underline{F}}\|_{HS}^2 \\ &\geq 0 . \end{aligned}$$

Thus clearly $\Gamma_\Theta \geq 0$ on $\mathcal{U}_2^{2\times 1}$, as claimed. ∎

LEMMA 8.2. *If Θ is a lossless chain scattering operator in $\mathcal{A}$, then* ran $\Gamma_\Theta = \mathcal{H}(\Theta)$, *as sets (where Γ_Θ is defined by (8.5)).*

PROOF. Suppose first that $\underline{\underline{F}} \in \mathcal{H}(\Theta)$. Then

$$\langle J\underline{\underline{F}}, \Theta \underline{\underline{G}}\rangle_{HS} = 0$$

for every choice of $\underline{\underline{G}} \in \mathcal{U}_2^{2\times 1}$. But this implies that $M_\Theta^* J \underline{\underline{F}} = 0$ and hence that

$$\underline{\underline{F}} = \Gamma_\Theta J \underline{\underline{F}} .$$

This proves that

$$\mathcal{H}(\Theta) \subset \text{ran } \Gamma_\Theta .$$

Suppose next that $\underline{\underline{F}}$ belongs to the indicated range. Then

$$\underline{\underline{F}} = \Gamma_\Theta \underline{\underline{G}}$$

for some choice of $\underline{\underline{G}} \in \mathcal{U}_2^{2\times 1}$ and therefore, for any $\underline{\underline{E}} \in \mathcal{U}_2^{2\times 1}$,

$$\begin{aligned}\langle J\underline{\underline{F}}, \Theta\underline{\underline{E}}\rangle_{HS} &= \langle J(J - M_\Theta J M_\Theta^*)\underline{\underline{G}}, \Theta\underline{\underline{E}}\rangle_{HS} \\ &= \langle \underline{\underline{G}}, \Theta\underline{\underline{E}}\rangle_{HS} - \langle J M_\Theta^* \underline{\underline{G}}, M_\Theta^* J\Theta\underline{\underline{E}}\rangle_{HS} \\ &= \langle \underline{\underline{G}}, \Theta\underline{\underline{E}}\rangle_{HS} - \langle J M_\Theta^* \underline{\underline{G}}, J\underline{\underline{E}}\rangle_{HS} ,\end{aligned}$$

since

$$M_\Theta^* J M_\Theta \underline{\underline{E}} = \underline{\underline{p}}\Theta^* J\Theta\underline{\underline{E}} = \underline{\underline{p}} J\underline{\underline{E}} = J\underline{\underline{E}}$$

for $\underline{\underline{E}} \in \mathcal{U}_2^{2\times 1}$. Thus

$$\langle J\underline{\underline{F}}, \Theta\underline{\underline{E}}\rangle_{HS} = 0$$

which serves to reverse the inclusion established earlier and hence to complete the proof.

∎

LEMMA 8.3. *If Θ is a lossless chain scattering operator in $\mathcal{A}$, then*

$$\text{ran } \Gamma_\Theta = \text{ran } \Gamma_\Theta^{\frac{1}{2}} ,$$

as sets (where both of the indicated operators act on $\mathcal{U}_2^{2\times 1}$).

PROOF. Lemma 8.2 serves to identify ran Γ_Θ as a closed subspace of $\mathcal{U}_2^{2\times 1}$. Therefore

$$\begin{aligned}\mathcal{U}_2^{2\times 1} &= \text{ran } \Gamma_\Theta \oplus \ker \Gamma_\Theta \\ &\subset \text{ran } \Gamma_\Theta^{\frac{1}{2}} \oplus \ker \Gamma_\Theta \\ &\subset \mathcal{U}_2^{2\times 1} .\end{aligned}$$

Therefore, equality prevails throughout.

∎

THEOREM 8.3. *Let Θ be a lossless chain scattering operator in $\mathcal{A}$. Then the Hilbert space $\mathcal{H}(\Theta)$ (with the J inner product in $\mathcal{X}_2^{2\times 1}$) is equal to the Hilbert space* ran $\Gamma_\Theta^{\frac{1}{2}}$ *equipped with the inner product based on* $\| \ \|_{\Gamma_\Theta}$.

PROOF. By Lemmas 8.2 and 8.3,

$$\mathcal{H}(\Theta) = \text{ran } \Gamma_\Theta = \text{ran } \Gamma_\Theta^{\frac{1}{2}}$$

as sets. Thus if $\underline{\underline{F}}_1$ and $\underline{\underline{F}}_2$ belong to $\mathcal{H}(\Theta)$, then

$$\underline{\underline{F}}_j = \Gamma_\Theta \underline{\underline{E}}_j \ , \qquad j = 1, 2 \ ,$$

for some choice of $\underline{\underline{E}}_j \in \mathcal{U}_2^{2\times 1}$. Moreover,

$$\begin{aligned}\langle J\underline{\underline{F}}_j, \underline{\underline{E}}_i\rangle_{HS} &= \langle J\underline{\underline{F}}_j, (J - M_\Theta J M_\Theta^*)\underline{\underline{E}}_i\rangle_{HS} \\ &= \langle \underline{\underline{F}}_j, \underline{\underline{E}}_i\rangle_{HS} \\ &= \langle \Gamma_\Theta \underline{\underline{E}}_j, \underline{\underline{E}}_i\rangle_{HS} \\ &= \langle \underline{\underline{F}}_j, \underline{\underline{F}}_i\rangle_{\Gamma_\Theta} \ ,\end{aligned}$$

as claimed.

∎

We now introduce the analogue of the generalized backwards shift in the present setting:

$$R_V F = \{F - F^\Delta(V)\}(Z - V)^{-1} \tag{8.6}$$

for $F \in \mathcal{U}$ and $V \in \mathcal{D}$ with $\ell_V < 1$.

THEOREM 8.4. *Let Θ be a lossless chain scattering operator in $\mathcal{A}$. Then $\mathcal{H}(\Theta)$ is R_V invariant for every choice of $V \in \mathcal{D}$ with $\ell_V < 1$.*

PROOF. Let $\underline{\underline{E}} = \begin{bmatrix} F \\ G \end{bmatrix}$ belong to $\mathcal{H}(\Theta)$. The claim is that

$$R_V \underline{\underline{E}} = \begin{bmatrix} R_V F \\ R_V G \end{bmatrix}$$

belong to $\mathcal{H}(\Theta)$.

By Theorem 3.4, the operator

$$R_V F = \{F - F^\Delta(V)\}(Z - V)^{-1}$$

belongs to $\mathcal{U}$. Moreover, since F and $F^\Delta(V)$ both belong to $\mathcal{U}_2$ while $(Z - V)^{-1}$ is bounded, $R_V F$ is also a Hilbert-Schmidt operator. Therefore it belongs to $\mathcal{U}_2$, as does $R_V G$ and hence

$$R_V \underline{\underline{E}} = \begin{bmatrix} R_V F \\ R_V G \end{bmatrix} (Z - V)^{-1}$$

belongs to $\mathcal{U}_2^{2\times 1}$.

Now, for any $\underline{\underline{H}} \in \mathcal{U}_2^{2\times 1}$,

$$\begin{aligned}
&\langle J \begin{bmatrix} F - F^{\Delta}(V) \\ G - G^{\Delta}(V) \end{bmatrix} (Z-V)^{-1}, \Theta \underline{\underline{H}} \rangle_{HS} \\
&= \langle J \begin{bmatrix} F - F^{\Delta}(V) \\ G - G^{\Delta}(V) \end{bmatrix}, \Theta \underline{\underline{H}} (Z^* - V^*)^{-1} \rangle_{HS} \\
&= \langle J \begin{bmatrix} F - F^{\Delta}(V) \\ G - G^{\Delta}(V) \end{bmatrix}, \Theta \underline{\underline{H}} \rho_V^{-1} Z \rangle_{HS} \\
&= \langle J \begin{bmatrix} F \\ G \end{bmatrix}, \Theta \underline{\underline{H}} \rho_V^{-1} Z \rangle_{HS} \ ,
\end{aligned}$$

since

$$J \begin{bmatrix} F^{\Delta}(V) \\ G^{\Delta}(V) \end{bmatrix} \in \mathcal{D}_2^{2\times 1}$$

whereas the right hand entry in the last inner product belongs to $Z\mathcal{U}_2^{2\times 1}$. In fact, the last inner product is equal to zero because its right hand side belongs to $\Theta \mathcal{U}_2^{2\times 1}$. The proof is complete.

∎

LEMMA 8.4. *Let $\Lambda \in \mathcal{X}$ [resp. $\mathcal{D}$] and let $\epsilon \geq 0$ be a given nonnegative constant. Then*

$$\langle \Lambda F, F \rangle_{HS} \geq \epsilon \langle F, F \rangle_{HS}$$

for every $F \in \mathcal{X}_2$ [resp. $\mathcal{D}_2$] if and only if

$$\langle \Lambda f, f \rangle_{\ell_{\mathcal{N}}^2} \geq \epsilon \langle f, f \rangle_{\ell_{\mathcal{N}}^2}$$

for every $f \in \ell_{\mathcal{N}}^2$.

PROOF. Suppose to begin with that the first inequality holds and let f be any element in $\ell_{\mathcal{N}}^2$ with unit norm. Let $f_1, f_2, \ldots$ be an orthonormal basis of $\ell_{\mathcal{N}}^2$ with $f_1 = f$. Then the operator F which is defined by the rule

$$F f_j = \begin{cases} f & \text{if } j = 1 \\ 0 & \text{if } j \neq 1 \end{cases}$$

clearly belongs to $\mathcal{X}_2$ and

$$\begin{aligned}\langle \Lambda f, f\rangle_{\ell^2_{\mathcal{N}}} &= \sum_{j=1}^{\infty} \langle \Lambda F f_j, F f_j\rangle_{\ell^2_{\mathcal{N}}} \\ &= \langle \Lambda F, F\rangle_{HS} \\ &\geq \epsilon \langle F, F\rangle_{HS} \\ &= \epsilon \langle f, f\rangle_{\ell^2_{\mathcal{N}}} \,.\end{aligned}$$

Thus the second inequality follows from the first. Conversely, if the second inequality holds and if $f_1, f_2, \ldots$ is any orthonormal basis of $\ell^2_{\mathcal{N}}$, then

$$\begin{aligned}\langle \Lambda F, F\rangle_{HS} &= \sum_{j=1}^{\infty} \langle \Lambda F f_j, F f_j\rangle_{\ell^2_{\mathcal{N}}} \\ &\geq \epsilon \sum_{j=1}^{\infty} \langle F f_j, F f_j\rangle_{\ell^2_{\mathcal{N}}} \\ &= \epsilon \langle F, F\rangle_{HS} \,.\end{aligned}$$

Finally, it remains to show that if $\Lambda \in \mathcal{D}$, then it suffices to take $F \in \mathcal{D}_2$ in the first inequality. To this end let $f \in \ell^2_{\mathcal{N}}$, let $\varphi_1, \varphi_2, \ldots$ be an orthonormal basis in $\mathcal{N}$ and let $D \in \mathcal{D}$ be the diagonal operator with block components $D_{tt} = \pi^* Z^t D Z^{*t} \pi$ which are defined by the rule

$$D_{tt}\varphi_j = \begin{cases} f_t & \text{if } j = 1 \\ 0 & \text{if } j \geq 1 \end{cases},$$

where f_t denotes the t-th component of the given $f \in \ell^2_{\mathcal{N}}$. Then

$$\begin{aligned}\langle \Lambda_{tt} D_{tt}, D_{tt}\rangle_{HS} &= \sum_{j=1}^{\infty} \langle \Lambda_{tt} D_{tt}\varphi_j, D_{tt}\varphi_j\rangle_{\mathcal{N}} \\ &= \langle \Lambda_{tt} f_t, f_t\rangle_{\mathcal{N}}\end{aligned}$$

and hence

$$\begin{aligned}\langle \Lambda D, D\rangle_{HS} &= \sum_{t=-\infty}^{\infty} \langle \Lambda_{tt} f_t, f_t\rangle_{\mathcal{N}} \\ &= \langle \Lambda f, f\rangle_{\ell^2_{\mathcal{N}}}\end{aligned}$$

and (as follows by taking $\Lambda = I$)

$$\langle D, D\rangle_{HS} = \langle f, f\rangle_{\ell^2_{\mathcal{N}}} \,.$$

Thus $D \in \mathcal{D}_2$. The rest goes through much as before.

∎

The proof of the preceding lemma can also be adapted to prove the following block version by invoking the identities established there term by term:

COROLLARY. *Let $\Lambda \in \mathcal{X}^{n\times n}$ [resp. $\mathcal{D}^{n\times n}$] and let $\epsilon \geq 0$ be a given non-negative constant. Then*

$$\langle \Lambda \underline{\underline{F}}, \underline{\underline{F}}\rangle_{HS} \geq \epsilon \langle \underline{\underline{F}}, \underline{\underline{F}}\rangle_{HS}$$

for every $\underline{\underline{F}} \in \mathcal{X}_2^{n\times 1}$ [resp. $\mathcal{D}_2^{n\times 1}$] if and only if

$$\langle \Lambda \underline{\underline{f}}, \underline{\underline{f}}\rangle_{\ell^2_{\mathcal{N}}} \geq \epsilon \langle \underline{\underline{f}}, \underline{\underline{f}}\rangle_{\ell^2_{\mathcal{N}}}$$

for every $\underline{\underline{f}} \in (\ell^2_{\mathcal{N}})^{n\times 1}$.

We return now to the definition of admissibility and give a new characterization in terms of multiplication operators: For

$$X = [A \quad B]$$

with A and B in $\mathcal{U}$, define the multiplication operator M_X from $\mathcal{U}_2^{2\times 1}$ to $\mathcal{U}_2$ by the rule

$$M_X \begin{bmatrix} F \\ G \end{bmatrix} = X \begin{bmatrix} F \\ G \end{bmatrix} = AF + BG \; .$$

THEOREM 8.5. *Let $X = [A \quad B]$ with A and B in $\mathcal{U}$. Then X is admissible if and only if the operator*

$$\Gamma_X = M_X J M_X^* \qquad (= A\underline{\underline{p}}A^* - B\underline{\underline{p}}B^*)$$

is positive semidefinite on $\mathcal{U}_2$.

PROOF. If $\Gamma_X \geq 0$ on $\mathcal{U}_2$, then

$$\sum_{i,j=1}^{n} \langle \Gamma_X \rho_{V_j}^{-1} D_j, \rho_{V_i}^{-1} D_i \rangle_{HS} \geq 0$$

for every choice of $D_j \in \mathcal{D}_2$ and $V_j \in \mathcal{D}$ with $\ell_{V_j} < 1$ for $j = 1, \ldots, n$. By Theorem 7.3 the given sum is equal to

$$\text{trace} \sum_{i,j=1}^{n} D_i^* \{\hat{X}(V_i) J \hat{X}(V_j)^* \rho_{V_j}^{-1}\}^\wedge (V_i) D_j$$

and therefore the asserted equivalence is immediate from the last corollary.

∎

We now have enough tools at our disposal to prove Theorem 5.1.

PROOF OF THEOREM 5.1. In view of Theorem 8.5, it suffices to show that $M_X J M_X^* \geq 0$ on $\mathcal{U}_2^{2\times 1}$. But this is immediate from the observation that

$$M_X J M_X^* = M_A\{I - M_S M_S^*\}M_A^*$$

and the inequality

$$\|M_S^* G\|_{HS} \leq \|S^* G\|_{HS} \leq \|S^*\| \; \|G\|_{HS} \leq \|G\|_{HS}$$

which is valid for every $G \in \mathcal{U}_2$.

∎

THEOREM 8.6. *If $X = [A \quad B]$ is admissible and if $\Gamma_X = M_X J M_X^*$, then the space*

$$\mathcal{B}(X) = \text{ran } \Gamma_X^{\frac{1}{2}}$$

(acting on $\mathcal{U}_2$) equipped with the inner product based on $\| \quad \|_{\Gamma_X}$ is a reproducing kernel Hilbert space with reproducing kernel

$$\Lambda_V = XJ\hat{X}(V)^* \rho_V^{-1}$$

for every $V \in \mathcal{D}$ with $\ell_V < 1$.

PROOF. By Theorem 8.1, $\mathcal{B}(X)$ is a Hilbert space and ran Γ_X is dense in $\mathcal{B}(X)$. Now, for any $D \in \mathcal{D}_2$ and $V \in \mathcal{D}$ with $\ell_V < 1$,

$$\begin{aligned} \Gamma_X \rho_V^{-1} D &= XJ\underline{\underline{p}}X^* \rho_V^{-1} D \\ &= XJ\hat{X}(V)^* \rho_V^{-1} D \ , \end{aligned}$$

by Theorem 7.3. Thus $\Lambda_V D \in \mathcal{B}(X)$. Moreover, for any $F \in \mathcal{B}(X)$,

$$\begin{aligned} \langle F, \Lambda_V D\rangle_{\Gamma_X} &= \langle F, \Gamma_X \rho_V^{-1} D\rangle_{\Gamma_X} \\ &= \langle F, \rho_V^{-1} D\rangle_{HS} \\ &= \text{trace}\{D^* \hat{F}(V)\} \ , \end{aligned}$$

thanks to Theorem 7.2.

∎

THEOREM 8.7. *Let* $X = [A \quad B]$ *be admissible, let* $V \in \mathcal{D}$ *with* $\ell_V < 1$ *and suppose further that the operator*

$$\hat{X}(V)J\hat{X}(V)^* = \hat{A}(V)\hat{A}(V)^* - \hat{B}(V)\hat{B}(V)^*$$

is positive definite on $\ell^2_{\mathcal{N}}$ *and invertible in* $\mathcal{X}$. *Let*

$$\Theta = H(-S_V)\begin{bmatrix} U_W & 0 \\ 0 & I \end{bmatrix},$$

where S_V *and* U_W *are defined as in Theorem 5.2. Then*

$$Y = X\Theta \qquad \textit{and} \qquad Y_1 = U_V^{-1}Y ,$$

are admissible.

PROOF. It was already shown in the proof of Theorem 5.2 that $Y \in \mathcal{U}^{1\times 2}$. Therefore, in view of Theorem 8.5, it remains only to show that

$$\Gamma_Y = M_Y J M_Y^* \geq 0$$

on $\mathcal{U}_2$. Clearly,

$$\Gamma_Y = M_X J M_X^* - M_X \Gamma_\Theta M_X^* .$$

The rest of the proof proceeds in steps.

STEP 1. *If* $F = \Sigma_{j=1}^n \rho_{\Omega_j}^{-1} D_j$ *for some choice of* $D_j \in \mathcal{D}_2$ *and* $\Omega_j \in \mathcal{D}$ *with* $\ell_{\Omega_j} < 1$, *then*

$$\langle \Gamma_X F, F\rangle_{HS} = \text{trace} \sum_{i,j=1}^{n} D_i^* \hat{\Lambda}_{\Omega_j}(\Omega_i) D_j .$$

PROOF OF STEP 1. This is a straightforward evaluation thanks to Theorems 7.2 and 7.3.

STEP 2. *If* $D \in \mathcal{D}_2$ *and* $\Omega \in \mathcal{D}$ *with* $\ell_\Omega < 1$, *then*

$$\Gamma_\Theta M_X^* \rho_\Omega^{-1} D = J\hat{X}(V)^* \rho_V^{-1} \hat{\Lambda}_V(V)^{-1} \hat{\Lambda}_\Omega(V) D .$$

PROOF OF STEP 2. Let

$$\underline{\underline{F}} = J\hat{X}(V)^* \rho_V^{-1} .$$

Then, by Theorems 5.2, 7.7 and 8.3,

$$\Gamma_\Theta M_X^* \rho_\Omega^{-1} D = \underline{\underline{F}} D'$$

for some choice of $D' \in \mathcal{D}_2$. It remains to show that

$$D' = \hat{\Lambda}_V(V)^{-1}\hat{\Lambda}_V(\Omega)D \ .$$

or equivalently, that

$$\langle J\Gamma_\Theta M_X^* \rho_\Omega^{-1} D, \underline{\underline{F}}E\rangle_{HS} = \langle J\underline{\underline{F}}\hat{\Lambda}_V(V)^{-1}\hat{\Lambda}_\Omega(V)D, \underline{\underline{F}}E\rangle_{HS}$$

for every $E \in \mathcal{D}_2$. But now as $\underline{\underline{F}}E$ is J orthogonal to $\Theta\mathcal{U}_2^{2\times 1}$, the left hand side of the last asserted identity is readily seen to reduce to

$$\begin{aligned}
\langle J(J - M_\Theta J M_\Theta^*)M_X^*\rho_\Omega^{-1}D, \underline{\underline{F}}E\rangle_{HS} &= \langle M_X^*\rho_\Omega^{-1}D, \underline{\underline{F}}E\rangle_{HS} \\
&= \langle \hat{X}(\Omega)^*\rho_\Omega^{-1}D, \underline{\underline{F}}E\rangle_{HS} \\
&= \langle J\hat{X}(\Omega)^*\rho_\Omega^{-1}D, M_X^*\rho_V^{-1}E\rangle_{HS} \\
&= \langle \Lambda_\Omega D, \rho_V^{-1}E\rangle_{HS} \\
&= \operatorname{trace}\{E^*\hat{\Lambda}_\Omega(V)D\} \ ,
\end{aligned}$$

whereas, the right hand side

$$\begin{aligned}
&= \langle \hat{X}(V)^*\rho_V^{-1}\hat{\Lambda}_V(V)^{-1}\hat{\Lambda}_\Omega(V)D, JM_X^*\rho_V^{-1}E\rangle_{HS} \\
&= \langle \Lambda_V\hat{\Lambda}_V(V)^{-1}\hat{\Lambda}_\Omega(V)D, \rho_V^{-1}E\rangle_{HS} \\
&= \operatorname{trace}\ E^*\{\Lambda_V\hat{\Lambda}_V(V)^{-1}\hat{\Lambda}_\Omega(V)D\}^\wedge(V) \\
&= \operatorname{trace}\ E^*\hat{\Lambda}_\Omega(V)D \ ,
\end{aligned}$$

by Theorem 7.2, which is applicable since $\hat{\Lambda}_V(V)^{-1}\hat{\Lambda}_\Omega(V)D \in \mathcal{D}_2$.

STEP 3. *If F is as in Step 1, then*

$$\langle M_X\Gamma_\Theta M_X^* F, F\rangle_{HS} = \operatorname{trace}\sum_{i,j=1}^n D_i^*\hat{\Lambda}_V(\Omega_i)\hat{\Lambda}_V(V)^{-1}\hat{\Lambda}_{\Omega_j}(V)D_j \ .$$

PROOF OF STEP 3. By Step 2,

$$\begin{aligned}
\langle M_X\Gamma_\Theta M_X^* F, F\rangle_{HS} &= \sum_{i,j=1}^n \langle M_X J\hat{X}(V)^*\rho_V^{-1}\hat{\Lambda}_V(V)^{-1}\hat{\Lambda}_{\Omega_j}(V)D_j, \rho_{\Omega_i}^{-1}D_i\rangle_{HS} \\
&= \sum_{i,j=1}^n \langle \Lambda_V\hat{\Lambda}_V(V)^{-1}\hat{\Lambda}_{\Omega_j}(V)D_j, \rho_{\Omega_i}^{-1}D_i\rangle_{HS}
\end{aligned}$$

which is readily seen to be equal to the asserted right hand side, thanks to Theorem 7.2.

STEP 4. *Y is admissible.*

PROOF OF STEP 4. It suffices to show that

$$\langle M_Y J M_Y^* F, F\rangle_{HS} \geq 0$$

for every finite sum F of the form given in Step 1, since these are dense in $\mathcal{U}_2$ by Theorem 7.2. By Steps 1 and 3

$$\begin{aligned}\langle M_Y J M_Y^* F, F\rangle_{HS} &= \langle M_X J M_X^* F, F\rangle_{HS} - \langle M_X \Gamma_\Theta M_X^* F, F\rangle_{HS} \\ &= \text{trace} \sum_{i,j=1}^{n} D_i^* \{\hat{\Lambda}_{\Omega_j}(\Omega_i) - \hat{\Lambda}_V(\Omega_i)\hat{\Lambda}_V(V)^{-1}\hat{\Lambda}_{\Omega_j}(V)\} D_j \\ &\geq 0\end{aligned}$$

since the term in curly brackets is the ij entry in the Schur complement of $\hat{\Lambda}_{\Omega_0}(\Omega_0)$ in the array

$$\left[\hat{\Lambda}_{\Omega_j}(\Omega_i)\right] \,, \qquad i,j = 0, \ldots, n$$

with $\Omega_0 = V$.

STEP 5. *$Y_1 = U_V^{-1} Y$ is admissible.*

PROOF. By Theorem 5.1, the block entries in Y_1 belong to $\mathcal{U}$. Therefore it remains only to check that

$$M_{Y_1} J M_{Y_1}^* \geq 0$$

on $\mathcal{U}_2$. But that is immediate from Step 4, since

$$M_{Y_1} = M_{U_V^{-1}} M_Y \;.$$

This completes the proof of both the step and the theorem.

∎

The next two theorems are analogues in the present setting of Theorems 4.1 and (part of) Theorem 5.5 of [AD2]. Factorizations of the latter kind appear first in de Branges [dB1].

THEOREM 8.8. *If $X = [A \quad B]$ is admissible and $\hat{X}(V)J\hat{X}(V)^*$ is positive definite and invertible in $\mathcal{X}$ for some $V \in \mathcal{D}$ with $\ell_V < 1$ and if S_V, W and*

$$\Theta = H(-S_V)\begin{bmatrix} U_W & 0 \\ 0 & I \end{bmatrix}$$

are defined as in Theorems 5.2 and 8.7, then:

(1) $X\mathcal{H}(\Theta)$ *is a closed subspace of* $\mathcal{B}(X)$ *and the inclusion is isometric:*

$$\|X\underline{\underline{F}}\|_{\mathcal{B}(X)} = \|\underline{\underline{F}}\|_{\mathcal{H}(\Theta)}$$

for every $\underline{\underline{F}} \in \mathcal{H}(\Theta)$.

(2) $\mathcal{B}(X\Theta)$ *is a closed subspace of* $\mathcal{B}(X)$ *and the inclusion is isometric:*

$$\|E\|^2_{\mathcal{B}(X\Theta)} = \|E\|^2_{\mathcal{B}(X)}$$

for every $E \in \mathcal{B}(X\Theta)$.

(3) $\mathcal{B}(X) = X\mathcal{H}(\Theta) \oplus \mathcal{B}(X\Theta)$.

PROOF. The proof is divided into steps.

STEP 1. $X\mathcal{H}(\Theta)$ *is a closed subspace of* $\mathcal{B}(X)$*; the inclusion is isometric.*

PROOF OF STEP 1. Let $\underline{\underline{F}} \in \mathcal{H}(\Theta)$. Then, by Theorem 7.7,

$$\underline{\underline{F}} = J\hat{X}(V)^*\rho_V^{-1}D$$

for some choice of $D \in \mathcal{D}_2$. Thus

$$X\underline{\underline{F}} = \Lambda_V D = \Gamma_X \rho_V^{-1} D$$

clearly belongs to

$$\text{ran } \Gamma_X \subset \text{ran } \Gamma_X^{\frac{1}{2}} = \mathcal{B}(X)\ .$$

Moreover,

$$\begin{aligned} \|\Gamma_X\rho_V^{-1}D\|^2_{\mathcal{B}(X)} : &= \|\Gamma_X\rho_V^{-1}D\|^2_{\Gamma_X} \\ &= \langle \Gamma_X\rho_V^{-1}D, \rho_V^{-1}D\rangle_{HS} \\ &= \langle JM_X^*\rho_V^{-1}D, M_X^*\rho_V^{-1}D\rangle_{HS} \\ &= \langle J\underline{\underline{F}}, \underline{\underline{F}}\rangle_{HS} \end{aligned}$$

as needed, thanks to Theorem 7.3. Finally, with the help of the exhibited isometry, it is readily checked that $X\mathcal{H}(\Theta)$ is a closed subspace of $\mathcal{B}(X)$.

STEP 2. *If* $\underline{\underline{F}} \in \mathcal{H}(\Theta)$, *then* $\underline{\underline{F}}$ *is* J *orthogonal to* ran $\Gamma_\Theta M_X^*$ *if and only if* $M_X\underline{\underline{F}} = 0$.

PROOF OF STEP 2. If $\underline{\underline{F}} \in \mathcal{H}(\Theta)$, then $\underline{\underline{F}} = \Gamma_\Theta \underline{\underline{E}}$ for some $\underline{\underline{E}} \in \mathcal{U}_2^{2\times 1}$ by Theorem 8.3 and Lemma 8.3. Thus if $\underline{\underline{F}}$ is also J orthogonal to ran $\Gamma_\Theta M_X^*$, then

$$\begin{aligned} 0 &= \langle J\underline{\underline{F}}, \Gamma_\Theta M_X^* G\rangle_{HS} \\ &= \langle J\Gamma_\Theta \underline{\underline{E}}, \Gamma_\Theta M_X^* G\rangle_{HS} \\ &= \langle M_X \Gamma_\Theta J \Gamma_\Theta \underline{\underline{E}}, G\rangle_{HS} \end{aligned}$$

for every $G \in \mathcal{U}_2$. Thus

$$M_X \Gamma_\Theta J \Gamma_\Theta \underline{\underline{E}} = 0 ,$$

which implies in turn that

$$M_X \underline{\underline{F}} = 0 ,$$

because

$$\Gamma_\Theta J \Gamma_\Theta = \Gamma_\Theta . \tag{8.7}$$

Since the argument can be reversed, the proof of this step is complete.

STEP 3. $\operatorname{ran}(M_X \Gamma_\Theta M_X^*)$ *is dense in* $M_X \mathcal{H}(\Theta)$.

PROOF OF STEP 3. Clearly

$$M_X \mathcal{H}(\Theta) = \operatorname{ran}(M_X \Gamma_\Theta) \supset \operatorname{ran}(M_X \Gamma_\Theta M_X^*) .$$

Suppose there exists an $\underline{\underline{E}} \in \mathcal{U}_2^{2\times 1}$ such that

$$\langle M_X \Gamma_\Theta \underline{\underline{E}}, M_X \Gamma_\Theta M_X^* G\rangle_{\mathcal{B}(X)} = 0$$

for every $G \in \mathcal{U}_2$. Then, by Step 1,

$$0 = \langle J\Gamma_\Theta \underline{\underline{E}}, \Gamma_\Theta M_X^* G\rangle_{HS}$$

which in turn implies, by Step 2, that

$$M_X \Gamma_\Theta \underline{\underline{E}} = 0 .$$

Thus $\operatorname{ran}(M_X \Gamma_\Theta M_X^*)$ is dense in $\operatorname{ran}(M_X \Gamma_\Theta)$.

STEP 4. ran $\Gamma_{X\Theta}$ *is a subspace of* $\mathcal{B}(X)$ *which is orthogonal to* $M_X \mathcal{H}(\Theta)$ *in* $\mathcal{B}(X)$.

PROOF OF STEP 4. In view of Step 1, the identity

$$\Gamma_X = M_X \Gamma_\Theta M_X^* + \Gamma_{X\Theta} \tag{8.8}$$

serves to exhibit ran $\Gamma_{X\Theta}$ as a subspace of $\mathcal{B}(X)$. Moreover, for every choice of F and G in $\mathcal{U}_2$,

$$\langle M_X\Gamma_\Theta M_X^* F, \Gamma_{X\Theta}G\rangle_{\mathcal{B}(X)} = ① + ② ,$$

where

$$① = -\langle M_X\Gamma_\Theta M_X^* F, M_X\Gamma_\Theta M_X^* G\rangle_{\mathcal{B}(X)}$$

and

$$② = \langle M_X\Gamma_\Theta M_X^* F, \Gamma_X G\rangle_{\mathcal{B}(X)} \cdot$$

By Step 1,

$$① = -\langle J\Gamma_\Theta M_X^* F, \Gamma_\Theta M_X^* G\rangle_{HS}$$

and

$$M_X\Gamma_\Theta M_X^* F = \Gamma_X^{\frac{1}{2}} E$$

for some $E \in \mathcal{U}_2$, since the left hand side belongs to $\mathcal{B}(X) = \text{ran } \Gamma_X^{\frac{1}{2}}$. Thus

$$\begin{aligned} ② &= \langle \Gamma_X^{\frac{1}{2}} E, \Gamma_X G\rangle_{\mathcal{B}(X)} \\ &= \langle \Gamma_X^{\frac{1}{2}} E, G\rangle_{HS} \\ &= \langle M_X\Gamma_\Theta M_X^* F, G\rangle_{HS} \ . \end{aligned}$$

It now follows readily from formula (8.7) that

$$① + ② = 0 \ .$$

This proves the asserted orthogonality, since ran$(M_X\Gamma_\Theta M_X^*)$ is dense in $X\mathcal{H}(\Theta)$ by Step 3.

STEP 5. $\|\Gamma_{X\Theta}F\|_{\mathcal{B}(X\Theta)} = \|\Gamma_{X\Theta}F\|_{\mathcal{B}(X)}$ *for every* $F \in \mathcal{U}_2$.

PROOF OF STEP 5. By successive applications of (8.8), (8.7) and Step 1,

$$\begin{aligned} \|\Gamma_{X\Theta}F\|^2_{\mathcal{B}(X\Theta)} &= \langle \Gamma_{X\Theta}F, F\rangle_{HS} \\ &= \langle \Gamma_X F, F\rangle_{HS} - \langle M_X\Gamma_\Theta M_X^* F, F\rangle_{HS} \\ &= \|\Gamma_X F\|^2_{\mathcal{B}(X)} - \langle J\Gamma_\Theta M_X^* F, \Gamma_\Theta M_X^* F\rangle_{HS} \\ &= \|\Gamma_X F\|^2_{\mathcal{B}(X)} - \|M_X\Gamma_\Theta M_X^* F\|^2_{\mathcal{B}(X)} \cdot \end{aligned}$$

On the other hand, by (8.8) and Step 4,

$$\begin{aligned}\|\Gamma_X F\|^2_{\mathcal{B}(X)} &= \|M_X\Gamma_\Theta M_X^* F + \Gamma_{X\Theta}F\|^2_{\mathcal{B}(X)}\\ &= \|M_X\Gamma_\Theta M_X^* F\|^2_{\mathcal{B}(X)} + \|\Gamma_{X\Theta}F\|^2_{\mathcal{B}(X)} ,\end{aligned}$$

which, together with the previous formula, leads immediately to the asserted isometry.

STEP 6. $\mathcal{B}(X\Theta)$ *is a closed subspace of* $\mathcal{B}(X)$. *The inclusion is isometric.*

PROOF OF STEP 6. Let $F \in \mathcal{B}(X\Theta)$. Then there exists a sequence $\{\Gamma_{X\Theta}E_n\}$ with $E_n \in \mathcal{U}_2$ such that

$$\lim_{n\uparrow\infty} \|F - \Gamma_{X\Theta}E_n\|_{\mathcal{B}(X\Theta)} = 0 .$$

Therefore $\{\Gamma_{X\Theta}E_n\}$ is a Cauchy sequence in $\mathcal{B}(X\Theta)$ and hence, by Step 5, it is also a Cauchy sequence in $\mathcal{B}(X)$. Thus there exists a $G \in \mathcal{B}(X)$ such that

$$\lim_{n\uparrow\infty} \|G - \Gamma_{X\Theta}E_n\|^2_{\mathcal{B}(X)} = 0 .$$

A standard reproducing kernel type argument serves to identify G with F: For every choice of $D \in \mathcal{D}_2$ and $V \in \mathcal{D}$ with $\ell_V < 1$,

$$\begin{aligned}\text{trace } D^*\hat{G}(V) &= \langle G, \Gamma_X\rho_V^{-1}D\rangle_{\mathcal{B}(X)}\\ &= \lim_{n\uparrow\infty}\langle \Gamma_{X\Theta}E_n, \Gamma_X\rho_V^{-1}D\rangle_{\mathcal{B}(X)}\\ &= \lim_{n\uparrow\infty}\langle \Gamma_{X\Theta}E_n, \Gamma_{X\Theta}\rho_V^{-1}D\rangle_{\mathcal{B}(X)}\\ &= \lim_{n\uparrow\infty}\langle \Gamma_{X\Theta}E_n, \Gamma_{X\Theta}\rho_V^{-1}D\rangle_{\mathcal{B}(X\Theta)} ,\end{aligned}$$

by successive applications of (8.8), Step 4 and Step 5. The last limit is however equal to

$$\langle F, \Gamma_{X\Theta}\rho_V^{-1}D\rangle_{\mathcal{B}(X\Theta)} = \text{trace}\, D^*\hat{F}(V) ,$$

since $\Gamma_{X\Theta}\rho_V^{-1}$ is the reproducing kernel for $\mathcal{B}(X\Theta)$. Thus, by Lemma 7.4 and Theorem 3.2,

$$F = G .$$

This proves that $\mathcal{B}(X\Theta) \subset \mathcal{B}(X)$ and with but a small modification yields the asserted isometry:

$$\begin{aligned}\|F\|^2_{\mathcal{B}(X\Theta)} &= \lim_{n\uparrow\infty} \|\Gamma_{X\Theta}E_n\|^2_{\mathcal{B}(X\Theta)}\\ &= \lim_{n\uparrow\infty} \|\Gamma_{X\Theta}E_n\|^2_{\mathcal{B}(X)} = \|F\|^2_{\mathcal{B}(X)} .\end{aligned}$$

Suppose next that $\{C_n\}$ is a sequence in $\mathcal{B}(X\Theta)$ such that

$$\lim_{n\uparrow\infty} \|C - C_n\|_{\mathcal{B}(X)} = 0 ,$$

for some $C \in \mathcal{B}(X)$. Then there exists an $E \in \mathcal{B}(X\Theta)$ such that

$$\lim_{n\uparrow\infty} \|E - C_n\|_{\mathcal{B}(X\Theta)} = 0 ,$$

and therefore, since $\mathcal{B}(X\Theta) \subset \mathcal{B}(X)$ and the inclusion is isometric,

$$\lim_{n\uparrow\infty} \|C - E_n\|_{\mathcal{B}(X)} = 0 .$$

This implies that $C = E$ and hence that $\mathcal{B}(X\Theta)$ is a closed subspace of $\mathcal{B}(X)$.

STEP 7. $\mathcal{B}(X) = X\mathcal{H}(\Theta) \oplus \mathcal{B}(X\Theta)$.

PROOF OF STEP 7. By (8.8) and Steps 4, 1 and 6

$$\begin{aligned} \text{ran } \Gamma_X &\subset X\mathcal{H}(\Theta) \oplus \text{ran } \Gamma_{X\Theta} \\ &\subset X\mathcal{H}(\Theta) \oplus \mathcal{B}(X\Theta) \\ &\subset \mathcal{B}(X) ; \end{aligned}$$

the orthogonality established in Step 4 propagates from the first line to the second because ran $\Gamma_{X\Theta}$ is dense in $\mathcal{B}(X\Theta)$ and, by Step 6, $\mathcal{B}(X\Theta)$ is included isometrically in $\mathcal{B}(X)$. The rest is straightforward since ran Γ_X is dense in $\mathcal{B}(X)$.

∎

THEOREM 8.9. *If Θ_1 and Θ_2 are lossless chain scattering operators in $\mathcal{A}$, then*

$$\mathcal{H}(\Theta_1\Theta_2) = \mathcal{H}(\Theta_1) \oplus \Theta_1\mathcal{H}(\Theta_2) . \tag{8.9}$$

PROOF. It follows readily from the identity

$$\Gamma_{\Theta_1\Theta_2} = \Gamma_{\Theta_1} + M_{\Theta_1}\Gamma_{\Theta_2}M_{\Theta_1}^*$$

and Theorems 8.1 and 8.3 that

$$\mathcal{H}(\Theta_1) = \text{ran } \Gamma_{\Theta_1}^{\frac{1}{2}} \subset \text{ran } \Gamma_{\Theta_1\Theta_2}^{\frac{1}{2}} = \mathcal{H}(\Theta_1\Theta_2) .$$

Moreover, the inclusion is isometric since both $\mathcal{H}(\Theta_1)$ and $\mathcal{H}(\Theta_1\Theta_2)$ are endowed with the J inner product in $\mathcal{X}_2^{2\times 1}$. Similarly

$$\text{ran } M_{\Theta_1}\Gamma_{\Theta_2}M_{\Theta_1}^* \subset \text{ran } (M_{\Theta_1}\Gamma_{\Theta_2}M_{\Theta_1}^*)^{\frac{1}{2}} \subset \mathcal{H}(\Theta_1\Theta_2)$$

and therefore, since

$$\langle JM_{\Theta_1}\Gamma_{\Theta_2}M^*_{\Theta_1}\underline{\underline{F}}, M_{\Theta_1}\Gamma_{\Theta_2}\underline{\underline{G}}\rangle_{HS} = \langle J\Gamma_{\Theta_2}M^*_{\Theta_1}\underline{\underline{F}}, \Gamma_{\Theta_2}\underline{\underline{G}}\rangle_{HS}$$
$$= \langle \underline{\underline{F}}, M_{\Theta_1}\Gamma_{\Theta_2}\underline{\underline{G}}\rangle_{HS}$$

for every choice of $\underline{\underline{F}}$ and $\underline{\underline{G}}$ in $\mathcal{U}_2^{2\times 1}$, it is readily seen that the the closure of the range of $M_{\Theta_1}\Gamma_{\Theta_2}M^*_{\Theta_1}$ is equal to the range of $M_{\Theta_1}\Gamma_{\Theta_2}$. Moreover, by Lemma 8.3 and Theorem 8.3,

$$\begin{aligned}\text{ran } M_{\Theta_1}\Gamma_{\Theta_2} &= M_{\Theta_1} \text{ ran } \Gamma_{\Theta_2} \\ &= M_{\Theta_1} \text{ ran}(\Gamma_{\Theta_2})^{\frac{1}{2}} \\ &= M_{\Theta_1}\mathcal{H}(\Theta_2) .\end{aligned}$$

Thus, since $\Gamma_{\Theta_1}JM_{\Theta_1} = 0$, it follows that $\mathcal{H}(\Theta_1)$ is J orthogonal to $M_{\Theta_1}\mathcal{H}(\Theta_2)$ and

$$\mathcal{H}(\Theta_1) \oplus M_{\Theta_1}\mathcal{H}(\Theta_2) \subset \mathcal{H}(\Theta_1\Theta_2) .$$

Finally, if $\underline{\underline{E}} \in \mathcal{H}(\Theta_1\Theta_2)$ is J orthogonal to the indicated subspace, then in particular

$$\langle J\Gamma_{\Theta_1\Theta_2}\underline{\underline{F}}, \ \underline{\underline{E}}\rangle_{HS} = \langle J\{\Gamma_{\Theta_1}\underline{\underline{F}} + M_{\Theta_1}\Gamma_{\Theta_2}M^*_{\Theta_1}\underline{\underline{F}}\}, \ \underline{\underline{E}}\rangle_{HS} \quad = \quad 0$$

for every choice of $\underline{\underline{F}} \in \mathcal{U}_2^{2\times 1}$. This proves that $\underline{\underline{E}} = 0$ and hence that the left hand side of the indicated inclusion is dense in the right. Therefore equality prevails, since the left hand side is also closed.

∎

9. REFERENCES

[ADD] D. Alpay, P. Dewilde and H. Dym, *On the existence and convergence of solutions to the partial lossless inverse scattering problem with applications to estimation theory*, IEEE Trans. Inform. Theory **35** (1989), 1184-1205.

[ADe] D. Alpay and P. Dewilde, *Time-varying signal approximation and estimation*, Signal Processing, Scattering and Operator Theory, and Numerical Methods. Proceedings of the International Symposium MTNS-89, Vol. 3, Birkhäuser, Boston, in press.

[AD1] D. Alpay and H. Dym, *Hilbert spaces of analytic functions, inverse scattering, and operator models I*, Integral Equations Operator Theory **7** (1984), 589–641.

[AD2] —, *On applications of reproducing kernel spaces to the Schur algorithm and rational J unitary factorization*, in: I. Schur Methods in Operator Theory and Signal Processing (I. Gohberg, ed.), Oper. Theory: Adv. Appl., **OT18**, Birkhäuser Verlag, Basel, 1986, pp. 89-159.

[dB1] L. de Branges, *Some Hilbert spaces of analytic functions I*, Trans. Amer. Math. Soc. **106** (1963), 445–468.

[dB2] —, *Complementation in Krein spaces*, Trans. Amer. Math. Soc. **305** (1988), 277-291.

[dBR1] L. de Branges and J. Rovnyak, *Square summable power series*, Holt, Rinehart and Winston, New York, 1966.

[dBR2] —, *Canonical models in quantum scattering theory*, in: Perturbation Theory and its Applications in Quantum Mechanics (C. Wilcox, ed.), Wiley, New York, 1966, pp. 295-392.

[Dep] E.F.A. Deprettere, *Mixed form time variant lattice recursions*, in: Outils et Modèles Mathématiques pour l'Automatique, l'Analyse des Systèmes et le Traitement du Signal, CNRS, Paris, 1981, pp. 545-562.

[DeDe] P. Dewilde and E.F.A. Deprettere, *The generalized Schur Algorithm: Approximation and hierarchy*, in: Oper. Theory: Adv. Appl., **OT29**, Birkhäuser Verlag, Basel, 1988, pp. 97-115.

[DD1] P. Dewilde and H. Dym, *Schur recursions, error formulas and convergence of rational estimators for stationary stochastic processes*, IEEE Trans. Inform. Theory, **27** (1981), 446-461.

[DD2] — *Lossless chain scattering matrices and optimum linear prediction: The vector case*, Circuit Theory and Applications **9** (1981), 135-175.

[DD3] —, *Lossless inverse scattering for digital filters*, IEEE Trans. Inform. Theory **30** (1984), 644-662.

[DVK] P. Dewilde, A.C. Vieira and T. Kailath, *On a generalized Szegö-Levinson realization algorithm for optimal linear predictors based on a network synthesis approach*, IEEE Trans. Circuits and Systems, **25** (1978), 663-675.

[Do] R.G. Douglas, *On majorization, factorization and range inclusion of operators in Hilbert spaces*, Proc. Amer. Math. Soc. **17** (1966), 413-416.

[D] H. Dym, *J-Contractive Matrix, Functions Reproducing Kernel Spaces and Interpolation*, CBMS Regional Conf. Ser. in Math., No. 71, Amer. Math. Soc., Providence, 1989.

[DG] H. Dym and I. Gohberg, *Extensions of band matrices with band inverses*, Linear Algebra Appl. **36** (1981), 1-24.

[FW] P.W. Fillmore and J.P. Williams, *On operator ranges*, Adv. in Math. **7** (1971), 254-281.

[GK] I.C. Gohberg and M.G. Krein, *Introduction to the Theory of Linear Nonselfadjoint Operators*, Transl. Math. Monographs, Vol. 18, Amer. Math. Soc., Providence, 1969.

[Ru] W. Rudin, *Real and Complex Analysis.* McGraw Hill, New York, 1966.

[Sc] L. Schwartz, *Sous espaces hilbertiens d'espaces vectoriels topologiques et noyaux associés (noyaux reproduisants)*, J. Analyse Math. **13** (1964), 115-256.

Daniel Alpay and Harry Dym
Department of Theoretical Mathematics
The Weizmann Institute of Science
Rehovot 76100, Israel

Patrick Dewilde
Department of Electrical Engineering
Delft University of Technology
2600GA Delft, The Netherlands

Operator Theory
Advances and Applications, Vol. 47

ZERO-POLE STRUCTURE OF NONREGULAR RATIONAL MATRIX FUNCTIONS

Joseph A. Ball[1] and Marek Rakowski[2]

DEDICATED TO THE MEMORY OF M.G. KREIN

We introduce and develop a notion of left zero-pole structure for a possibly nonregular rational matrix function analogous to notions introduced recently for the regular case. We then solve the inverse problem of constructing a rational matrix function with a given left zero-pole structure.

0. INTRODUCTION

In recent work (see [BGR1, BGR3] and the references there) a notion of left null-pole structure over an arbitrary subset σ of $\mathbf{C}$ of a regular rational matrix function has been introduced and studied, where by definition a rational matrix function W (i.e. matrix W over the field of scalar rational functions) is regular if W is square and det W does not vanish identically. The pole structure of a rational matrix function W involves a pair of matrices (C_π, A_π) which encode the data about poles and partial pole multiplicities of W at points in σ and the information concerning the principal parts of the Laurent expansions of W at these poles. Similarly, the null structure is encoded in a pair of matrices (A_ζ, B_ζ), called a left null pair for W, which gives the zeros and partial multiplicities of zeros of W in σ together with directional information about these zeros. These notions have proved to be basic in a multitude of applications in factorization, interplolation, and control theory (see [BGR2, BGR5, BGR6, BGR3]). With a few exceptions ([Ro, BCR, BR]), all the development of this approach has been done exclusively for the regular case. In this paper we develop systematically from first principles analogous notions of zero and pole structure over a subset σ of $\mathbf{C}_\infty$ for the nonregular case. We also solve the inverse problem of constructing a rational matrix function having a prescribed zero-pole structure over a subset σ of $\mathbf{C}$.

We will frequently use below the well-known fact that every rational matrix func-

---

[1] The research of this author was partially supported by National Science Foundation Grant DMS 8701615-02.

[2] The research of this author was partially supported by the Institute of Electron Technology CEMI, Warszawa, Poland.

tion W admits a factorization EDF such that E and F are square matrix polynomials with constant nonzero determinant and

$$D = \begin{bmatrix} d_1 & & & & \\ & d_2 & & & 0 \\ & & \ddots & & \\ & & & d_k & \\ & 0 & & & \end{bmatrix},$$

where each d_i is a quotient of monic polynomials and d_{i+1}/d_i is a polynomial for $i = 1, 2, ..., k-1$. Such a factorization is called a Smith-McMillan factorization of W. The middle factor in a Smith-McMillan factorization of W is unique; it is called the Smith-McMillan form of W (see e.g. [K]). We note that a Smith-McMillan factorization of a rational matrix function W can be systematically obtained by means of elementary column and row operations. It can be practically found on a computer using a symbolic manipulator program such as MACSYMA.

Our notion of zero-pole structure of a nonregular rational matrix function W over σ involves the following collections of finite matrices: a pole pair (C_π, A_π) which encodes information about singular part of W at the poles of W in σ, a null pair (A_ζ, B_ζ) which encodes directional information about discrete zeros of W in σ, and a kernel triple $(A_\kappa, B_\kappa, D_\kappa)$ from which one gets a realization for a left kernel polynomial of W. Computation of these objects can be obtained easily from the Smith-McMillan factorization of W. Progress on developing methods for finding these objects directly from a realization of W, in analogy to what can be done in the regular case (see [BGR1]), has recently been made by R. Faulkenberry of the University of Maryland (private communication).

Our solution of the inverse problem, unlike that for the regular case (see [GK1, GK2]), does not give an explicit formula for a realization for the solution, but is an iterative construction involving several steps. Each step, however, involves only standard easily implementable linear algebra operations. We will illustrate particular steps of the construction with specific examples.

A new feature in the exposition in this paper is the use of orthogonality with respect to a valuation on the field of scalar rational functions. The general theory was developed by Monna (see [M2] and the references there) in the context of non-Archimedean Banach or locally convex spaces, and has been applied to the study of rational matrix functions by Forney [F] and Verghese-Kailath [VK]. For our work here this theory provides a convenient tool for embedding a rectangular matrix function W into a regular rational matrix function Q so that W and Q have the same local pole and discrete zero structure. In the context of rational matrix functions which we have here, orthogonal decomposition of a subspace Λ can be easily obtained from the Smith-McMillan factorization of a function formed by a basis for Λ. The language of orthogonality versus that of a Smith-McMillan factorization smooths out the exposition in roughly the same way as the notion of subspace versus basis for a subspace does in the standard treatment of linear algebra.

We mention that in the regular case, whenever a zero and a pole of a function W occur at the same point, in addition to a null pair and a pole pair an additional invariant, called the null-pole coupling matrix, is needed for a complete description of null-pole structure (see e.g. [BGR4]). In the nonregular case as well one needs a null-pole coupling matrix, in addition to the left null pair, right pole pair, and a left kernel triple, to characterize the set

$$\{Wr : \quad r \text{ is a rational vector function with no poles in } \sigma\}.$$

This setup, together with the associated inverse problem, is developed and solved in the second author's dissertation [Ra]; we shall return to this in a future pupblication.

In future work we plan to develop applications of the theory of zero-pole structure of nonregular rational matrix functions developed here to factorization and interpolation problems analogous to those mentioned above for the regular case.

The paper is organized as follows. In Section 1 we describe the right and left pole structure of a rational matrix function $W \in \mathcal{R}^{m\times n}$ in terms of pairs of matrices called right and left pole pairs. In Section 2 we recall basic facts about orthogonality in $\mathcal{R}^n$ induced by a standard valuation on $\mathcal{R}$. In Section 3 we describe the continuous and discrete zero structure of a function $W \in \mathcal{R}^{m\times n}$ in terms of kernel triples and null pairs. In Section 4 we establish a necessary and sufficient condition for existence of a rational matrix function W with a given left kernel triple $(A_\kappa, B_\kappa, D_\kappa)$, a left null pair over a subset σ of $\mathbf{C}$ equal to (A_ζ, B_ζ), and a right pole pair over σ equal to (C_π, A_π). We also describe there an algorithm for constructing a function $W \in \mathcal{R}^{m\times n}$ with the left kernel triple $(A_\kappa, B_\kappa, D_\kappa)$, and the spectral data over $\sigma \subset \mathbf{C}$ equal to (C_π, A_π) and (A_ζ, B_ζ), from a regular rational matrix function with the same spectral data.

Finally, we mention that many of the results of this paper are presented in the second author's dissertation written under the direction of the first author.

1. POLE PAIRS

Let $\mathcal{R}^{m\times n}$ denote the set of all $m \times n$ rational matrix functions, that is $m \times n$ matrices over the field $\mathcal{R}$ of scalar rational functions. We say that $W \in \mathcal{R}^{m\times n}$ has a pole at a point λ of the extended complex plane $\mathbf{C}_\infty$ (or λ is a pole of W) if some entry of W has a pole at λ. Suppose that the first l diagonal entries in the Smith-McMillan form D of W have a pole at λ and all other entries of D are analytic at λ. Then we say that the geometric multiplicity of the pole of W at λ equals l. The multiplicities of poles at λ of the diagonal entries of D are called partial multiplicities of the pole of W at λ. The geometric multiplicity and the partial multiplicities of a pole of W at infinity are defined to be the geometric multiplicity and the partial multiplicities of the pole at 0 of the rational matrix function H such that $H(z) = W(z^{-1})$. The sum of partial multiplicities of the pole of W at $\lambda \in \mathbf{C}_\infty$ is called the (total) multiplicity of the pole of W at λ. The sum of multiplicities of all poles of W in $\mathbf{C}_\infty$ is called the McMillan degree of W.

We note that if a rational (row or column) vector function W has a pole at a point $\lambda \in \mathbf{C}_\infty$, then the geometric multiplicity of the pole of W at λ equals 1 and the total multiplicity of the pole of W at λ equals k where k is a positive integer such that in a deleted neighborhood of λ

$$W(z) = \sum_{i=-k}^{\infty} (z-\lambda)^i W_i$$

with all W_i's constant vectors and $W_{-k} \neq 0$. Thus, the multiplicity of a pole of a rational vector function W at λ equals the largest multiplicity of a pole at λ of some entry of W.

Let W be an $m \times n$ rational matrix function and let $\lambda \in \mathbf{C}_\infty$. A function $\psi \in \mathcal{R}^{m\times 1}$ is called a right pole function for W at λ if

(i) ψ is analytic at λ and $\psi(\lambda) \neq 0$,

(ii) there is a positive integer k and a function $\phi \in \mathcal{R}^{n\times 1}$ such that ϕ is analytic at λ and

$$\psi(z) = \begin{cases} (z-\lambda)^k W(z)\phi(z), & \text{if } \lambda \in \mathbf{C} \\ z^{-k} W(z)\phi(z), & \text{if } \lambda = \infty. \end{cases} \tag{1.1}$$

The maximal integer k such that (1.1) holds for an appropriate ϕ is called the order of the right pole function ψ.

We note that $W \in \mathcal{R}^{m\times n}$ has a right pole function at λ if and only if W has a pole at λ.

Suppose $W \in \mathcal{R}^{m\times n}$ has a pole at $\lambda \in \mathbf{C}_\infty$. Then the values at λ of right pole functions for W at λ form, together with the 0 vector, a subspace V of $\mathbf{C}^m$. It can be seen from a Smith-McMillan factorization of W that dim V is equal to the geometric multiplicity of the pole of W at λ. We say that right pole functions $\psi_1, \psi_2, \ldots, \psi_\eta$ for W at λ of orders $k_1, k_2, \ldots, k_\eta$ respectively form a canonical set of right pole functions for W at λ if

(i) $\{\psi_1(\lambda), \psi_2(\lambda), \ldots, \psi_k(\lambda)\}$ is a basis for V,

(ii) $\sum_{i=1}^{\eta} k_i$ is maximal subject to condition (i).

It follows from a Smith-McMillan factorization of W that the orders of the pole functions in any canonical set of right pole functions for W at λ coincide with the partial pole multiplicities for W at λ.

A canonical set of right pole functions for a rational matrix function W at a point $\lambda \in \mathbf{C}_\infty$ can be found as follows (cf. Lemma 2.4 in [BR]). Let $w_{i_1}, w_{i_2}, ..., w_{i_k}$ be the columns of W that have a pole at λ and denote by $[\phi]_{z=\lambda}$ the leading coefficient in the Laurent expansion at λ of a rational column (resp. row) vector function ϕ. If $[w_{i_1}]_{z=\lambda}, [w_{i_2}]_{z=\lambda}, ..., [w_{i_k}]_{z=\lambda}$ are linearly independent,[1] then a canonical set of right pole functions for W at λ is formed by $\tilde{w}_{i_1}, \tilde{w}_{i_2}, ..., \tilde{w}_{i_k}$, where

$$\tilde{w}_{i_j} = \begin{cases} (z-\lambda)^{\delta(w_{i_j},\lambda)} w_{i_j}, & \text{if } \lambda \in \mathbf{C} \\ z^{-\delta(w_{i_j},\lambda)} w_{i_j}, & \text{if } \lambda = \infty. \end{cases}$$

If $[w_{i_1}]_{z=\lambda}, [w_{i_2}]_{z=\lambda}, ..., [w_{i_k}]_{z=\lambda}$ are linearly dependent, we can find a finite chain of rational matrix functions $W_1, W_2, ..., W_\eta$ such that the leading coefficients in the Laurent expansion at λ of the columns of W_η that have a pole at λ are linearly independent and every W_j is equal to WQ_j for some regular rational matrix function Q_j with neither a pole nor a zero at λ (that is, the functions $W, W_1, W_2, ..., W_\eta$ are right equivalent at λ; cf. [BR]). Clearly, W and W_η have the same canonical sets of right pole functions at λ. One such set can be easily found from the columns of W_η.

Example 1.1 Find canonical sets of right pole functions at $z = 1$ and $z = \infty$ for a function

$$W(z) = \begin{bmatrix} (z-2)/(z-1)^3 & z/(z-1)^2 \\ (z-1)/z & z/(z-1) \\ z^2 & z \end{bmatrix}.$$

Call the columns of W w_1 and w_2. Since

$$[w_1]_{z=\infty} = \begin{bmatrix} 0 \\ 0 \\ 1 \end{bmatrix} \quad \text{and} \quad [w_2]_{z=\infty} = \begin{bmatrix} 0 \\ 1 \\ 1 \end{bmatrix}$$

are linearly independent, $\{z^{-2}w_1, z^{-1}w_2\}$ is a canonical set of right pole functions for W at infinity. The orders of these functions are 2 and 1.

Now

$$[w_1]_{z=1} = \begin{bmatrix} -1 \\ 0 \\ 0 \end{bmatrix} \quad \text{and} \quad [w_2]_{z=1} = \begin{bmatrix} 1 \\ 0 \\ 0 \end{bmatrix}$$

are lineraly dependent, and $\delta(w_2, 1) \leq \delta(w_1, 1)$. Multiplying W on the right by the function

$$\begin{bmatrix} 1 & z-1 \\ 0 & 1 \end{bmatrix}$$

[1] In the terminology of Section 2 below (see also [F, K]), the rational vector functions $w_{i_1}, w_{i_2}, ..., w_{i_k}$ are then orthogonal at λ.

(cf. Procedure 1 in [VK]) we get

$$W_1(z) = \begin{bmatrix} (z-2)/(z-1)^3 & 2/(z-1) \\ (z-1)/z & z/(z-1)+(z-1)^2/z \\ z^2 & z+(z-1)z^2 \end{bmatrix}.$$

since the leading coefficients in the expansions at $z = 1$ of the columns of W

$$\begin{bmatrix} 1 \\ 0 \\ 0 \end{bmatrix} \quad \text{and} \quad \begin{bmatrix} 2 \\ 1 \\ 1 \end{bmatrix}$$

are linearly independent,

$$\left\{ \begin{bmatrix} z-2 \\ (z-1)^4/z \\ z^2(z-1)^3 \end{bmatrix}, \begin{bmatrix} 2 \\ z+(z-1)^3/z \\ z(z-1)(z^2-z+1) \end{bmatrix} \right\} \tag{1.2}$$

is a canonical set of right pole functions for W at $z = 1$. The orders of the functions in (1.2) are 3 and 1.

□

Let $W \in \mathcal{R}^{m \times n}$ and let $\lambda \in \mathbf{C}_\infty$ be a pole of W. Choose a canonical set of right pole functions for W at λ, $\{\psi_1, \psi_2, ..., \psi_\eta\}$. Suppose the order of ψ_i $(i = 1, 2, ..., \eta)$ as a right pole function for W at λ is k_i and let $\psi_{i,j} \in \mathbf{C}^{m \times 1}$ $(1 \le i \le \eta, j \ge 0)$ be such that

$$\psi_i(z) = \begin{cases} \sum_{j=0}^{\infty} (z-\lambda)^j \psi_{i,j}, & \text{if } \lambda \in \mathbf{C} \\ \sum_{j=0}^{\infty} z^{-j} \psi_{i,j}, & \text{if } \lambda = \infty \end{cases} \tag{1.3}$$

For a positive integer k and a point $\lambda \in \mathbf{C}$, let $J_k(\lambda)$ denote the $k \times k$ Jordan block matrix with λ on the diagonal. Any ordered pair of matrices (C, A), where C equals

$$\begin{bmatrix} \psi_{1,0} & \psi_{1,1} & \cdots & \psi_{1,k_1-1} & \psi_{2,0} & \psi_{2,1} & \cdots & \psi_{2,k_2-1} & \cdots & \psi_{\eta,0} & \psi_{\eta,1} & \cdots & \psi_{\eta,k_\eta-1} \end{bmatrix} S,$$

and

$$A = \begin{cases} S^{-1} \begin{bmatrix} J_{k_1}(\lambda) & & & \\ & J_{k_2}(\lambda) & & \\ & & \ddots & \\ & & & J_{k_\eta}(\lambda) \end{bmatrix} S, & \text{if } \lambda \in \mathbf{C} \\ S^{-1} \begin{bmatrix} J_{k_1}(0) & & & \\ & J_{k_2}(0) & & \\ & & \ddots & \\ & & & J_{k_\eta}(0) \end{bmatrix} S, & \text{if } \lambda = \infty \end{cases}$$

for some invertible matrix S of an appropriate size, is called a <u>right pole pair</u> for W at λ (see [BGR1, GK1]).

Let W be an $m \times n$ rational matrix function and let $\sigma \subset \mathbf{C}$. Let $\lambda_1, \lambda_2, ..., \lambda_p$ be the poles of W in σ and let (C_i, A_i) be a right pole pair for W at λ_i $(i = 1, 2, ..., p)$. Any ordered pair of matrices (C, A), where

$$C = \begin{bmatrix} C_1 & C_2 & \dots & C_p \end{bmatrix} S,$$

and

$$A = S^{-1} \begin{bmatrix} A_1 & & & \\ & A_2 & & \\ & & \ddots & \\ & & & A_p \end{bmatrix} S$$

for some invertible matrix S of an appropriate size, is called a <u>right pole pair</u> for W over σ (or a σ-right pole pair for W). We note that if W is analytic on σ, the right pole pair for W over σ is vacuous. Also, if λ is the only pole of W in σ, the right pole pair for W at λ is the right pole pair for W over σ.

Example 1.2 Find a right pole pair over the open right half-plane for the function W defined in Example 1.1.

We can rewrite (1.2) as

$$\left\{ \begin{bmatrix} -1 \\ 0 \\ 0 \end{bmatrix} + (z-1) \begin{bmatrix} 1 \\ 0 \\ 0 \end{bmatrix} + (z-1)^2 \begin{bmatrix} 0 \\ 0 \\ 0 \end{bmatrix} + (z-1)^3 v_1(z), \quad \begin{bmatrix} 2 \\ 1 \\ 0 \end{bmatrix} + (z-1) v_2(z) \right\}$$

where v_1 and v_2 are functions analytic at $z = 1$. Hence

$$(C, A) = \left(\begin{bmatrix} -1 & 1 & 0 & 2 \\ 0 & 0 & 0 & 1 \\ 0 & 0 & 0 & 0 \end{bmatrix}, \begin{bmatrix} 1 & 1 & 0 & 0 \\ 0 & 1 & 1 & 0 \\ 0 & 0 & 1 & 0 \\ 0 & 0 & 0 & 1 \end{bmatrix} \right)$$

is a right pole pair for W at $z = 1$. Since the point $z = 1$ is the only pole of W in the right half-plane, (C, A) is a right pole pair for W over $\{z : \ Re z > 0\}$.

□

Let W be an $m \times n$ rational matrix function and choose a Smith-McMillan factorization EDF of W. Let $\tilde{D}$ be a regular rational matrix function such that

$$D = \begin{bmatrix} \tilde{D} & 0 \\ 0 & 0 \end{bmatrix}.$$

For $\lambda \in \mathbf{C}$, $S = \sum_{i=-\eta}^{-1} (z-\lambda)^i c_i$ is a singular part at λ of a function $W\phi$ if and only if S is a singular part at λ of a function

$$E \begin{bmatrix} \tilde{D} & 0 \\ 0 & I \end{bmatrix} \tilde{\phi}$$

where $\tilde{\phi}$ is such that the first k components of $F\phi$ and $\tilde{\phi}$ coincide. It follows that (C, A) is a right pole pair for W over a subset σ of $\mathbf{C}$ if and only if (C, A) is a right pole pair over σ of a regular rational matrix function

$$E\begin{bmatrix} \tilde{D} & \\ & I \end{bmatrix}.$$

Thus, all the properties of right pole pairs of regular rational matrix functions carry over to nonregular rational matrix functions. In particular (see Section 2 in [BGR1]), right pole pairs have the following properties.

Proposition 1.3 *Let $W \in \mathcal{R}^{m\times n}$ and let (C, A) be a pole pair for W over a subset σ of* $\mathbf{C}$. *Then*

(i)

$$ker \begin{bmatrix} C \\ CA \\ \vdots \\ CA^{l-1} \end{bmatrix} = (0) \tag{1.4}$$

where l is an integer such that the size of the matrix A is $l \times l$;

(ii) $\sigma(A) \subset \sigma$;

(iii) $(\tilde{C}, \tilde{A})$ *is a right pole pair for W over σ if and only if*

$$\tilde{C} = CS \quad \text{and} \quad \tilde{A} = S^{-1}AS \tag{1.5}$$

for some invertible matrix S.

We note that the pair (C, A) such that (1.4) holds for an appropriate l is said to be <u>observable</u>. Observable pairs are also called null-kernel pairs in the literature. The pairs (C, A) and $(\tilde{C}, \tilde{A})$ such that (1.5) holds for some invertible matrix S are said to be <u>right-similar</u>.

Let W be an $m \times n$ rational matrix function and let $\lambda \in \mathbf{C}_\infty$. A function $\psi \in \mathcal{R}^{1\times m}$ is called a <u>left pole function</u> for W at λ if

(i) ψ is analytic at λ and $\psi(\lambda) \neq 0$,

(ii) there is a positive integer k and a function $\phi \in \mathcal{R}^{1\times m}$ such that ϕ is analytic at λ and

$$\psi(z) = \begin{cases} (z-\lambda)^k \phi(z) W(z), & \text{if } \lambda \neq \infty \\ z^{-k}\phi(z)W(z), & \text{if } \lambda = \infty. \end{cases} \tag{1.6}$$

The maximal integer k such that (1.6) holds for an appropriate ϕ is called the <u>order</u> of the left pole function ψ.

The left pole functions $\psi_1, \psi_2, ..., \psi_\eta$ for W at λ of orders $k_1, k_2, ..., k_\eta$ respectively form a <u>canonical set of left pole functions</u> for W at λ if

(i) $\psi_1(\lambda), \psi_2(\lambda), ..., \psi_k(\lambda)$ are linearly independent,

(ii) $\sum_{i=1}^{\eta} k_i$ is maximal subject to condition (i).

We note that the number of functions in any canonical set of left pole functions for W at λ is equal to the geometric multiplicity of the pole of W at λ. Also, the orders of the functions in a canonical set of left pole functions for W are equal to the partial multiplicities of the pole of W at λ.

Let $\{\psi_1, \psi_2, ..., \psi_\eta\}$ be a canonical set of left pole functions for $W \in \mathcal{R}^{m\times n}$ at $\lambda \in \mathbf{C}_\infty$ of orders $k_1, k_2, ..., k_\eta$ respectively and let $\psi_{i,j}$ be such that (1.3) holds. Let S be an invertible matrix of the same size as A. An ordered pair of matrices (A, B), where

$$A = \begin{cases} S \begin{bmatrix} J_{k_1}(\lambda) & & & \\ & J_{k_2}(\lambda) & & \\ & & \ddots & \\ & & & J_{k_\eta}(\lambda) \end{bmatrix} S^{-1}, & \text{if } \lambda \in \mathbf{C} \\ \\ S \begin{bmatrix} J_{k_1}(0) & & & \\ & J_{k_2}(0) & & \\ & & \ddots & \\ & & & J_{k_\eta}(0) \end{bmatrix} S^{-1}, & \text{if } \lambda = \infty \end{cases}$$

and

$$B = S \begin{bmatrix} \psi_{1,k_1-1} \\ \psi_{1,k_1-2} \\ \vdots \\ \psi_{1,0} \\ \psi_{2,k_2-1} \\ \psi_{2,k_2-2} \\ \vdots \\ \psi_{2,0} \\ \vdots \\ \psi_{\eta,k_\eta-1} \\ \psi_{\eta,k_\eta-2} \\ \vdots \\ \psi_{\eta,0} \end{bmatrix},$$

is called a <u>left pole pair</u> for W at λ.

As an illustration of the above definition,

$$\left(\begin{bmatrix} 0 & 1 \\ 0 & 0 \end{bmatrix}, \begin{bmatrix} 0 & 1 \\ 1 & 0 \end{bmatrix} \right)$$

is a left pole pair at infinity for a rational matrix function W in Example 1.1.

Let $W \in \mathcal{R}^{m\times n}$ and let $\sigma \subset \mathbf{C}$. Let $\lambda_1, \lambda_2, ..., \lambda_p$ be the poles of W in σ and let (A_i, B_i) be a left pole pair for W at λ_i $(i = 1, 2, ..., p)$. Any ordered pair of matrices (A, B), where

$$A = S\begin{bmatrix} A_1 & & & \\ & A_2 & & \\ & & \ddots & \\ & & & A_p \end{bmatrix} S^{-1}, \qquad B = S\begin{bmatrix} B_1 \\ B_2 \\ \vdots \\ B_p \end{bmatrix}$$

for some invertible matrix S of an appropriate size, is called a <u>left pole pair</u> for W over σ (or a σ-left pole pair for W).

If (A, B) is a left pole pair for a rational matrix function W over $\sigma \subset \mathbf{C}$ and W is analytic on $\mathbf{C}_\infty \backslash \sigma$, (A, B) is called the <u>global left pole pair</u> for W.

We note that, similarly as in the regular case, any two σ-left pole pairs (A_1, B_1) and (A_2, B_2) for a rational matrix function W over $\sigma \subset \mathbf{C}$ (resp. at infinity) are <u>left-similar</u>, that is

$$A_2 = SA_1S^{-1} \qquad \text{and} \qquad B_2 = SB_1$$

for some matrix S (see [BGR1]). Also, the similarity matrix S is unique. It is shown in [BGR1] (Theorem 3.4) that a left pole pair (A, B) for W over σ is <u>controllable</u>, that is for sufficiently large integers l the matrix

$$\begin{bmatrix} B & AB & \dots & A^{l-1}B \end{bmatrix}$$

has full row rank. Controllable pairs are also called full-range pairs in the literature.

The following lemma connects the realization of a rational matrix function with its pole pairs.

Lemma 1.4 *Let (C, A) be a right pole pair for $W \in \mathcal{R}^{m\times n}$ at $\lambda \in \mathbf{C}_\infty$. Then there exists a unique matrix B such that the rational matrix function*

$$\begin{cases} W(z) - C(z-A)^{-1}B, & \text{if } \lambda \in \mathbf{C} \\ W(z) - C(z^{-1}-A)^{-1}B, & \text{if } \lambda = \infty \end{cases} \tag{1.7}$$

is analytic at λ. Moreover, (A, B) is a left pole pair for W at λ.

Lemma 1.3 can be proved in the same way as Theorem 5.1 in [BGR1].

We note that the left version of Lemma 1.3 also holds; that is, if the left pole pair (A, B) for W at λ is given, then there exists a unique matrix C such that the rational matrix function (1.7) is analytic at λ. Moreover, (C, A) is a right pole pair for W at λ.

In view of Liouville's Theorem, Lemma 1.1 implies that each rational matrix function has the following representation (cf. [BGK, C]).

Theorem 1.5 *Let W be a rational matrix function. Let (C, A) and (C_∞, A_∞) be right pole pairs for W over $\mathbf{C}$ and at infinity. Then there exist unique matrices B, B_∞ and D such* *that*

$$W(z) = C(z - A)^{-1}B + D + C_\infty(z^{-1} - A_\infty)^{-1}B_\infty. \tag{1.8}$$

The right-hand side of formula (1.8) is usually written down as (A, B, C, D, E, F, G) where $E = C_\infty$, $F = B_\infty$ and $G = A_\infty$. Any representation of a rational matrix function W of the form

$$C(\phi(z) - A)^{-1}B + d + E(\psi(z) - G)^{-1}F, \tag{1.9}$$

where A, B, C, D, E, F, G are matrices and ϕ, ψ are möbius transformations, is called a <u>realization</u> of W. If the sizes of matrices A and G add up to the McMillan degree of W, the realization (1.9) is said to be <u>minimal</u>. Since the orders of pole functions in a canonical set of right pole functions for a rational matrix function W at $\lambda \in \mathbf{C}_\infty$ are equal to partial multiplicities of the pole of W at λ, the sum of sizes of matrices A and A_∞ in (1.8) is equal to the McMillan degree of W. Thus, the realization (1.8) of W is minimal. In fact, the following holds.

Theorem 1.6 *Suppose*

$$W(z) = C(z - A)^{-1}B + D + E(z^{-1} - G)^{-1}F$$

and $\sigma(G) \subset \{0\}$. *Then the following are equivalent:*

(i) (C, A) *is a right pole pair for* W *over* $\mathbf{C}$ *and* (E, G) *is a right pole pair for* W *at infinity;*

(ii) (A, B) *is a left pole pair for* W *over* $\mathbf{C}$ *and* (G, F) *is a left pole pair for* W *at infinity;*

(iii) (A, B, C, D, E, F, G) *is a minimal realization of* W.

Theorem 1.6 follows immediately from Theorem 1.5 and the fact that if

$$(A_1, B_1, C_1, D_1, E_1, F_1, G_1) \quad \text{and} \quad (A_2, B_2, C_2, D_2, E_2, F_2, G_2)$$

are two minimal realizations of a rational matrix function W such that $\sigma(G_1) = \sigma(G_2) \subset \{0\}$, then

$$A_2 = SA_1S^{-1}, \quad B_2 = SB_1, \quad C_2 = C_1S$$
$$G_2 = TG_1T^{-1}, \quad F_2 = TF_1, \quad E_2 = E_1T$$

for some unique invertible matrices S and T (see Theorem I.7 in [C] or Chapter 5 in [BGR3]).

We note that if a rational matrix function W is proper (that is, if W is analytic at infinity), then the pair (C_∞, A_∞) is vacuous and (1.8) reduces to the usual realization

(A, B, C, D). On the other hand, if W is a matrix polynomial (that is, if W is analytic on $\mathbf{C}$), then the first term on the right hand side of (1.8) does not occur.

2. ORTHOGONALITY IN $\mathcal{R}^n$

Let λ be a point of the extended complex plane. We define a function $|\cdot|_{z=\lambda}$ from $\mathcal{R}$ into the set of real numbers by putting

$$|r|_{z=\lambda} = \begin{cases} 0, & \text{if } r = 0 \\ e^{-\eta}, & \text{if } r \neq 0 \end{cases}$$

where η is the unique integer such that

$$r(z) = \begin{cases} (z-\lambda)^{\eta}\tilde{r}(z), & \text{if } \lambda \in \mathbf{C} \\ z^{-\eta}\tilde{r}(z), & \text{if } \lambda = \infty \end{cases}$$

with $\tilde{r}$ analytic and nonzero at λ. The function $|\cdot|_{z=\lambda}$ is a real valuation of $\mathcal{R}$. Since $|n|_{z=\lambda} \leq 1$ for every integer n, the valuation $|\cdot|_{z=\lambda}$ is non-Archimedean and the stronger triangle inequality

$$|r_1 + r_2|_{z=\lambda} \leq \max\,\{|r_1|_{z=\lambda}, |r_2|_{z=\lambda}\}$$

holds for all $r_1, r_2 \in \mathcal{R}$ (see [VdW]).

Let n be a positive integer and let $\lambda \in \mathbf{C}_\infty$. We define a function $\|\cdot\|_{z=\lambda}$ on $\mathcal{R}^n$, the product of n copies of $\mathcal{R}$, by putting

$$\|(r_1, r_2, ..., r_n)\|_{z=\lambda} = \max\,\{|r_1|_{z=\lambda}, |r_2|_{z=\lambda}, ..., |r_n|_{z=\lambda}\}.$$

In this way $\mathcal{R}^n$ becomes a normed vector space over the real valued field $(\mathcal{R}, |\cdot|_{z=\lambda})$. We note that $(\mathcal{R}^n, \|\cdot\|_{z=\lambda})$ is not a Banach space. Indeed, exp (z) is an example of a non-rational function which is in the completion of $(\mathcal{R}, \|\cdot\|_{z=\lambda})$ for any $\lambda \in \mathbf{C}$.

Example 2.1 We have

$$|z^2-1|_{z=1} = \frac{1}{e}, \qquad \left\| \begin{bmatrix} z^2-1 & \frac{1}{(z+1)^2} \end{bmatrix} \right\|_{z=-1} = e^2, \qquad \left\| \begin{bmatrix} z^2-1 & \frac{1}{(z+1)^2} \end{bmatrix} \right\|_{z=0} = 1.$$

□

Following the definition of orthogonality in a non-Archimedean normed space, we shall say that two subspaces Λ and Ω of $\mathcal{R}^n$ are <u>orthogonal</u> at $\lambda \in \mathbf{C}_\infty$ if

$$\|x+y\|_{z=\lambda} = \max\,\{\|x\|_{z=\lambda}, \|y\|_{z=\lambda}\} \tag{2.1}$$

for each $x \in \Lambda$, $y \in \Omega$. We shall say that Λ and Ω are orthogonal on $\sigma \subset \mathbf{C}_\infty$ if they are orthogonal at every point of σ. We shall say that vectors $x_1, x_2, ..., x_k$ are orthogonal at $\lambda \in \mathbf{C}_\infty$ (resp. on a subset σ of $\mathbf{C}_\infty$) if the spans over $\mathcal{R}$ of $\{x_j\}$ and $\{x_i : i \neq j\}$ are orthogonal at λ (resp. on σ) for $j = 1, 2, ..., k$.

We say that a point $\lambda \in \mathbf{C}$ is a zero of a rational matrix function W (or, equivalently, that W has a zero at λ) if λ is a zero of some nonzero diagonal entry in the Smith-McMillan form of W (see eg. [Ros]). W has a zero at infinity if $H(z) = W(z^{-1})$ has a zero at 0. We shall need below the following lemma.

Lemma 2.2 *Let Λ be a subspace of $\mathcal{R}^{n\times 1}$ and let $\lambda \in \mathbf{C}_\infty$. Then there exists a basis $\{v_1, v_2, ..., v_k\}$ for Λ such that the rational matrix function $[v_1 \quad v_2 \quad \ldots \quad v_k]$ has neither a pole nor a zero at λ.*

Proof Choose an algebraic basis $\{w_1, w_2, ..., w_k\}$ for Λ and let

$$W = [w_1 \quad w_2 \quad \ldots \quad w_k].$$

Considering, if necessary, $H(z) = W(z^{-1})$ we may assume $\lambda \in \mathbf{C}$. Choose a Smith-McMillan factorization EDF of W and let $v_1, v_2, ..., v_k$ be the first k columns of E. Clearly $\{v_1, v_2, ..., v_k\}$ is a basis for Λ and the rational matrix function $V = [v_1 \quad v_2 \quad \ldots \quad v_k]$ is analytic at λ. Since

$$E \begin{bmatrix} I \\ 0 \end{bmatrix} I$$

is a Smith-McMillan factorization of V, V does not have a zero at λ.

□

We note that an analogous result holds if Λ is a subspace of $\mathcal{R}^{1\times n}$.

If Λ is a subspace of $\mathcal{R}^n$ and $\lambda \in \mathbf{C}_\infty$, we will denote by $\Lambda(\lambda)$ the set of values at λ of those functions in Λ which are analytic at λ. Plainly, $\Lambda(\lambda)$ is a subspace of $\mathbf{C}^n$. The space $\Lambda(\lambda)$ can be characterized equivalently as the linear span over $\mathbf{C}$ of the leading coefficients in the expansions at λ of the functions in Λ. Using this notation, we can characterize orthogonality in $\mathcal{R}^n$ as follows.

Proposition 2.3 *Let Λ and Ω be two subspaces of $\mathcal{R}^n$ and let $\lambda \in \mathbf{C}_\infty$. Then Λ and Ω are orthogonal at λ if and only if $\Lambda(\lambda) \cap \Omega(\lambda) = (0)$.*

Proof Suppose Λ and Ω are orthogonal at λ and let $\{v_1, v_2, ..., v_k\}$ and $\{w_1, w_2, ..., w_l\}$ be algebraic bases for Λ and Ω respectively. In view of Lemma 2.2, we may assume that $v_1, v_2, ..., v_k, w_1, w_2, ..., w_l$ are analytic at λ and $\{v_1(\lambda), v_2(\lambda), ..., v_k(\lambda)\}$ and $\{w_1(\lambda), w_2(\lambda), ...,$

$w_l(\lambda)\}$ are linearly independent sets. Suppose that

$$\sum_{i=1}^{k} \alpha_i v_i(\lambda) + \sum_{j=1}^{l} \beta_j w_j(\lambda) = 0$$

for some numbers α_i, β_j. Then $\| \sum_{i=1}^{k} \alpha_i v_i + \sum_{j=1}^{l} \beta_j w_j \|_{z=\lambda} < 1$. So, by the orthogonality of Λ and Ω at λ, $\| \sum_{i=1}^{k} \alpha_i v_i \|_{z=\lambda} < 1$ and $\| \sum_{j=1}^{l} \beta_j w_j \|_{z=\lambda} < 1$. Hence $\sum_{i=1}^{k} \alpha_i v_i(\lambda) = 0$ and $\sum_{j=1}^{l} \beta_j w_j(\lambda) = 0$. Since $\{v_1(\lambda), v_2(\lambda), ..., v_k(\lambda)\}$ and $\{w_1(\lambda), w_2(\lambda), ..., w_l(\lambda)\}$ are linearly independent sets, $\alpha_1 = \alpha_2 = ... = \alpha_k = \beta_1 = \beta_2 = ... = \beta_l = 0$. Thus, $v_1(\lambda), v_2(\lambda), ..., v_k(\lambda), w_1(\lambda), w_2(\lambda), ..., w_l(\lambda)$ are linearly independent and $\Lambda(\lambda) \cap \Omega(\lambda) = (0)$.

Conversely, suppose $\Lambda(\lambda) \cap \Omega(\lambda) = (0)$. Choose a basis $\{v_1, v_2, ..., v_k\}$ for Λ and a basis $\{w_1, w_2, ..., w_l\}$ for Ω such that $v_1, v_2, ..., v_k, w_1, w_2, ..., w_l$ are analytic at λ and $\{v_1(\lambda), v_2(\lambda), ..., v_k(\lambda)\}$ and $\{w_1(\lambda), w_2(\lambda), w_l(\lambda)\}$ are linearly independent sets. Then $v_1(\lambda)$, $v_2(\lambda), ..., v_k(\lambda), w_1(\lambda), w_2(\lambda), ..., w_l(\lambda)$ are linearly independent. Let $r_1, r_2, ..., r_{k+l}$ be scalar rational functions, not all equal to 0. Choose κ such that

$$|r_\kappa|_{z=\lambda} = \max \{|r_1|_{z=\lambda}, |r_2|_{z=\lambda}, ..., |r_{k+l}|_{z=\lambda}\}.$$

We assume without loss of generality $1 \leq \kappa \leq k$. The linear independence of $v_1(\lambda)$, $v_2(\lambda), ..., v_k(\lambda), w_1(\lambda), w_2(\lambda), ..., w_l(\lambda)$ implies then the equalities

$$\| \sum_{i=1}^{k} r_i v_i + \sum_{j=1}^{l} r_{k+j} w_j \|_{z=\lambda} = |r_\kappa|_{z=\lambda}$$

and

$$\| \sum_{i=1}^{k} r_i v_i \|_{z=\lambda} = |r_\kappa|_{z=\lambda}.$$

Since $\| \sum_{j=1}^{l} r_{k+j} w_j \|_{z=\lambda} \leq |r_\kappa|_{z=\lambda}$,

$$\| \sum_{i=1}^{k} r_i v_i + \sum_{j=1}^{l} r_{k+j} w_j \|_{z=\lambda} = \max \{\| \sum_{i=1}^{k} r_i v_i \|_{z=\lambda}, \| \sum_{j=1}^{l} r_{k+j} w_j \|_{z=\lambda}\}.$$

Thus, Λ and Ω are orthogonal at λ.

□

Corollary 2.4 *Nonzero functions $x_1, x_2, ..., x_k \in \mathcal{R}^n$ are orthogonal at $\lambda \subset \mathbf{C}_\infty$ if and only if the vectors $[x_1]_{z=\lambda}, [x_2]_{z=\lambda}, ..., [x_k]_{z=\lambda}$ are linearly independent.*

It follows from the definition of orthogonality (see equality (2.1)) that two subspaces of $\mathcal{R}^n$ orthogonal at a single point $\lambda \in \mathbf{C}_\infty$ necessarily have the trivial intersection.

Let Λ, Ω and Σ be subspaces of $\mathcal{R}^n$. We say that the subspace Ω is an orthogonal complement of the subspace Λ in (Σ, σ) if Λ and Ω are orthogonal on σ and $\Lambda + \Omega = \Sigma$. The existence of orthogonal complements of a subspace Λ in $(\mathcal{R}^n, \sigma)$ follows from the next proposition.

Proposition 2.5 *Let Λ and Ω be subspaces of $\mathcal{R}^n$ which are orthogonal on a proper subset σ of $\mathbf{C}_\infty$. Then Ω has an extension to an orthogonal complement of Λ in $(\mathcal{R}^n, \sigma)$.*

Proof We identify $\mathcal{R}^n$ with $\mathcal{R}^{n\times 1}$. After applying a suitable Möbius transformation, we may assume $\sigma \subset \mathbf{C}$. Find a basis $\{w_1, w_2, ..., w_k\}$ for $\Lambda + \Omega$ and let EDF be a Smith-McMillan factorization of the rational matrix function $[\,w_1 \quad w_2 \quad \ldots \quad w_k\,]$. We show that the subspace V spanned by the last $n-k$ columns of E is orthogonal to $\Lambda + \Omega$ on σ. Then the subspace $\Omega + V$ is an orthogonal complement of Λ in $(\mathcal{R}^n, \sigma)$. In fact, it suffices to show that the spaces spanned by the first k and the last $n-k$ columns of E are orthogonal on σ. But this follows from Proposition 2.3 and the properties of a Smith-McMillan factorization of a rational matrix function.

□

Corollary 2.6 *If σ is a proper subset of $\mathbf{C}_\infty$, then every subspace Λ of $\mathcal{R}^n$ has an orthogonal complement in $(\mathcal{R}^n, \sigma)$.*

We note that unlike in a Hilbert space, an orthogonal complement of Λ in $(\mathcal{R}^n, \sigma)$ is generally not unique. We also note that there exist subspaces of $\mathcal{R}^n$ which are orthogonal on the whole extended complex plane. Indeed, if $S = \{c_1, c_2, ..., c_n\}$ is a basis for $\mathbf{C}^n$ and $S_1 \cup S_2$ is a partition of S, the subspaces of $\mathcal{R}^n$ spanned by the (constant) rational vector functions contained in S_1 and S_2, respectively, are orthogonal on $\mathbf{C}_\infty$. In general, however, as can be seen from the following example, the hypothesis $\sigma \neq \mathbf{C}_\infty$ in Proposition 2.5 and Corollary 2.6 is necessary. In Example 2.7, we will utilize the concept of a minimal polynomial basis for a subspace Λ of $\mathcal{R}^{1\times n}$, defined by Forney in [F] as follows. Let the degree of a row vector polynomial $g = [\,g_1 \quad g_2 \quad \cdots \quad g_n\,]$ be

$$\deg g = \max\,\{\deg g_1, \deg g_2, \ldots, \deg g_n\}.$$

A polynomial basis $\{v_1, v_2, ..., v_k\}$ for Λ is said to be minimal if $\sum_{i=1}^{k} \deg v_i$ is as small as possible. Forney showed that a polynomial basis $\{v_1, v_2, ..., v_k\}$ for Λ is minimal if and only if the coefficients of the highest degree terms in $v_1, v_2, ..., v_k$ are linearly independent and $v_1(\lambda), v_2(\lambda), ..., v_k(\lambda)$ are linearly independent for every $\lambda \in \mathbf{C}$. This characterization of a minimal polynomial basis underlies, in fact, the algorithm in [F] of reduction of a basis for Λ to a minimal polynomial basis.

Example 2.7 Let Λ be a subspace of $\mathcal{R}^2$ spanned by $[1 \quad z]^T$ and let Ω be a subspace of $\mathcal{R}^2$ which is orthogonal to Λ on $\mathbf{C}_\infty$. Then dim $\Omega \leq 1$. Suppose dim $\Omega = 1$ and let $[p_1 \quad p_2]^T$ be a minimal polynomial basis for Ω. In view of Proposition 2.3, orthogonality of Λ and Ω on $\mathbf{C}$ implies that

$$\det \begin{bmatrix} 1 & p_1(z) \\ z & p_2(z) \end{bmatrix}$$

does not vanish in $\mathbf{C}$. By the fundamental theorem of algebra, $zp_1(z) - p_2(z) = c$ for some constant c. If $p_1 \neq 0$ then

$$\begin{aligned} \| \, p_1(z)[1 \quad z] - [p_1(z) \quad p_2(z)] \, \|_{z=\lambda} &= \| \, [0 \quad c] \, \|_{z=\infty} = 1 \\ &< e \leq \| \, p_1(z)[1 \quad z] \, \|_{z=\infty} \\ &\leq \max \{\| \, p_1(z)[1 \quad z] \, \|_{z=\infty}, \| \, [p_1 \quad p_2] \, \|_{z=\infty}\}, \end{aligned}$$

a contradiction. Thus, $p_1 = 0$ and Ω is spanned by $[0 \quad p_2]^T$. If $p_2 \neq 0$,

$$\begin{aligned} \| \, [1 \quad z] - (z/p_2(z))[0 \quad p_2(z)] \, \|_{z=\infty} &= 1 \\ &< e = \max \{\| \, [1 \quad z] \, \|_{z=\infty}, \| \, (z/p_2(z))[0 \quad p_2(z)] \, \|_{z=\infty}\} \end{aligned}$$

which is again a contradiction. Thus, the only subspace of $\mathcal{R}^2$ orthogonal to Λ on $\mathbf{C}_\infty$ is (0).

□

Let Λ be a subspace of $\mathcal{R}^{n\times 1}$ (resp. $\mathcal{R}^{1\times n}$). We identify the dual of $\mathcal{R}^{n\times 1}$ (resp. $\mathcal{R}^{1\times n}$), i.e. the space of $\mathcal{R}$-valued linear functional on $\mathcal{R}^{n\times 1}$ (or $\mathcal{R}^{1\times n}$), with $\mathcal{R}^{1\times n}$ (resp. $\mathcal{R}^{n\times 1}$) via the usual matrix multiplication. We will denote by Λ° the subspace of the dual of $\mathcal{R}^{n\times 1}$ which annihilates Λ. We shall need below the following fact.

Theorem 2.8 *Let Λ, Ω be algebraic complements in $\mathcal{R}^n$ and let $\sigma \subset \mathbf{C}_\infty$. Then Λ and Ω are orthogonal on σ if and only if Λ° and Ω° are orthogonal on σ.*

Proof We identify $\mathcal{R}^n$ with $\mathcal{R}^{n\times 1}$ and show that Λ° and Ω° are orthogonal on σ whenever Λ and Ω are. Choose any $\lambda \in \sigma$. In view of Lemma 2.2, we can find bases $\{v_1, v_2, ..., v_k\}$ and $\{w_1, w_2, ..., w_l\}$ for Λ and Ω such that the rational matrix functions $V = [v_1 \; v_2 \; ... \; v_k]$ and $W = [w_1 \; w_2 \; ... \; w_l]$ have neither a pole nor a zero at λ. Let $R = [\, V \quad W \,]$. Plainly, R is a regular rational matrix function without a pole at λ. By Proposition 2.3, the matrix $R(\lambda)$ is nonsingular. Thus R and, consequently, R^{-1} have neither a pole nor a zero at λ. Invoking once again Proposition 2.3, we see that the subspaces of $\mathcal{R}^{1\times n}$ spanned by the first k and last l rows of R^{-1}, respectively, are orthogonal at λ. But these subspaces, in turn, can be identified with Ω° and Λ°.

□

We conclude this section with the following stronger version of Lemma 2.2.

Proposition 2.9 *Let Λ be a subspace of $\mathcal{R}^{n\times 1}$ and let σ be a proper subset of $\mathbf{C}_\infty$. Then there exists a basis $\{v_1, v_2, ..., v_k\}$ for Λ such that the rational matrix function $[v_1 \quad v_2 \quad \dots \quad v_k]$ has no poles nor zeros in σ.*

Proof Choose a minimal polynomial basis $\{w_1, w_2, ..., w_k\}$ for Λ. If $\sigma \subset \mathbf{C}$, let $v_i = w_i$ $(i = 1, 2, ..., k)$. Otherwise, choose a point λ in $\mathbf{C}\backslash\sigma$ and let $v_i(z) = (z-\lambda)^{-\delta(w_i,\infty)}w_i(z)$, where $\delta(w_i, \infty)$ denotes the multiplicity of a pole of the function w_i at infinity $(i = 1, 2, ..., k)$.

□

3. ZERO STRUCTURE

Recall that a point $\lambda \in \mathbf{C}$ (resp. $\lambda = \infty$) is a zero of a rational matrix function W if λ is a zero of some nonzero diagonal entry in the Smith-McMillan form of W (resp. $H(z) = W(z^{-1})$). The number of nonzero diagonal entries in the Smith-McMillan form D of W which vanish at $\lambda \in \mathbf{C}$ is called the geometric multiplicity of the zero of W at λ. The orders of zeros at λ of the nonzero diagonal entries in the Smith-McMillan form of W are called the partial multiplicities of the zero of W at λ. The geometric multiplicity and the partial multiplicities of the zero of W at infinity are defined to be the geometric multiplicity and the partial multiplicities of the zero of $H(z) = W(z^{-1})$ at 0. The sum of partial multiplicities of a zero of W at $\lambda \in \mathbf{C}_\infty$ is called the (total) multiplicity of the zero of W at λ. In the literature, partial multiplicities and the total multiplicity of a zero of a rational matrix function W at $\lambda \in \mathbf{C}_\infty$ are also called the partial and total zero multiplicities of W at λ.

As it is well known, the sum of multiplicities of all zeros of a rational matrix function W is at most equal to the McMillan degree of W (see [VDK, VK]; cf. [WSCP]).

If W is a rational (row or column) vector function which has a zero at $\lambda \in \mathbf{C}_\infty$, the geometric multiplicity of the zero of W at λ equals 1. In this case the multiplicity of the zero of W at λ is an integer k such that the rational vector function

$$\begin{cases} (z-\lambda)^{-k}W(z), & \text{if } \lambda \in \mathbf{C} \\ z^k W(z), & \text{if } \lambda = \infty \end{cases}$$

is analytic and nonzero at λ.

If $W \in \mathcal{R}^{m\times n}$ is right invertible, $\phi W \neq 0$ whenever $\phi \neq 0$. Then, for all ϕ with $\phi(\lambda) \neq 0$, ϕW can vanish to a finite order equal to the largest partial multiplicity of a zero of W at λ. If W is not right invertible, there exist functions $\phi \in \mathcal{R}^{1\times m}\backslash\{0\}$ for which ϕW vanishes identically. Such functions form, together with a zero function, a subspace of $\mathcal{R}^{1\times m}$

which is called the left annihilator of W. Following ther terminology in [Ro] we shall call the part of the left zero structure related to the left annihilator of W the continuous left zero structure of W. We shall call the part of the zero structure of W related to the zeros of W the discrete left zero structure of W.

We begin with the discription of the continuous left zero structure of an $m \times n$ rational matrix function W. We will denote by W^{ol} the left annihilator of W. Thus,

$$W^{ol} = \{r \in \mathcal{R}^{1 \times m} : \ rW = 0\}.$$

We note that if W is right invertible, then $W^{ol} = (0)$. If $W^{ol} \neq (0)$, a matrix polynomial whose rows form a minimal polynomial basis for W^{ol} is called a left kernel polynomial for W (see [BCR]). We will assume that the rows in any left kernel polynomial for W are ordered according to decreasing degrees.

In view of Theorem 1.5, any matrix polynomial P has a realization of the form

$$D + E(z^{-1} - G)^{-1}F$$

where $D = P(0)$ and $(E, G), (G, F)$ are right and left pole pairs for P at infinity. We now describe one canonical form for such a realization of a left kernel polynomial.

Proposition 3.1 *Let $P \in \mathcal{R}^{k \times n}$ be a matrix polynomial with the ith row*

$$P_i(z) = P_{i,\eta_i} + zP_{i,\eta_i - 1} + \ldots + z^{\eta_i} P_{i,0}$$

($P_{i,0} \neq 0$ if $P_i \neq 0$). Suppose that the first l rows of P have a pole at infinity and the last $k - l$ rows of P are constant. Then

$$P(z) = P(0) + [E_1 \quad E_2 \quad \ldots \quad E_l] \cdot \left(z^{-1} - \begin{bmatrix} J_{\eta_1}(0) & & & \\ & J_{\eta_2}(0) & & \\ & & \ddots & \\ & & & J_{\eta_l}(0) \end{bmatrix} \right)^{-1} \begin{bmatrix} P_{1,\eta_1 - 1} \\ P_{1,\eta_1 - 2} \\ \vdots \\ P_{1,0} \\ P_{2,\eta_2 - 1} \\ P_{2,\eta_2 - 2} \\ \vdots \\ P_{2,0} \\ \vdots \\ P_{l,\eta_l - 1} \\ P_{l,\eta_l - 2} \\ \vdots \\ P_{l,0} \end{bmatrix} \tag{3.1}$$

where E_i is the $k \times \eta_i$ matrix with 1 at the position $(i, 1)$ and 0's elsewhere.

Proof Since

$$\left(z^{-1} - \begin{bmatrix} J_{\eta_1}(0) & & & \\ & J_{\eta_2}(0) & & \\ & & \ddots & \\ & & & J_{\eta_l}(0) \end{bmatrix} \right)^{-1} =$$

$$= \begin{bmatrix} (z^{-1} - J_{\eta_1}(0))^{-1} & & & \\ & (z^{-1} - J_{\eta_2}(0))^{-1} & & \\ & & \ddots & \\ & & & (z^{-1} - J_{\eta_l}(0))^{-1} \end{bmatrix}$$

and

$$\left(z^{-1} - J_{\eta_i}(0) \right)^{-1} = \begin{bmatrix} z & z^2 & \dots & z^{\eta_i} \\ 0 & z & \dots & z^{\eta_i - 1} \\ \vdots & \vdots & \ddots & \vdots \\ 0 & 0 & \dots & z \end{bmatrix},$$

the second term on the right hand side in (3.1) equals

$$\sum_{i=1}^{l} E_i \begin{bmatrix} zP_{i,\eta_i-1} + z^2 P_{i,\eta_i-2} + \dots + z^{\eta_i} P_{i,0} \\ * \\ \vdots \\ * \end{bmatrix}.$$

So equality (3.1) follows.

□

Proposition 3.2 *The realization* (3.1) *of a matrix polynomial P is minimal if and only if the columns of P are orthogonal at infinity.*

Proof Since

$$\left(\begin{bmatrix} J_{\eta_1}(0) & & & \\ & J_{\eta_2}(0) & & \\ & & \ddots & \\ & & & J_{\eta_l}(0) \end{bmatrix}, \begin{bmatrix} P_{1,\eta_1-1} \\ P_{1,\eta_1-2} \\ \vdots \\ P_{1,0} \\ P_{2,\eta_2-1} \\ P_{2,\eta_2-1} \\ \vdots \\ P_{2,0} \\ \vdots \\ P_{l,\eta_l-1} \\ P_{l,\eta_l-2} \\ \vdots \\ P_{l,0} \end{bmatrix} \right)$$

is a left pole pair at infinity for P if and only if $P_{1,0}, P_{2,0}, \dots, P_{l,0}$ are linearly independent, the assertion follows by Theorem 1.5 and Corollary 2.4.

□

Corollary 3.3 *If P is a left kernel polynomial for some rational matrix function then the realization* (3.1) *of P is minimal.*

The matrices $E_1, E_2, ..., E_l$ in the realization (3.1) of P are determined by the matrices $P(0)$ and $J_{\eta_1}(0), J_{\eta_2}(0), ..., J_{\eta_l}(0)$. Indeed, the number of rows of any E_i equals the number of rows of $P(0)$, and the number of columns of E_i equals the number of columns of $J_{\eta_i}(0)$. Consequently, the ordered triple of matrices

$$\left(\begin{bmatrix} J_{\eta_1}(0) & & & \\ & J_{\eta_2}(0) & & \\ & & \ddots & \\ & & & J_{\eta_l}(0) \end{bmatrix}, \begin{bmatrix} P_{1,\eta_1-1} \\ P_{1,\eta_1-2} \\ \vdots \\ P_{1,0} \\ P_{2,\eta_2-1} \\ P_{2,\eta_2-1} \\ \vdots \\ P_{2,0} \\ \vdots \\ P_{l,\eta_l-1} \\ P_{l,\eta_l-2} \\ \vdots \\ P_{l,0} \end{bmatrix}, P(0) \right) \tag{3.2}$$

determines a matrix polynomial P uniquely. A triple of matrices as in (3.2) which describes a left kernel polynomial for a rational matrix function W will be called a left kernel triple for W. Such collection determines a matrix polynomial P whose rows form a minimal polynomial basis for their span if and only if the set $\{P_{1,0}, P_{2,0}, ..., P_{l,0}\}$ is linearly independent and P has no zeros in $\mathbf{C}$. If $(A_\kappa, B_\kappa, D_\kappa)$ is a left kernel triple, we shall call the corresponding matrix polynomial the associated matrix polynomial. Also, we shall call the matrix C_κ such that the associated matrix polynomial is given by the formula

$$D_\kappa + C_\kappa(z^{-1} - A_\kappa)^{-1}B_\kappa$$

the associated state-space-output operator.

Example 3.4 Find a left kernel triple of a rational matrix function

$$W(z) = \begin{bmatrix} z/(z^2-1) \\ 1/z \\ (z^2-2)/(z+1) \end{bmatrix}.$$

Since W has a Smith-McMillan factorization EDF where

$$D = \begin{bmatrix} 1/(z(z-1)(z+1)) \\ 0 \\ 0 \end{bmatrix}$$

and E is a matrix polynomial with the inverse

$$E^{-1}(z) = \begin{bmatrix} z^2 & -1 & 0 \\ -z^2+1 & 1 & 0 \\ -z^4+z^3+2z^2-2z & 0 & 1 \end{bmatrix},$$

the rows of a matrix polynomial

$$P(z) = \begin{bmatrix} -z^2+1 & 1 & 0 \\ -z^4+z^3+2z^2-2z & 0 & 1 \end{bmatrix},$$

form a basis for W^{ol}. Since $E^{-1}(z)$ has no zeros in $\mathbf{C}$, $P(z)$ has no zeros in $\mathbf{C}$. Hence we can find a left kernel polynomial for W by "row-reducing" P at infinity (cf. Step 3 in [F, Section 4]). We put

$$\begin{aligned} P_1(z) &= \begin{bmatrix} 1 & 0 \\ -z^2 & 1 \end{bmatrix} P(z) \\ &= \begin{bmatrix} -z^2+1 & 1 & 0 \\ z^3+z^2-2z & -z^2 & 1 \end{bmatrix}, \end{aligned}$$

then

$$\begin{aligned} P_2(z) &= \begin{bmatrix} 1 & 0 \\ z & 1 \end{bmatrix} P_1(z) \\ &= \begin{bmatrix} -z^2+1 & 1 & 0 \\ z^2-z & -z^2+z & 1 \end{bmatrix}, \end{aligned}$$

and finally

$$\begin{aligned} P_3(z) &= \begin{bmatrix} 1 & 0 \\ 1 & 1 \end{bmatrix} P_2(z) \\ &= \begin{bmatrix} -z^2+1 & 1 & 0 \\ -z+1 & -z^2+z+1 & 1 \end{bmatrix}. \end{aligned}$$

Since the rows of the function P_3 are orthogonal at infinity, P_3 is a left kernel polynomial for W. The corresponding left kernel triple is

$$\left(\begin{bmatrix} 0 & 1 & 0 & 0 \\ 0 & 0 & 0 & 0 \\ 0 & 0 & 0 & 1 \\ 0 & 0 & 0 & 0 \end{bmatrix}, \begin{bmatrix} 0 & 0 & 0 \\ -1 & 0 & 0 \\ -1 & 1 & 0 \\ 0 & -1 & 0 \end{bmatrix}, \begin{bmatrix} 1 & 1 & 0 \\ 1 & 1 & 1 \end{bmatrix} \right).$$

□

The following proposition summarizes conditions under which a given triple of matrices is a left kernel triple for some rational matrix function.

Proposition 3.5 *A triple of matrices $(A_\kappa, B_\kappa, D_\kappa)$ is a left kernel triple for some rational matrix function W if and only if the following hold:*

(i) *the matrix* A_κ *is in Jordan form,* $\sigma(A_\kappa) \subset \{0\}$*, and the number of blocks of* A_κ *is at most equal to the number of rows of* D_κ*; the number of columns of* B_κ *equals the number of columns of* D_κ*;*

(ii) *the pair* (A_κ, B_κ) *is controllable;*

(iii) *the Smith-McMillan form of the associated matrix polynomial corresponding is* $[I \;\; 0]$.

Proof (i) holds if and only if there exists a matrix polynomial P corresponding to the triple $(A_\kappa, B_\kappa, D_\kappa)$. Since conditions (ii) and (iii) are equivalent to conditions 2.(b) and 2.(a) in the main theorem in [F], (ii) and (iii) hold if and only if the rows of P form a minimal polynomial basis for the row space of P.

□

Let Λ be a subspace of $\mathcal{R}^{1\times n}$ and let P be a matrix polynomial such that the rows $P_1, P_2, ..., P_k$ of P form a minimal polynomial basis for Λ. Suppose first $\lambda \in \mathbf{C}$. Plainly,

$$\text{span}\,\{P_1(\lambda), P_2(\lambda), ..., P_k(\lambda)\} \subset \Lambda(\lambda). \tag{3.3}$$

Since $P(\lambda)$ has full row rank, $P_1(\lambda), P_2(\lambda), ..., P_k(\lambda)$ are linearly independent. Hence, if ψ is a linear combination over scalar rational functions of $P_1, P_2, ..., P_k$, then the leading coefficient in the Laurent expansion of ψ at λ is contained in span $\{P_1(\lambda),\, P_2(\lambda), ...,\, P_k(\lambda)\}$. It follows that inclusion (3.3) is an equality. Thus, $\{P_1(\lambda), P_2(\lambda), ..., P_k(\lambda)\}$ is a basis for $\Lambda(\lambda)$ for every $\lambda \in \mathbf{C}$. Suppose now $\lambda = \infty$ and let $\tilde{P}_i(z) = z^{-\alpha_i} P_i(z)$ where α_i is such that $\tilde{P}_i$ is analytic and nonzero at infinity ($i = 1, 2, ..., k$). Since the leading coefficients in the Laurent expansions of $P_1, P_2, ..., P_k$ at infinity are linearly independent, $\tilde{P}_1(\infty), \tilde{P}_2(\infty), ..., \tilde{P}_k(\infty)$ are linearly independent. Hence, by the same reasoning as above, $\Lambda(\infty) = \text{span}\{\tilde{P}_1(\infty), \tilde{P}_2(\infty), ..., \tilde{P}_k(\infty)\}$. So the leading coefficients of the rows of P form a basis for $\Lambda(\infty)$. In particular, if P and Λ are a left kernel polynomial and the left annihilator of an $m \times n$ rational matrix function W, then, for each $\lambda \in \mathbf{C}$, $\{P_1(\lambda), P_2(\lambda), ..., P_k(\lambda)\}$ is a basis for $W^{ol}(\lambda)$ and $\{\tilde{P}_1(\infty), \tilde{P}_2(\infty), ..., \tilde{P}_k(\infty)\}$ is a basis for $W^{ol}(\infty)$. As a consequence, we have the following fact.

Proposition 3.6 *Let*

$$\left(\begin{bmatrix} J_{\eta_1}(0) & & & \\ & J_{\eta_2}(0) & & \\ & & \ddots & \\ & & & J_{\eta_l}(0) \end{bmatrix}, \begin{bmatrix} b_{1,\eta_1-1} \\ b_{1,\eta_1-2} \\ \vdots \\ b_{1,0} \\ b_{2,\eta_2-1} \\ b_{2,\eta_2-1} \\ \vdots \\ b_{2,0} \\ \vdots \\ b_{l,\eta_l-1} \\ b_{l,\eta_l-2} \\ \vdots \\ b_{l,0} \end{bmatrix}, \begin{bmatrix} d_1 \\ d_2 \\ \vdots \\ d_l \\ d_{l+1} \\ \vdots \\ d_k \end{bmatrix} \right) \tag{3.4}$$

be a left kernel triple for an $m \times n$ rational matrix function W. Then

$$\{b_{1,0}, b_{2,0}, ..., b_{l,0}, d_{l+1}, d_{l+2}, ..., d_k\} \tag{3.5}$$

is a basis for $W^{ol}(\infty)$ and, if $\lambda \in \mathbf{C}$, the set of vectors $\{v_1, v_2, ..., v_k\}$ such that

$$v_i = \begin{cases} d_i + \lambda b_{i,\eta_i-1} + ... + \lambda^{\eta_i} b_{i,0}, & \text{if } i \leq l \\ d_i, & \text{if } i > l \end{cases} \tag{3.6}$$

is a basis for $W^{ol}(\lambda)$.

We note that the implication in Proposition 3.6 can be reversed: the triple of matrices (3.4) is a left kernel triple for W if the set (3.5) is a basis for $W^{ol}(\infty)$ and the vectors (3.6) form a basis for $W^{ol}(\lambda)$ for every $\lambda \in \mathbf{C}$. Indeed, the set (3.5) is a basis for $W^{ol}(\infty)$ if and only if condition (ii) in Proposition 3.5 holds, and the vectors (3.6) form a basis for $W^{ol}(\lambda)$ for every $\lambda \in \mathbf{C}$ if and only if condition (iii) in Proposition 3.5 is satisfied. Conditions (ii) and (iii) in Proposition 3.5 are more useful as they can be verified on a computer.

Another property of a triple of matrices $(A_\kappa, B_\kappa, D_\kappa)$, which is equivalent to condition (iii) in Proposition 3.5, is that the principal minors of the associated matrix polynomial are relatively prime. This condition was, in fact, used in [F]. Verification of this condition involves computation of roots of a polynomial.

Still another formulation of condition (iii) in Proposition 3.5 is via the "completion" approach. Suppose a triple of matrices $(A_\kappa, B_\kappa, D_\kappa)$ satisfies conditions (i) and (ii). Then the associated matrix polynomial P has the Smith-McMillan form $[\,I \quad 0\,]$ if and only if there exists a matrix polynomial $\tilde{P}$ such that the square matrix polynomial

$$Q = \begin{bmatrix} P \\ \tilde{P} \end{bmatrix}$$

has a constant nonzero determinant. Let $P(z) = D_\kappa + C_\kappa(z^{-1} - A_\kappa)^{-1}B_\kappa$ and let $\tilde{D}_\kappa + \tilde{C}_\kappa(z^{-1} - \tilde{A}_\kappa)^{-1}\tilde{B}_\kappa$ be a minimal realization of $\tilde{P}$. Let $H(z) = Q(z^{-1})$. Thus,

$$H(z) = \begin{bmatrix} D \\ \tilde{D} \end{bmatrix} + \begin{bmatrix} z - A_\kappa & \\ & z - \tilde{A}_\kappa \end{bmatrix}^{-1} \begin{bmatrix} B_\kappa \\ \tilde{B}_\kappa \end{bmatrix}$$

is a rational matrix function with all the poles at $z = 0$. Now Q has all zeros in $\mathbf{C}$ if and only if W has no zeros in $\mathbf{C}_\infty \backslash \{0\}$. The latter condition holds if and only if $W^{-1}(z)$ has all poles at $z = 0$, that is if and only if

$$\sigma\left(\begin{bmatrix} A_\kappa & \\ & \tilde{A}_\kappa \end{bmatrix} - \begin{bmatrix} B_\kappa \\ \tilde{B}_\kappa \end{bmatrix} \begin{bmatrix} D_\kappa \\ \tilde{D}_\kappa \end{bmatrix}^{-1} \begin{bmatrix} C_\kappa & \\ & \tilde{C}_\kappa \end{bmatrix}\right) = \{0\}. \tag{3.7}$$

As a consequence, Proposition 3.5 has the following analogue.

Proposition 3.7 *A triple of matrices $(A_\kappa, B_\kappa, D_\kappa)$ is a left kernel triple for some rational matrix function W if and only if $A_\kappa, B_\kappa, D_\kappa$ satisfy conditions (i) and (ii) in Proposition 3.5 and there exist matrices $\tilde{A}_\kappa, \tilde{B}_\kappa, \tilde{C}_\kappa \tilde{D}_\kappa$ such that $\tilde{A}_\kappa, \tilde{B}_\kappa, \tilde{D}_\kappa$ satisfy conditions (i) and (ii) in Proposition 3.5 and (3.7) holds with C_κ equal to the associated state-space-output operator of $(A_\kappa, B_\kappa, D_\kappa)$.*

We describe now the discrete left zero structure of a rational matrix function W. Let W be an $m \times n$ rational matrix function and let $\lambda \in \mathbf{C}_\infty$. We say that a function $\phi \in \mathcal{R}^{1 \times m}$ is a <u>left null function</u> for W at λ of <u>order</u> k, k a positive integer, if

(i) ϕ is analytic and nonzero at λ,

(ii) ϕW has a zero at λ of order k,

(iii) ϕ is orthogonal to W^{ol} at λ.

In view of Proposition 2.3 above conditions (i)-(iii) above are equivalent to (i), (ii) and

(iii') $\phi(\lambda) \notin W^{ol}(\lambda)$.

A set of left null functions $\{\phi_1, \phi_2, ..., \phi_\eta\}$ for an $m \times n$ rational matrix function W at $\lambda \in \mathbf{C}_\infty$ of orders $k_1, k_2, ..., k_\eta$, respectively, is called a <u>canonical set of left null functions</u> for W at λ if

(i) $\phi_1(\lambda), \phi_2(\lambda), ..., \phi_\eta(\lambda)$ are linearly independent,

(ii) $\phi_1, \phi_2, ..., \phi_\eta$ are contained in an orthogonal complement of W^{ol} in $(\mathcal{R}^{1 \times m}, \{\lambda\})$,

(iii) $\sum_{i=1}^{\eta} k_i$ is maximal subject to conditions (i) and (ii).

We note that if $\tau = (A_\kappa, B_\kappa, D_\kappa)$ is a left kernel triple for W, then, in view of Proposition 2.3 and Corollary 2.4, conditions (i) and (ii) above are equivalent to the linear independence of the rows of the matrix

$$\begin{bmatrix} \pi(\lambda) \\ \phi_1(\lambda) \\ \phi_2(\lambda) \\ \vdots \\ \phi_\eta(\lambda) \end{bmatrix}$$

where $\pi(\lambda)$ is a basis for $W^{ol}(\lambda)$ obtained from τ.

We show first that the canonical sets of left null functions for $W \in \mathcal{R}^{m\times n}$ at $\lambda \in \mathbf{C}_\infty$ can be projected along certain subspaces of $\mathcal{R}^{1\times m}$.

Proposition 3.8 *Let $\{\phi_1, \phi_2, ..., \phi_\eta\}$ be a canonical set of left null functions for an $m \times n$ rational matrix function W at $\lambda \in \mathbf{C}_\infty$ of orders $k_1, k_2, ..., k_\eta$ respectively and let Λ be an orthogonal complement of W^{ol} in $(\mathcal{R}^{1\times m}, \{\lambda\})$. Let $\tilde{\phi}_1, \tilde{\phi}_2, ..., \tilde{\phi}_\eta$ be the projections of $\phi_1, \phi_2, ..., \phi_\eta$ onto Λ along W^{ol}. Then $\{\tilde{\phi}_1, \tilde{\phi}_2, ..., \tilde{\phi}_\eta\}$ is a canonical set of left null functions for W at λ of orders $k_1, k_2, ..., k_\eta$.*

Proof Choose orthonormal bases $\{v_1, v_2, ..., v_k\}$ and $\{w_1, w_2, ..., w_l\}$ for $(\Lambda, \{\lambda\})$ and $(W^{ol}, \{\lambda\})$. Then the rational matrix function

$$Q = \begin{bmatrix} v_1 \\ v_2 \\ \vdots \\ v_k \\ w_1 \\ w_2 \\ \vdots \\ w_l \end{bmatrix}$$

is regular and analytic at λ. By Proposition 2.3, Q does not have a zero at λ and hence Q^{-1} is analytic at λ. Consequently, the functions $\psi_1 = \phi_1 Q^{-1}, \psi_2 = \phi_2 Q^{-1}, ..., \psi_\eta = \phi_\eta Q^{-1}$ are analytic at λ. If $\psi_i = [\,\psi_{i1} \quad \psi_{i2} \quad \cdots \quad \psi_{im}\,]$ ($\psi_{ij} \in \mathcal{R}$), let $\tilde{\phi}_i = \sum_{j=1}^{k} \psi_{ij} v_j$ ($i = 1, 2, ..., \eta$). Then $\tilde{\phi}_1, \tilde{\phi}_1, ..., \tilde{\phi}_\eta$ are the projections of $\phi_1, \phi_2, ..., \phi_\eta$ onto Λ along W^{ol} and $\tilde{\phi}_1, \tilde{\phi}_2, ..., \tilde{\phi}_\eta$ are analytic at λ. Since $\text{span}\{\tilde{\phi}_1(\lambda), \tilde{\phi}_2(\lambda), ..., \tilde{\phi}_\eta(\lambda)\} = \text{span}\{\phi_1(\lambda), \phi_2(\lambda), ..., \phi_\eta(\lambda)\}$ modulo $W^{ol}(\lambda)$, the vectors $\tilde{\phi}_1(\lambda), \tilde{\phi}_2(\lambda), ..., \tilde{\phi}_\eta(\lambda)$ are linearly independent. Since $\tilde{\phi}_i W = \phi_i W$ ($i = 1, 2, ..., \eta$), the orders of $\tilde{\phi}_i$ and ϕ_i as left null functions for W at λ are equal. Finally, suppose that there exists a left null function $\tilde{\phi}_{\eta+1} \in \Lambda$ for W at λ such that $\tilde{\phi}_{\eta+1} \notin$ span $\{\tilde{\phi}_1(\lambda), \tilde{\phi}_2(\lambda), ..., \tilde{\phi}_\eta(\lambda)\}$ or such that $\tilde{\phi}_{\eta+1}$ should replace some function ϕ_i in the canonical set of left null functions $\{\phi_1, \phi_2, ..., \phi_\eta\}$ because of order considerations. Then, reversing the argument above, we can find a left null function $\phi_{\eta+1}$ for W at λ such that $\phi_1, \phi_2, ..., \phi_{\eta+1}$ are contained in a subspace orthogonal to W^{ol} at λ and either $\phi_{\eta+1}(\lambda) \notin$ span $\{\phi_1(\lambda), \phi_2(\lambda), ..., \phi_\eta(\lambda)\}$ or $\phi_{\eta+1}$ should replace some functions in $\{\phi_1, \phi_2, ..., \phi_\eta\}$ because of order considerations. Since $\{\phi_1, \phi_2, ..., \phi_\eta\}$

is a canonical set of left null functions for W at λ, this is a contradiction. It follows that $\{\tilde{\phi}_1, \tilde{\phi}_2, ..., \tilde{\phi}_\eta\}$ is a canonical set of left null functions for W at λ and the proof is complete. □

Proposition 3.9 *Let $\lambda \in \mathbf{C}_\infty$ be a zero of an $m \times n$ rational matrix function W. Then the orders of the functions in any canonical set of left null functions for W at λ are equal to the partial multiplicities of the zero of W at λ.*

Proof We assume without loss of generality that $\lambda \in \mathbf{C}$. Let EDF be a Smith-McMillan factorization of W and suppose that the first k diagonal entries of D are nonzero and all the other entries of D are zero. Then W^{ol} is spanned by the last $m-k$ rows of E^{-1}. By Proposition 2.3 the space Λ spanned by the first k rows of E^{-1} is orthogonal to W^{ol} at λ. Clearly, the rows of E^{-1} corresponding to the nonzero entries of D which vanish at λ are the left null functions for W at λ of orders equal to the partial zero multipicities of W at λ. Hence, by Proposition 3.8, the orders of functions in any canonical set of left null functions for W at λ are equal to the partial multiplicities of the zero of W at λ. □

Proposition 3.8 implies also the following fact.

Proposition 3.10 *Let W_1 and W_2 be $m \times n_1$ and $m \times n_2$ rational matrix functions, let $\lambda \in \mathbf{C}_\infty$, and suppose that the column space of W_1 is orthogonal at λ to the column space of W_2. Then there exist canonical sets Φ_1 and Φ_2 of left null functions at λ for W_1 and W_2, respectively, such that every left null function $\phi \in \Phi_1$ for W_1 (resp. $\phi \in \Phi_2$ for W_2) at λ of order k is a left null function at λ of order k for the rational matrix function $W = [\, W_1 \quad W_2 \,]$, and $\Phi = \Phi_1 \cup \Phi_2$ is a canonical set of left null functions at λ for W.*

Proof Let Λ_i be the column space of W_i $(i = 1, 2)$ and let Λ_3 be an orthogonal complement of $\Lambda_1 + \Lambda_2$ in $\mathcal{R}^m$. Let

$$\begin{aligned} \Omega_1 &= (\Lambda_2 \oplus \Lambda_3)^\circ, \\ \Omega_2 &= (\Lambda_1 \oplus \Lambda_3)^\circ, \\ \Omega_3 &= (\Lambda_1 \oplus \Lambda_2)^\circ. \end{aligned}$$

Then, by Theorem 2.8, Ω_i is orthogonal to Ω_j at λ whenever $i \neq j$ and $\Omega_1 \oplus \Omega_2 \oplus \Omega_3 = \mathcal{R}^{1 \times m}$. Also,

$$\begin{aligned} W_1^{ol} &= \Omega_2 \oplus \Omega_3, \\ W_2^{ol} &= \Omega_1 \oplus \Omega_3, \\ W^{ol} &= \Omega_3. \end{aligned}$$

Choose a canonical set $\tilde{\Phi}_i$ of left null functions at λ for W_i and project functions in $\tilde{\Phi}_i$ onto Ω_i along W_i^{ol} to get Φ_i $(i = 1,2)$. Let $\Phi = \Phi_1 \cup \Phi_2$. If $\phi \in \Phi_1$ is a left null function for W_1 at λ of order k, then ϕW_1 has a zero at λ of order k and ϕW_2 vanishes identically. Since Ω_1 is orthogonal to W^{ol} at λ, ϕ is a left null function for W at λ of order k. Similarly, every left null function ϕ for W_2 at λ of order k is a left null function for W at λ of order k. Let ϕ be a left null function at λ for W of order k. We may assume $\phi \in \Omega_1 \oplus \Omega_2$. Let $\phi = \phi_{\Omega_1} + \phi_{\Omega_2}$ with $\phi_{\Omega_1} \in \Omega_1$. Then

$$\begin{aligned}(\phi_{\Omega_1} + \phi_{\Omega_2})\,[\,W_1 \quad W_2\,] &= [\,(\phi_{\Omega_1} + \phi_{\Omega_2})W_1 \quad (\phi_{\Omega_1} + \phi_{\Omega_2})W_2\,] \\ &= [\,\phi_{\Omega_1} W_1 \quad \phi_{\Omega_2} W_2\,]\end{aligned}$$

vanishes to order k at λ. So each of $\phi_{\Omega_1} W_1$ and $\phi_{\Omega_2} W_2$ vanish to order at least k at λ and $\phi(\lambda)$ is contained in the span of values at λ of null functions in Φ of order at least k. Since, by Proposition 2.3, the values at λ of functions in Φ are linearly independent, Φ is a canonical set of left null functions for W at λ.

□

A canonical set of left null functions for an $m \times n$ rational matrix function W at λ can be found similarly as in the regular case (see [BGR1]) with the additional constraint that the functions have to be contained in a subspace Λ orthogonal to W^{ol} at λ. That is, choose an orthogonal complement Λ of W^{ol} in $(\mathcal{R}^{1\times m}, \{\lambda\})$ and find a left null function $\phi_1 \in \Lambda$ for W of maximal possible order. Inductively, given left null functions $\phi_1, \phi_2, ..., \phi_\gamma \in \Lambda$ for W at λ, find a left null function $\phi_{\gamma+1}$ for W at λ of maximal possible order such that $\phi_{\gamma+1} \in \Lambda$ and $\phi_{\gamma+1}(\lambda) \notin \text{span}\ \{\phi_1(\lambda), \phi_2(\lambda), ..., \phi_\gamma(\lambda)\}$. Continue until the span of values at λ of left null functions for W at λ which are contained in Λ has been exhausted.

An alternative way to find a canonical set of left null functions for a rational matrix function W at λ is to find a Smith-McMillan factorization of W, compute the inverse of the matrix polynomial E, and select those rows of E^{-1} which correspond to the nonzero diagonal entries of D that vanish at λ.

A criterion whether a set of functions $\{\phi_1, \phi_2, ..., \phi_k\}$ is a canonical set of left null functions for a function $W \in \mathcal{R}^{m\times n}$ at a point $\lambda \in \mathbf{C}_\infty$ which does not depend explicitly on W^{ol} is as follows.

Proposition 3.11 *Let W be an $m \times n$ rational matrix function, let V be the linear span over $\mathcal{R}$ of the columns of W and let $\lambda \in \mathbf{C}_\infty$. Suppose $\Xi_1, \Xi_2, ..., \Xi_l \in \mathcal{R}^{1\times m}$ have no poles at λ and, for $i = 1,2,...,l$, $\Xi_i W$ has a zero at λ of order k_i, $k_i > 0$. Then Ξ_i is a left null function for W at λ of order k_i $(i = 1,2,...,l)$ and $\{\Xi_1, \Xi_2, ..., \Xi_l\}$ is a canonical set of left null functions for W at λ if and only if the following two conditions hold:*

(i) total multiplicity of the zero of W at λ equals $\sum_{i=1}^{l} k_i$,

(ii) $dim\ (\Xi_1(\lambda)V(\lambda) + \Xi_2(\lambda)V(\lambda) + ... + \Xi_l(\lambda)V(\lambda)) = l.$

Proof We show one of the asserted implications; the other one is clear. Suppose conditions (i) and (ii) hold. Condition (ii) implies that $\Xi_1(\lambda), \Xi_2(\lambda), ..., \Xi_l(\lambda)$ are linearly independent and the subspace of $\mathcal{R}^{1\times m}$ spanned by $\Xi_1, \Xi_2, ..., \Xi_l$ is orthogonal to W^{ol} at λ. Hence each Ξ_i is a left null function for W at λ of order k_i and, by (i), $\{\Xi_1, \Xi_2, ..., \Xi_l\}$ is a canonical set of left null functions for W at λ.

□

Example 3.12 Let

$$W(z) = \begin{bmatrix} (z-1)^2/z^2 & z^2/(z-1) \\ z^2-1 & 0 \\ z^3-z & z^2/(z+1) \end{bmatrix}, \tag{3.8}$$

let $\phi_1(z) = [0 \quad z \quad -1]$, and let $\phi_2(z) = [0 \quad 1 \quad 0]$. Since

$$\begin{aligned} W(z) &= \begin{bmatrix} z-1 & z+1 \\ z^2(z+1) & 0 \\ z^3(z+1) & z-1 \end{bmatrix} \begin{bmatrix} (z-1)/z^2 & \\ & z^2/(z^2-1) \end{bmatrix} \\ &\equiv \tilde{W}(z) \begin{bmatrix} z-1 & \\ & z \end{bmatrix} \end{aligned}$$

and

$$\tilde{W}(1) = \begin{bmatrix} 0 & 2 \\ 2 & 0 \\ 2 & 0 \end{bmatrix}, \qquad \tilde{W}(0) = \begin{bmatrix} -1 & 1 \\ 0 & 0 \\ 0 & -1 \end{bmatrix},$$

W has zeros at $z = 0$ and $z = 1$ of multiplicities 2 and 1 respectively (cf. Result 2 in [VK]). Since

$$\phi_1(z)W(z) = [0 \quad -z^2/(z+1)]$$

vanishees at 0 to the order 2 and

$$\begin{aligned} \phi_1(0)\tilde{W}(0) &= [0 \quad 1] \\ &\neq 0, \end{aligned}$$

$\{\phi_1\}$ is a canonical set of left null functions for W at 0. Similarly,

$$\phi_2(z)W(z) = [(z-1)(z+1) \quad 0]$$

vanishes at 1 and

$$\begin{aligned} \phi_2(1)\tilde{W}(1) &= [2 \quad 0] \\ &\neq 0, \end{aligned}$$

so $\{\phi_2\}$ is a canonical set of left null functions for W at 1.

□

Let $\lambda \in \mathbf{C}_\infty$ be a zero of an $m \times n$ rational matrix function W and let $\{\phi_1, \phi_2, ..., \phi_\eta\}$ be a canonical set of left null functions for W at λ of orders $k_1, k_2, ..., k_\eta$, respectively. Let $\phi_{i,j} \in \mathbf{C}^{1\times n}$ be such that

$$\phi(z) = \begin{cases} \sum_{j=0}^{\infty} (z-\lambda)^j \phi_{i,j}, & \text{if } \lambda \in \mathbf{C} \\ \sum_{j=0}^{\infty} z^{-j} \phi_{i,j}, & \text{if } \lambda = \infty. \end{cases}$$

Any ordered pair of matrices (A, B), where

$$A = \begin{cases} S \begin{bmatrix} J_{k_1}(\lambda) & & & \\ & J_{k_2}(\lambda) & & \\ & & \ddots & \\ & & & J_{k_\eta}(\lambda) \end{bmatrix} S^{-1}, & \text{if } \lambda \in \mathbf{C} \\ S \begin{bmatrix} J_{k_1}(0) & & & \\ & J_{k_2}(0) & & \\ & & \ddots & \\ & & & J_{k_\eta}(0) \end{bmatrix} S^{-1}, & \text{if } \lambda = \infty \end{cases}$$

and

$$B = S \begin{bmatrix} \phi_{1,k_1-1} \\ \phi_{1,k_1-2} \\ \vdots \\ \phi_{1,0} \\ \phi_{2,k_2-1} \\ \phi_{2,k_2-2} \\ \vdots \\ \phi_{2,0} \\ \vdots \\ \phi_{\eta,k_\eta-1} \\ \phi_{\eta,k_\eta-2} \\ \vdots \\ \phi_{\eta,0} \end{bmatrix}$$

for some invertible matrix S of an appropriate size, is called a <u>left null pair</u> for W at λ.

Let $\sigma \subset \mathbf{C}$ and let $\lambda_1, \lambda_2, ..., \lambda_r$ be the zeros of a rational matrix function W in σ. Let (A_i, B_i) be a left null pair for W at λ_i $(i = 1, 2, ..., r)$. Any ordered pair of matrices (A, B), where

$$A = S \begin{bmatrix} A_1 & & & \\ & A_2 & & \\ & & \ddots & \\ & & & A_p \end{bmatrix} S^{-1}, \qquad B = S \begin{bmatrix} B_1 \\ B_2 \\ \vdots \\ B_p \end{bmatrix}$$

for some invertible matrix S of an appropriate size, is called a left null pair for W over σ (or a σ-left null pair for W).

We note that since the values at $\lambda \in \mathbf{C}$ of functions in any canonical set of left null functions for $W \in \mathcal{R}^{m\times n}$ at λ are linearly independent, a left null pair (A,B) for W over a $\sigma \subset \mathbf{C}$, or at infinity, is controllable (see Theorem 3.4 in [BGR1]), i.e. the matrix $\begin{bmatrix} B & AB & \dots & A^lB \end{bmatrix}$ has full column rank for sufficiently large integers l.

Example 3.13 The function W in Example 3.12 has

$$(A_\zeta, B_\zeta) = \left(\begin{bmatrix} 0 & 1 & 0 \\ 0 & 0 & 0 \\ 0 & 0 & 1 \end{bmatrix}, \begin{bmatrix} 0 & 1 & 0 \\ 0 & 0 & -1 \\ 0 & 1 & 0 \end{bmatrix} \right)$$

as a left null pair over $\{0,1\}$.

□

In the next lemma we show, among other things, that there is a collection of functions $\Phi = \{\phi_1, \phi_2, ..., \phi_\kappa\}$, where κ is the largest geometric multiplicity of a zero of W, so that a canonical set of left null functions for W at any zero of W can be chosen from among the members of Φ.

Lemma 3.14 *Let $\sigma \subset \mathbf{C}$, let (A,B) be a left null pair for a rational matrix function W over σ, and let $\{\lambda_1, \lambda_2, ..., \lambda_s\}$ be a finite set of points in $\mathbf{C}_\infty$. Suppose that the largest geometric multiplicity of a zero of W in σ equals κ. Then there exist rational row vector functions $\phi_1, \phi_2, ..., \phi_\kappa$ such that the pair (A,B) is left-similar to the pair constructed from the Taylor coefficients of $\phi_1, \phi_2, ..., \phi_\kappa$ at the eigenvalues of A and span$\{\phi_j\}$ is orthogonal to $W^{ol} +$ span$\{\phi_i : i \neq j\}$ on $\{\lambda_1, \lambda_2, ..., \lambda_s\}$. Moreover, if the geometric multiplicity of the zero of W at a point $\lambda \in \sigma$ equals k, then $\{\phi_1, \phi_2, ..., \phi_k\}$ is a canonical set of left null functions for W at λ.*

Proof After augmenting the set $\{\lambda_1, \lambda_2, ..., \lambda_s\}$ if necessary, we may assume that $s > r$, the eigenvalues of A are $\{\lambda_1, \lambda_2, ..., \lambda_r\}$, and $\lambda_s = \infty$. Let $\{\psi_{i,1}, \psi_{i,2}, ..., \psi_{i,\eta_i}\}$ $(i = 1,2,...,r)$ be a canonical set of left null functions for W at λ_i such that the pair (A,B) is left-similar to the pair constructed from the coefficients in the Taylor expansions of $\psi_{i,j}$ at λ_i. We assume $\kappa = \eta_1 \geq \eta_2 \geq ... \geq \eta_r$. Let $k_{i,j}$ be the order of $\psi_{i,j}$ as a left null function for W at λ_i $(i = 1,2,...,r,\ j = 1,2,...,\eta_i)$, and let

$$\begin{aligned} \psi_j(z) = {} & (z-\lambda_2)^{k_{2,j}}(z-\lambda_3)^{k_{3,j}}...(z-\lambda_{\gamma_j})^{k_{\gamma_j,j}}\psi_{1,j} + \\ & + (z-\lambda_1)^{k_{1,j}}(z-\lambda_3)^{k_{3,j}}...(z-\lambda_{\gamma_j})^{k_{\gamma_j,j}}\psi_{2,j} + ... + \\ & + (z-\lambda_1)^{k_{1,j}}(z-\lambda_2)^{k_{2,j}}...(z-\lambda_{\gamma_j-1})^{k_{\gamma_j-1,j}}\psi_{\gamma_j,j} \end{aligned} \tag{3.9}$$

where γ_j is the largest integer such that $\eta_{\gamma_j} \geq j$ $(j = 1, 2, ..., \kappa)$. Then ψ_j is a left null function for W at the points $\lambda_1, \lambda_2, ..., \lambda_{\gamma_j}$ of orders $k_{1,j}, k_{2,j}, ..., k_{\gamma_j,j}$ respectively.

We claim that the pair (A, B) is left-similar to the pair constructed from the canonical sets $\{\psi_1, \psi_2, ..., \psi_{\eta_1}\}$, $\{\psi_1, \psi_2, ..., \psi_{\eta_2}\}, ...,$ $\{\psi_1, \psi_2, ..., \psi_{\eta_r}\}$ of left null functions for W at $\lambda_1, \lambda_2, ..., \lambda_r$. To prove the claim it suffices to show that if $\psi_{i,j}^{\gamma}$ and ψ_j^{γ} are the γth coefficients in the Taylor expansions of $\psi_{i,j}$ and ψ_j at λ_i $(1 \leq i \leq r,\ 1 \leq j \leq \eta_i)$, then the pairs of matrices

$$\left(J_{k_{i,j}}(\lambda_i), \begin{bmatrix} \psi_{i,j}^{k_{i,j}-1} \\ \psi_{i,j}^{k_{i,j}-2} \\ \vdots \\ \psi_{i,j}^{0} \end{bmatrix} \right) \quad \text{and} \quad \left(J_{k_{i,j}}(\lambda_i), \begin{bmatrix} \psi_{j}^{k_{i,j}-1} \\ \psi_{j}^{k_{i,j}-2} \\ \vdots \\ \psi_{j}^{0} \end{bmatrix} \right)$$

are left-similar. Now it follows from (3.9) that

$$\psi_j(z) = p_{i,j}(z)\psi_{i,j}(z) + (z - \lambda_j)^{k_{i,j}} q_{i,j}(z)$$

where $p_{i,j}(z)$ is a scalar polynomial with $p_{i,j}(\lambda_i) \neq 0$ and where $q_{i,j}(z)$ is a vector function analytic at λ_i. Hence for $\gamma \in \{0, 1, ..., k_{i,j} - 1\}$ we have $\psi_j^{\gamma} = \sum_{\mu+\nu=\gamma} a_\mu \psi_{i,j}^{\nu}$ where $a_0, a_1, ..., a_{k_{i,j}-1}$ are numbers such that $p_{i,j}(z) = \sum_{\mu=0}^{L} a_\mu (\lambda - \lambda_i)^\mu$ and $a_0 \neq 0$. Hence

$$\begin{bmatrix} \psi_{j}^{k_{i,j}-1} \\ \psi_{j}^{k_{i,j}-2} \\ \vdots \\ \psi_{j}^{0} \end{bmatrix} = \begin{bmatrix} a_0 & a_1 & \cdots & a_{k_{i,j}-1} \\ & a_0 & \cdots & a_{k_{i,j}-2} \\ & & \ddots & \vdots \\ & & & a_0 \end{bmatrix} \begin{bmatrix} \psi_{i,j}^{k_{i,j}-1} \\ \psi_{i,j}^{k_{i,j}-2} \\ \vdots \\ \psi_{i,j}^{0} \end{bmatrix} \equiv S \begin{bmatrix} \psi_{i,j}^{k_{i,j}-1} \\ \psi_{i,j}^{k_{i,j}-2} \\ \vdots \\ \psi_{i,j}^{0} \end{bmatrix}.$$

Since the inverse of S is the upper triangular matrix

$$\begin{bmatrix} b_0 & b_1 & \cdots & b_{k_{i,j}-1} \\ & b_0 & \cdots & a_{k_{i,j}-2} \\ & & \ddots & \vdots \\ & & & b_0 \end{bmatrix}$$

such that

$$\sum_{\mu+\nu=\gamma,\ \mu,\nu\geq 0} a_\mu b_\nu = \begin{cases} 1, & \text{if } \gamma = 0 \\ 0, & \text{if } \gamma = 1, 2, ..., k_{i,j} - 1, \end{cases}$$

we have

$$SJ_{k_{i,j}}(\lambda_i)S^{-1} = S\left(\lambda_i + \begin{bmatrix} 0 & 1 & 0 & \dots & 0 \\ & 0 & 1 & \dots & 0 \\ & & \ddots & \ddots & \vdots \\ & & & \ddots & 1 \\ & & & & 0 \end{bmatrix}\right)S^{-1}$$

$$= \lambda_i + \begin{bmatrix} a_0 & a_1 & a_2 & \dots & a_{k_{i,j}-1} \\ & a_0 & a_1 & \dots & a_{k_{i,j}-2} \\ & & \ddots & \ddots & \vdots \\ & & & \ddots & a_1 \\ & & & & a_0 \end{bmatrix} \begin{bmatrix} 0 & b_0 & b_1 & \dots & b_{k_{i,j}-2} \\ & 0 & b_0 & \dots & \vdots \\ & & \ddots & \ddots & \vdots \\ & & & \ddots & b_0 \\ & & & & 0 \end{bmatrix}$$

$$= \lambda_i + \begin{bmatrix} 0 & 1 & 0 & \dots & 0 \\ & 0 & 1 & \dots & 0 \\ & & \ddots & \ddots & \vdots \\ & & & \ddots & 1 \\ & & & & 0 \end{bmatrix} = J_{k_{i,j}}(\lambda)$$

and the claim is established.

We now describe the *orthogonalization process* which transforms $\psi_1, \psi_2, ..., \psi_\kappa$ into functions $\phi_1, \phi_2, ..., \phi_\kappa$ with the required properties (cf. Proposition 2.3). We begin with ψ_1. Choose constant vectors $c_{1,r+1}, c_{1,r+2}, c_{1,s}$ such that $c_{1,j} \notin W^{ol}(\lambda_j)$ $(j = r+1, r+2, ..., s)$ and suppose that the integers $\mu_{1,j}$ are such that the rational vector function

$$\begin{cases} (z-\lambda_j)^{-\mu_{1,j}}\psi_1(z), & \text{if } j = r+1, r+2, ..., s-1 \\ z^{\mu_{1,j}}\psi_1(z), & \text{if } j = s \end{cases}$$

is analytic and nonzero at λ_j $(j = r+1, r+2, ..., s)$. We put

$$\begin{aligned} \phi_1(z) =& \psi_1(z) + \prod_{i=1}^{r}(z-\lambda_i)^{k_{i,1}+1}\cdot \\ & \cdot\left(\sum_{j=r+1}^{s-1}(z-\lambda_j)^{\mu_{1,j}}\left(c_{i,j} - [\psi_1(z)]_{\lambda_j}\Big/\prod_{i=1}^{r}(\lambda_j-\lambda_i)^{k_{i,1}+1}\right)+\right. \\ & \left.+ z^{-\mu_{1,s}-\sum_{i=1}^{r}(k_{i,1}+1)}\left(c_{1,s} - [\psi_1(z)]_\infty\right)\right). \end{aligned}$$

Then ϕ_1 is a left null function for W at λ_i of order $k_{i,1}$ $(i = 1, 2, ..., r)$, the first $k_{i,1}$ coefficients in Taylor expansions of ϕ_1 and ψ_1 at λ_i coincide $(i = 1, 2, ..., r)$, and the leading coefficients in the Laurent expansions of ϕ_1 at $\lambda_{r+1}, \lambda_{r+2}, ..., \lambda_s$ are $c_{1,r+1}, c_{1,r+2}, ..., c_{1,s}$. Inductively, suppose $\phi_1, \phi_2, ..., \phi_{j-1}$ with $j \le \kappa$ are given and let $\gamma_j \le r$ be the largest integer such that $\eta_{\gamma_j} \ge j$. Choose nonzero constant vectors $c_{j,\gamma_j+1}, c_{j,\gamma_j+2}, ..., c_{j,s}$ such that the span of $c_{j,\nu}$ and the leading coefficients in the Laurent expansions of $\phi_1, \phi_2, ..., \phi_{j-1}$ at λ_ν intersects trivially with $W^{ol}(\lambda_\nu)$ $(\nu = \gamma_{j+1}, \gamma_{j+2}, ..., \gamma_s)$, and suppose that the integers $\mu_{j,\nu}$ are such that the rational vector function

$$\begin{cases} (z-\lambda_\nu)^{-\mu_{j,\nu}}\psi_j(z), & \text{if } \nu = j+1, j+2, ..., s-1 \\ z^{\mu_{j,\nu}}\psi_j(z), & \text{if } \nu = s \end{cases}$$

is analytic and nonzero at λ_ν $(\nu = j+1, j+2, ..., s)$. Let

$$\begin{aligned} \phi_j(z) = & \psi_j(z) + \prod_{i=1}^{\gamma_j}(z-\lambda_i)^{k_{i,j}+1} \cdot \\ & \cdot \Bigg(\sum_{\nu=\gamma_j+1}^{s-1} (z-\lambda_\nu)^{\mu_{j,\nu}} \Big(c_{j,\nu} - [\psi_j(z)]_{\lambda_j} \Big/ \prod_{i=1}^{\gamma_j} (\lambda_\nu - \lambda_i)^{k_{i,j}+1} \Big) + \\ & \qquad + z^{-\mu_{j,s} - \sum_{i=1}^{\gamma_j}(k_{i,j}+1)} \Big(c_{j,s} - [\psi_j(z)]_\infty \Big) \Bigg). \end{aligned}$$

The functions $\phi_1, \phi_2, ..., \phi_\kappa$ have the required properties.

□

We note that by choosing constant vectors $c_{i,j}$ in the orthogonalization process in the proof of Lemma 3.14, we have actually chosen the values of $\phi_1, \phi_2, ..., \phi_\kappa$ at $\alpha_{r+1}, \alpha_{r+2}, ...,$ α_s. Also, in general the values of ϕ_j's at $\alpha_2, \alpha_3, ..., \alpha_r$ were not completely determined and depended only on the values of $\psi_{i,1}, \psi_{i,2}, ..., \psi_{i,\eta_i}$ at λ_i. For future reference, we state this in the following corollary.

Corollary 3.15 *Let $\sigma \subset \mathbf{C}$, let (A, B) be a left null pair for an $m \times n$ rational matrix function W over σ, and let κ be the largest geometric multiplicity of a zero of W in σ. Suppose that the pair (A, B) is left-similar to the pair constructed from canonical sets of left null functions*

$$\{\{\phi_{i_j}\} : \ z_j \in \sigma \text{ is a zero of } W \text{ of geometric multiplicity } \eta_j, \ 1 \le i \le \eta_j\},$$

and, for each distinct eigenvalue z_j of A, let

$$\Omega_j = span\ \{\phi_{i_j}(\lambda) : \ 1 \le i \le \eta_j\}.$$

Let k be an integer such that $\kappa \le k \le m$. Let $\{\lambda_1, \lambda_2, ..., \lambda_s\}$ be a finite set of points in $\mathbf{C}_\infty$, and suppose that to every integer $i = 1, 2, ..., s$ there corresponds a k-dimensional subspace Λ_i of $\mathbf{C}^{1\times m}$ such that Λ_i contains Ω_j whenever $\lambda_i = z_j$ is an eigenvalue of A. Then we can find a subspace Ξ of $\mathcal{R}^{1\times m}$ and functions $\phi_1, \phi_2, ..., \phi_\kappa \in \Xi$ such that $\phi_1, \phi_2, ..., \phi_\kappa$ are analytic on $\sigma(A_\zeta) \cup \{\lambda_1, \lambda_2, ..., \lambda_s\}$, $\Xi(\lambda_i) = \Lambda_i$ $(i = 1, 2, ..., s)$, and the pair (A, B) is left-similar to *the left null pair constructed from Taylor coefficients of* $\phi_1, \phi_2, ..., \phi_{\eta_j}$ *at every zero* z_j *of* W *in* σ.

Lemma 3.14 implies also another fact.

Lemma 3.16 *Let* (A, B) *be a left null pair of an* $m \times n$ *rational matrix function* W *over a subset* σ *of* $\mathbf{C}$. *Then there exists a subspace* Λ *of* $\mathcal{R}^{1\times m}$ *such that* Λ *is orthogonal to* W^{ol} *on* $\mathbf{C}$ *and the pair* (A, B) *is left-similar to a left null pair for* W *over* σ *constructed from functions in* Λ.

Proof Let $\phi_1, \phi_2, ..., \phi_k$ be as in Lemma 3.14. After multiplying each ϕ_i by a scalar polynomial which vanishes only at the poles of ϕ_i, we may assume that $\phi_1, \phi_2, ..., \phi_k$ are polynomial row vector functions. Let a matrix polynomial $P(z)$ with rows $p_1, p_2, ..., p_\kappa$ be a left kernel polynomial for W and let

$$R(z) = \begin{bmatrix} P(z) \\ \phi_1(z) \\ \phi_2(z) \\ \vdots \\ \phi_k(z) \end{bmatrix}.$$

By Lemma 3.14 and Proposition 2.3, R has no zeros in $\sigma(A)$. Let $\lambda \in \mathbf{C}\backslash\sigma(A)$ be a zero of R. Then the rows of the matrix $R(\lambda)$ are linearly dependent. Suppose the set $\{p_{\mu_i}(\lambda) : i = 1, 2, ..., r\} \cup \{\phi_{\nu_j}(\lambda) : j = 1, 2, ..., s\}$ is linearly dependent and

$$\phi_{\nu_s}(\lambda) = \alpha_1 p_{\mu_1}(\lambda) + \alpha_2 p_{\mu_2}(\lambda) + ... + \alpha_r p_{\mu_r}(\lambda) + \beta_1 \phi\nu_1(\lambda) + \beta_2 \phi\nu_2(\lambda) + ...\beta_{s-1}\phi\nu_{s-1}(\lambda).$$

Let $q(z)$ be the minimal degree monic scalar polynomial which vanishes at z_0 to the order k_0 whenever $\phi\nu_s$ is a left null function for W at z_0 of oreder k_0. Replace the row ϕ_{ν_s} in R by

$$\tilde{\phi}_{\nu_s}(z) = \frac{q(z)}{q(\lambda)(z-\lambda)}[\phi_{\nu_s}(z) - \alpha_1 p_{\mu_1}(z) - ... - \alpha_r p_{\mu_r}(z) - \beta_1\phi\nu_1(z) - ...\beta_{s-1}\phi\nu_{s-1}(z)].$$

Clearly, $\tilde{\phi}_{\nu_s}$ is a left null function for W at z_0 of order k_0 whenever ϕ_{ν_s} is such a function. Also, a finite number of such operations yields row polynomials $\psi_1, \psi_2, ..., \psi_k$ such that the matrix polynomial

$$\begin{bmatrix} P(z) \\ \psi_1(z) \\ \psi_2(z) \\ \vdots \\ \psi_k(z) \end{bmatrix}$$

has no finite zeros. The subspace $\Lambda = \text{span}\{\psi_1, \psi_2, ..., \psi_k\}$ has the required properties.

□

Lemma 3.16 has the following consequence.

Theorem 3.17 *Let (A, B) be a left null pair for a function $W \in \mathcal{R}^{m\times n}$ over a subset σ of* **C**. *Then there exists an $m \times m$ regular rational matrix function H with the following properties:*

(i) (A, B) is a left null pair for H over σ;

(ii) (C_π, A_π) is a right pole pair for H over σ if and only if (C_π, A_π) is a right pole pair for W over σ.

Proof Let EDF be s Smith-McMillan factorization of W and let

$$D = \begin{bmatrix} \tilde{D} & 0 \\ 0 & 0 \end{bmatrix}$$

with $\tilde{D}$ a regular rational matrix function. Considering, if necessary, the function

$$E\begin{bmatrix} \tilde{D} \\ 0 \end{bmatrix},$$

we may assume that $m \geq n$ and W is left invertible.

Let Λ be as in Lemma 3.16. By Proposition 2.9, there exists a basis $\{v_1, v_2, ..., v_k\}$ for Λ^{or} such that the rational matrix function $[\,v_1 \quad v_2 \quad \dots \quad v_k\,]$ has no zeros nor poles in σ. In view or Theorem 2.8 and Proposition 2.3, the rational matrix function

$$H = [\,W \quad v_1 \quad v_2 \quad \dots \quad v_k\,]$$

has the required propertics.

□

Let us note here that the extension of W to a regular rational matrix function is not unique. In fact the column space of the added columns depends on the similarity class of the left null pair (A, B). The reason for this is that the functions $\phi_1, \phi_2, ..., \phi_\kappa$ arising in the construction in Lemma 3.3 depend on the choice of canonical sets of left null functions at each point of $\sigma(A)$.

Theorem 3.17 connects the discrete left zero structure of a nonregular rational matrix function with the left zero structure of a regular rational matrix function. As a consequence, many properties of left null pairs of regular rational matrix functions carry over to the nonregular case.

Proposition 3.18 *Let $W \in \mathcal{R}^{m\times n}$, let $\sigma \subset$ **C**, and let (A, B) be a left null pair for W over σ. Then*

(i) the pair (A, B) is controllable;

(ii) $\sigma(A_\zeta) \subset \sigma$;

(iii) for every $k \in \mathcal{R}^{n\times 1}$ such that k and Wk are analytic on σ, the function

$$(z-A)^{-1}BW(z)k(z)$$

is analytic on σ.

Proof Property (ii) follows immediately from the construction. Property (i) follows from Theorem 3.17 above and Theorem 3.3 in [BGR1]. Property (iii) can be proved using Theorem 3.17 above and the argument in the proof of Theorem 2.42 in [BGR2]. □

In particular, it follows from part (iii) of Proposition 3.18 that if (A, B) is a left null pair for a function $W \in \mathcal{R}^{m\times n}$ over a subset σ of $\mathbf{C}$ and W is analytic on σ, then the function $(z-A)^{-1}BW(z)$ is analytic on σ.

We emphasize that left null pairs for a nonregular rational matrix function W over a subset $\sigma \subset \mathbf{C}$ need not be left-similar. E.g. if

$$W(z) = \begin{bmatrix} z-1 \\ z-1 \end{bmatrix},$$

$(A_1, B_1) = ([1],\ [1\ \ 0])$ and $(A_2, B_2) = ([1],\ [0\ \ 1])$ are left null pairs for W over $\mathbf{C}$ which are not left-similar. A characterization of when two controllable pairs (A_1, B_1) and (A_2, B_2) are left null pairs for the same matrix polynomial W is given in [BCR].

We conclude this section with basic definitions for the right zero structure of a rational matrix function W.

Let W be an $m \times n$ rational matrix function. We will denote the right annihilator of W by W^{or}. The set of values at $\lambda \subset \mathbf{C}_\infty$ of functions in W^{or} which are analytic at λ will be denoted by $W^{or}(\lambda)$. A matrix polynomial whose columns form a minimal polynomial basis for W^{or} is called a right kernel polynomial for W. We will assume that the columns of a right kernel polynomial P are ordered according to decreasing degrees.

Proposition 3.19 *Let $P \in \mathcal{R}^{n\times k}$ be a matrix polynomial with columns orthogonal at infinity and suppose that the ith column of P equals $\sum_{j=0}^{\eta_i} z^{\eta_i - j}P_{i,j}$ ($P_{i,j} \in \mathbf{C}^{n\times 1}$, $P_{i,0} \neq 0$ if $P_i \neq 0$, $i = 1, 2, ..., k$). Suppose that the first l columns of P have a pole at infinity and the last $k-l$*

columns are constant. Then a minimal realization of P is given by the formula

$$P(z) = P(0) + \begin{bmatrix} P_{1,0} & P_{1,1} & \dots & P_{1,\eta_1-1} & P_{2,0} & P_{2,1} & \dots & P_{2,\eta_2-1} & \dots & P_{l,0} & P_{l,1} & \dots & P_{l,\eta_l-1} \end{bmatrix} \cdot \left(z^{-1} - \begin{bmatrix} J_{\eta_1}(0) & & & \\ & J_{\eta_2}(0) & & \\ & & \ddots & \\ & & & J_{\eta_l}(0) \end{bmatrix} \right)^{-1} \begin{bmatrix} E_1 \\ E_2 \\ \vdots \\ E_l \end{bmatrix} \tag{3.10}$$

where E_i is the $\eta_i \times k$ matrix with 1 at the position $(i, 1)$ and zeros elsewhere.

We note that the matrix $\begin{bmatrix} E_1^T & E_2^T & \dots & E_l^T \end{bmatrix}^T$ in the realization (3.10) of P can be easily reproduced from other matrices in (3.10). If P is a right kernel polynomial for a rational matrix function W, the ordered triple of matrices $(A_\kappa, C_\kappa, D_\kappa)$ where

$$C_\kappa = \begin{bmatrix} P_{1,0} & P_{1,1} & \dots & P_{1,\eta_1-1} & P_{2,0} & P_{2,1} & \dots & P_{l,0} & P_{l,1} & \dots & P_{l,\eta_l-1} \end{bmatrix},$$

$$A_\kappa = \begin{bmatrix} J_{\eta_1}(0) & & & \\ & J_{\eta_2}(0) & & \\ & & \ddots & \\ & & & J_{\eta_l}(0) \end{bmatrix} \quad \text{and} \quad D_\kappa = P(0)$$

will be called a <u>right</u> <u>kernel</u> <u>triple</u> for W.

We note that if W is a rational matrix function and $\lambda \in \mathbf{C}$, a basis for $W^{or}(\lambda)$ can be easily computed from a right kernel triple $(A_\kappa, C_\kappa, D_\kappa)$ for W; a basis for $W^{or}(\infty)$ can be read off directly from $(A_\kappa, C_\kappa, D_\kappa)$.

Let W be an $m \times n$ rational matrix function and let $\lambda \in \mathbf{C}_\infty$. A function $\phi \in \mathcal{R}^{n \times 1}$ is a <u>right</u> <u>null</u> <u>function</u> for W at λ of <u>order</u> k, k a positive integer, if

(i) ϕ is analytic and nonzero at λ,

(ii) $W\phi$ has a zero at λ of order k,

(iii) ϕ is orthogonal to W^{or} at λ.

Condition (iii) above can be replaced by

(iii') $\phi(\lambda) \notin W^{or}(\lambda)$.

A set of right null functions $\{\phi_1, \phi_2, ..., \phi_\eta\}$ for an $m \times n$ rational matrix function W at $\lambda \in \mathbf{C}_\infty$ of orders $k_1, k_2, ..., k_\eta$, respectively, is called a <u>canonical</u> <u>set</u> of <u>right</u> <u>null</u> <u>functions</u> for W at λ if

(i) $\phi_1(\lambda), \phi_2(\lambda), ..., \phi_\eta(\lambda)$ are linearly independent,

(ii) $\phi_1, \phi_2, ..., \phi_\eta$ are contained in an orthogonal complement of W^{or} in $(\mathcal{R}^{n \times 1}, \{\lambda\})$,

(iii) $\sum_{i=1}^{\eta} k_i$ is maximal subject to conditions (i) and (ii).

Condition (ii) above can be replaced with

(ii') span $\{\phi_1(\lambda), \phi_2(\lambda), ..., \phi_\eta(\lambda) \cap W^{or}(\lambda) = (0)$.

We note that the number of functions in a canonical set of right null functions for W at λ is equal to the geometric multiplicity of the zero of W at λ. The orders of functions in a canonical set of right null functions for W at λ coincide with the partial multiplicities of the zero of W at λ.

Let $\lambda \in \mathbf{C}_\infty$ be a zero of an $m \times n$ rational matrix function W and let $\{\phi_1, \phi_2, ..., \phi_\eta\}$ be a canonical set of right null functions for W at λ, of orders $k_1, k_2, ..., k_\eta$, respectively. Let $\phi_{i,j} \in \mathbf{C}^{n \times 1}$ be the jth coefficient in the Taylor expansion of ϕ_i at λ. Any ordered pair of matrices (C, A), where C equals

$$\begin{bmatrix} \phi_{1,0} & \phi_{1,1} & \cdots & \phi_{1,k_1-1} & \phi_{2,0} & \phi_{2,1} & \cdots & \phi_{2,k_2-1} & \cdots & \phi_{\eta,0} & \phi_{\eta,1} & \cdots & \phi_{\eta,k_\eta-1} \end{bmatrix} S$$

and

$$A = \begin{cases} S^{-1} \begin{bmatrix} J_{k_1}(\lambda) & & & \\ & J_{k_2}(\lambda) & & \\ & & \ddots & \\ & & & J_{k_\eta}(\lambda) \end{bmatrix} S, & \text{if } \lambda \in \mathbf{C} \\ \\ S^{-1} \begin{bmatrix} J_{k_1}(0) & & & \\ & J_{k_2}(0) & & \\ & & \ddots & \\ & & & J_{k_\eta}(0) \end{bmatrix} S, & \text{if } \lambda = \infty \end{cases}$$

for some invertible matrix S of an appropriate size, is called a <u>right null pair</u> for W at λ.

Let $\sigma \subset \mathbf{C}$ and let $\lambda_1, \lambda_2, ..., \lambda_r$ be the zeros of a rational matrix function W in σ. Let (C_i, A_i) be a right null pair for W at λ_i $(i = 1, 2, ..., r)$. Any ordered pair of matrices (C, A), where

$$C = [\, C_1 \quad C_2 \quad \ldots \quad C_r \,] S \quad \text{and} \quad A = S^{-1} \begin{bmatrix} A_1 & & & \\ & A_2 & & \\ & & \ddots & \\ & & & A_p \end{bmatrix} S$$

for some invertible matrix S of an appropriate size, is called a <u>right null pair</u> for W over σ (or a σ-right null pair for W).

Similarly as in the regular case, right null pairs are controllable. Right null pairs also have the following property.

Proposition 3.20 *Let $\sigma \subset \mathbf{C}$, let (C, A) be a right null pair for a rational matrix function W over σ, and let $\{\lambda_1, \lambda_2, ..., \lambda_s\}$ be a finite set of points in $\mathbf{C}_\infty$. Suppose that the largest geometric multiplicity of a zero of W in σ equals κ. Then there exists a subspace Λ orthogonal* *to* W^{or} *on* $\{\lambda_1, \lambda_2, ..., \lambda_s\}$ *and functions* $\phi_1, \phi_2, ..., \phi_\kappa \in \Lambda$ *such that the pair* (C, A) *is right-similar to the pair constructed from the Taylor coefficients of* $\phi_1, \phi_2, ..., \phi_\kappa$ *at the eigenvalues of* A.

Proposition 3.20 allows us, in particular, to associate with each right null pair (C, A) for a rational matrix function W over $\sigma \subset \mathbf{C}$ a subspace Λ of $\mathcal{R}^{m\times 1}$ which is orthogonal to W^{or} on $\sigma(A)$ such that (C, A) is right-similar to a pair constructed from right null functions for W which are contained in Λ. Right null pairs constructed from functions in a fixed orthogonal complement of W^{or} in $(\mathcal{R}^{n\times 1}, \{\lambda_1, \lambda_2, ..., \lambda_r\})$, where $\lambda_1, \lambda_2, ..., \lambda_r$ are all the zeros of W in σ, are right-similar.

4. FUNCTIONS WITH A GIVEN LOCAL NULL-POLE STRUCTURE

Right pole and left null pairs together with a left kernel triple for a rational matrix function have to satisfy certain obvious conditions. We summarize these conditions in the following proposition.

Proposition 4.1 *Suppose* (C_π, A_π) *and* (A_ζ, B_ζ) *are right pole and left null pairs over* $\sigma \subset \mathbf{C}$ *for a rational matrix function* W *and* $(A_\kappa, B_\kappa, D_\kappa)$ *is a left kernel triple for* W *with the associated matrix polynomial* P. *Then*

(i) the pair (C_π, A_π) *is observable and* $\sigma(A_\pi) \subset \sigma$,

(ii) the pair (A_ζ, B_ζ) *is controllable and* $\sigma(A_\zeta) \subset \sigma$,

(iii) the pair (A_κ, B_κ) *is controllable,* A_κ *is in Jordan form,* $\sigma(A_\kappa) \subset \{0\}$, *and the matrix polynomial* P *has the Smith-McMillan form* $[\,I \quad 0\,]$,

(iv) the rational matrix function $P(z)C_\pi(z - A_\pi)^{-1}$ *is analytic on* $\mathbf{C}$,

(v) if $\sigma(A_\zeta) = \{\lambda_1, \lambda_2, ..., \lambda_r\}$ *and* I_k *is the identity matrix with the same number of rows as* D_κ, *the pair*

$$\left(\begin{bmatrix} A_\zeta & & & & \\ & \lambda_1 \mathrm{I}_k & & & \\ & & \lambda_2 \mathrm{I}_k & & \\ & & & \ddots & \\ & & & & \lambda_r \mathrm{I}_k \end{bmatrix}, \begin{bmatrix} B_\zeta \\ P(\lambda_1) \\ P(\lambda_2) \\ \vdots \\ P(\lambda_r) \end{bmatrix}\right)$$

is controllable.

Proof Assertions (i)-(iii) follow from Propositions 1.2, 3.18, and 3.5. To show (iv) note that after multiplying $P(z)C_\pi(z - A_\pi)^{-1}$ on the right by a constant invertible matrix, we may assume that A_π is in Jordan form and the columns of C_π contain coefficients of the Taylor expansions of right pole functions for W at the corresponding eigenvalues of A_π.

We may also assume without loss of generality that A_π consists of a single Jordan cell of size $k \times k$ with λ on the diagonal and $P(z) = \sum_{i=0}^{l}(z-\lambda)^i P_i$ is a polynomial row vector function. Let the columns of C_π be $c_0, c_1, ..., c_{k-1}$. Then there exist vectors $c_k, c_{k+1}, ...$ such that $\psi(z) = \sum_{j=0}^{\infty}(z-\lambda)^j c_j$ is a right pole function for W at λ. Since P annihilates ψ,

$$\sum_{\substack{i+j=\nu \\ 0\leq i\leq l \\ j\geq 0}} P_i c_j = 0 \qquad (\nu = 0, 1, ...).$$

Hence

$$P(z)C_\pi(z-A_\pi)^{-1} = \left(\sum_{i=1}^{l}(z-\lambda)^i P_i\right)\left[(z-\lambda)^{-1}c_0 \quad (z-\lambda)^{-2}c_0 + (z-\lambda)^{-1}c_1 \quad \dots \quad (z-\lambda)^{-k}c_0 + (z-\lambda)^{-k+1}c_1 + \dots + (z-\lambda)^{-1}c_{k-1}\right]$$

is analytic at λ and (iv) is established. Finally, assertion (v) follows from condition (iii) in the definition of a canonical set of left null functions for W at λ.

□

We can now state the solution of the local inverse spectral problem. The proof will be completed with the proof of Theorem 4.12 below.

Theorem 4.2 *Let $\sigma \subset \mathbf{C}$ and let (C_π, A_π), (A_ζ, B_ζ) and $(A_\kappa, B_\kappa, D_\kappa)$ satisfy conditions (i)-(v) in Proposition 4.1. Then there exists a rational matrix function W with the right pole and left null pairs over σ (C_π, A_π) and (A_ζ, B_ζ) and with the left kernel triple $(A_\kappa, B_\kappa, D_\kappa)$ if and only if the matrix equation in Γ*

$$\Gamma A_\pi - A_\zeta \Gamma = B_\zeta C_\pi \tag{4.1}$$

has a solution.

Proof If (C_π, A_π) is a right pole pair for a rational matrix function W over σ and (A_ζ, B_ζ) is a left null pair for W over σ, then, by Theorem 3.17, there exists a regular rational matrix function H having (C_π, A_π) and (A_ζ, B_ζ) as right pole and left null pairs over $\sigma(A_\pi) \cup \sigma(A_\zeta)$. After multiplying H on the right by a regular rational matrix function Q without poles or zeros in $\sigma(A_\pi) \cup \sigma(A_\zeta)$ we may assume $H(\infty) = \mathrm{I}$. Then it follows by Theorem 2.3 in [GK2] that equation (4.1) has a solution.

Conversely, suppose equation (4.1) has a solution. Then, since (C_π, A_π) and (A_ζ, B_ζ) satisfy conditions (i) and (ii) in Proposition 4.1, by Theorem 2.3 in [GK2], there exists a regular rational matrix function H with the right pole and left null pairs over σ equal to (C_π, A_π) and (A_ζ, B_ζ). Since (C_π, A_π), (A_ζ, B_ζ) and $(A_\kappa, B_\kappa, D_\kappa)$ satisfy conditions (iii)-(v) in Proposition 4.1, by Theorem 4.12 below, there exists a rational matrix function W with a left kernel triple equal to $(A_\kappa, B_\kappa, D_\kappa)$ and with right pole and left null pairs over σ equal to (C_π, A_π) and (A_ζ, B_ζ).

□

We shall need below the following properties of orthogonal projections in $\mathcal{R}^n$.

Proposition 4.3 *Let σ be a subset of $\mathbf{C}_\infty$ and let Λ, Ω be orthogonal complements in $(\mathcal{R}^n, \sigma)$. If $h \in \mathcal{R}^n$ is analytic on σ then the projection of h onto Λ along Ω is analytic on σ.*

Proof Let $h_\Lambda \in \Lambda$ and $h_\Omega \in \Omega$ be such that $h = h_\Lambda + h_\Omega$ and suppose that h_Λ has a pole at $\lambda \in \sigma$. Then, by Proposition 2.3, $[h_\Lambda]_{z=\lambda}$ and $[h_\Omega]_{z=\lambda}$ are linearly independent. Hence h has a pole at λ, a contradiction.

□

Proposition 4.4 *Let $\lambda \in \mathbf{C}_\infty$ and let Λ, Ω be orthogonal complements in $(\mathcal{R}^n, \{\lambda\})$. Let $h \in \mathcal{R}^n$ have the Laurent expansion at λ*

$$\begin{cases} \sum_{j=k}^{\infty}(z-\lambda)^j h_j, & \text{if } \lambda \in \mathbf{C} \\ \sum_{j=-\infty}^{k} z^j h_j, & \text{if } \lambda = \infty \end{cases}$$

with $h_k \neq 0$. If h_Λ is the projection of h onto Λ along Ω, then

$$h_\Lambda(z) = \begin{cases} \sum_{j=k}^{\infty}(z-\lambda)^j h_{\Lambda j}, & \text{if } \lambda \in \mathbf{C} \\ \sum_{j=-\infty}^{k} z^j h_{\Lambda j}, & \text{if } \lambda = \infty \end{cases} \tag{4.2}$$

where the coefficient $h_{\Lambda k}$ is the projection of the leading coefficient h_k of h along $\Omega(\lambda)$ onto $\Lambda(\lambda)$.

Proof In view of Proposition 4.3, h_Λ can be represented as in (4.2). Similarly, the projection h_Ω of h along Λ onto Ω has the Laurent expansion at λ

$$h_\Omega(z) = \begin{cases} \sum_{j=k}^{\infty}(z-\lambda)^j h_{\Omega j}, & \text{if } \lambda \in \mathbf{C}\ , \\ \sum_{j=-\infty}^{k} z^j h_{\Omega j}, & \text{if } \lambda = \infty\ . \end{cases}$$

Since $h_\Lambda \in \Lambda$ and $h_\Omega \in \Omega$, $h_{\Lambda k} \in \Lambda(\lambda)$ and $h_{\Omega k} \in \Omega(\lambda)$. (We note that $h_{\Lambda k}$ or $h_{\Omega k}$ may be equal to 0.) Since $h_k = h_{\Lambda k} + h_{\Omega k}$, the assertion follows.

□

Corollary 4.5 *Let $\lambda \in \mathbf{C}_\infty$ and let Λ, Ω be orthogonal complements in $(\mathcal{R}^n, \{\lambda\})$. Let $h \in \mathcal{R}^n$ and let h_Λ be the projection of h along Ω onto Λ. If $[h]_{z=\lambda} \notin \Omega(\lambda)$, then*

$$\|h\|_{z=\lambda} = \|h_\Lambda\|_{z=\lambda}.$$

Let $\lambda \in \mathbf{C}_\infty$, let Λ and Ω be orthogonal complements in $(\mathcal{R}^{1\times m}, \{\lambda\})$, and let $h \in \mathcal{R}^{m\times 1}$. In view of Theorem 2.8, if h has a pole of multiplicity 1 at λ and $[h]_\lambda \in \Lambda^\circ(\lambda)$, then Proposition 4.4 implies that the projection of h along Ω° onto Λ° does not affect the singular part at λ of h. This observation can be generalized as follows.

Proposition 4.6 *Let $\lambda \in \mathbf{C}_\infty$ and suppose that the rows of a rational matrix function P form an orthonormal basis for a subspace Λ of $(\mathcal{R}^{1\times m}, \{\lambda\})$. Let $h \in \mathcal{R}^{m\times 1}$ and let h_{Λ° be a projection of h onto Λ° along some orthogonal complement Ω° of Λ° in $(\mathcal{R}^{m\times 1}, \{\lambda\})$. Then $h - h_{\Lambda^\circ}$ is analytic at λ whenever Ph is analytic at λ.*

Proof We have $h = h_{\Lambda^\circ} + h_{\Omega^\circ}$ with $h_{\Omega^\circ} \in \Omega^\circ$. Suppose h_{Ω° has a pole at λ. Then $[h_{\Omega^\circ}]_\lambda \in \Omega^\circ(\lambda)$, so $P(\lambda)[h_{\Omega^\circ}]_\lambda \neq 0$. Hence

$$P(h_{\Lambda^\circ} + h_{\Omega^\circ}) = Ph_{\Omega^\circ}$$

has a pole at λ. Since Ph is analytic at λ, this is a contradiction.

□

We show now how to construct from a regular rational matrix function H a left invertible rational matrix function W so that W inherits from H left null and right pole pairs over $\sigma \subset \mathbf{C}$ and has a given left kernel triple $(A_\kappa, B_\kappa, D_\kappa)$. The construction consists of several steps, which are of independent interest. We analyze these steps in separate lemmas.

In the first step we construct from a given $m \times m$ regular rational matrix function H, with (C_π, A_π) as a right pole pair and (A_ζ, B_ζ) as a left null pair over a subset σ of C, a "tall" $m \times \eta$ function W with (C_π, A_π) and (A_ζ, B_ζ) as right pole and left null pairs over σ. Here the integer η is as small as possible; it is equal to the largest sum of the geometric multiplicity of a pole and the geometric multiplicity of a zero of H at any single point of σ.

Lemma 4.7 *Let H be a regular $m \times m$ rational matrix function with (C_π, A_π) as a right pole pair and (A_ζ, B_ζ) as a left null pair over a subset σ of $\mathbf{C}$, and let η be the largest sum of the geometric multiplicity of a pole and the geometric multiplicity of a zero of H at any single point of σ. Then we can construct an $m \times \eta$ rational matrix function W with (C_π, A_π) as a right pole pair and (A_ζ, B_ζ) as a left null pair over σ.*

Proof Find a Smith-McMillan factorization EDF of H and let $W_1 = ED$. Let d_i denote the ith diagonal entry of D $(i = 1, 2, ..., m)$ and let ν and μ be the largest geometric multiplicity of a pole and of a zero of W_1 in σ, respectively. For $i = \eta - \mu + 1, \eta - \mu + 2, ..., \nu$, let p_i be the minimal degree monic polynomial such that if $d_{m-\eta+i}$ has a zero at a point $\lambda \in \sigma$ of order k then $p_i d_i$ has a zero at λ of order k. Define an $m \times \eta$ matrix polynomial $Q = [q_{ij}]$ by

$$q_{ij} = \begin{cases} 1, & \text{if } i = j \leq \eta - \mu \\ p_i, & \text{if } \eta - \mu < i = j \leq \nu \\ 1, & \text{if } i = j - \eta + m > m - \mu \\ 0, & \text{otherwise,} \end{cases}$$

and put

$$W = W_1 Q.$$

Since W_1 and H are right equivalent on $\mathbf{C}$, W_1 has (C_π, A_π) as a right pole pair and (A_ζ, B_ζ) as a left null pair over σ. After applying appropriate similarity transformations to the pairs (C_π, A_π) and (A_ζ, B_ζ), we assume that

- A_π and A_ζ are in Jordan forms,
- (C_π, A_π) has been read off from the first ν columns of E, namely $E_1, E_2, ..., E_\nu$,
- (A_ζ, B_ζ) has been read off from the last μ rows of E^{-1}, namely $\Xi_{m-\mu+1}, \Xi_{m-\mu+2}, ...,$ Ξ_m.

We claim that W has the following properties:

(i) W has no zero columns;

(ii) the columns of W are orthogonal on σ;

(iii) the ith column of W has the same poles in σ with the same multiplicities as the ith column of W_1 $(1 \leq i \leq \nu)$ and the $(\eta - j)^{th}$ column of W has the same zeros in σ with the same multiplicities as the $(m-j)^{th}$ column of W_1 $(1 \leq j \leq \mu)$.

Property (i) follows from the choice of η. We show that (iii) holds. Define two $m \times \eta$ rational matrix functions V_L and V_R by

$$V_L = [\, d_1 E_1 \quad d_2 E_2 \quad \ldots \quad d_\nu E_\nu \quad 0 \quad \ldots \quad 0 \,],$$

$$V_R = \begin{bmatrix} 0 & \ldots & 0 & d_{m-\mu+1} E_{m-\mu+1} & d_{m-\mu+2} E_{m-\mu+2} & \ldots & d_m E_m \end{bmatrix}.$$

Then

$$W = V_L \begin{bmatrix} 1 & & & & & & & & \\ & \ddots & & & & & & & \\ & & p_{\eta-\mu+1} & & & & & & \\ & & & p_{\eta-\mu+2} & & & & & \\ & & & & \ddots & & & & \\ & & & & & p_\nu & & & \\ & & & & & & 1 & & \\ & & & & & & & \ddots & \\ & & & & & & & & 1 \end{bmatrix} + V_R. \tag{4.3}$$

Suppose that the first l columns of V_L have a pole at $\lambda \in \sigma$. Then, by the choice of η, at most $\eta - l$ columns of V_R have a zero at λ and, by construction, these columns are the last columns of V_R. Thus, it cannot happen that the ith column of V_L has a pole at λ while the ith column of V_R has a zero at λ. Consequently, the ith column of W has the same poles in σ with the same multiplicities as the ith column of V_L and hence as the ith column of W_1. By construction, the $(\eta - j)^{th}$ column of W has the same zeros in σ with the same multiplicities as the $(\eta - j)^{th}$ column of V_R and hence as the $(m-j)^{th}$ column of W_1. This establishes (iii).

We show now (ii). For each column w_i of W we can find a scalar rational function $\tilde{d}_i$ such that $w_i = \tilde{d}_i \tilde{w}_i$ and $\tilde{w}_i$ has no poles nor zeros in σ. Let $\tilde{W} = [\,\tilde{w}_1 \quad \tilde{w}_2 \quad \dots \quad \tilde{w}_1\,]$ so that

$$W = \tilde{W} \begin{bmatrix} \tilde{d}_1 & & & \\ & \tilde{d}_2 & & \\ & & \ddots & \\ & & & \tilde{d}_\eta \end{bmatrix}.$$

Since for each $\lambda \in \sigma$ the set $\{E_1(\lambda), E_2(\lambda), ..., E_m(\lambda)\}$ is linearly independent, the set

$$\{E_1(\lambda), E_2(\lambda), ..., E_{\eta-\mu}(\lambda), E_{\eta-\mu+1}(\lambda) + \alpha_1 E_{m-\mu+1}(\lambda), \dots, \\ E_\nu(\lambda) + \alpha_{\mu+\nu-\eta} E_{m-\nu+\eta}(\lambda), E_{m-\eta+\nu+1}(\lambda), \dots, E_m(\alpha)\}$$

is linearly independent for all constants $\alpha_1, \alpha_2, ..., \alpha_{\mu+\nu-\eta}$. Therefore it follows from (4.3) that $\tilde{W}(\lambda)$ has full column rank for all $\lambda \in \sigma$. By Corollary 2.4, the columns of W are orthogonal on σ as asserted.

In view of (iii), in order to prove that (C_π, A_π) is a right pole pair for W over σ it suffices to show that if $\tilde{d}_i$ has a pole at $\lambda \in \sigma$ of order l then we can find a right pole function ψ_i for W at λ of order l such that the first l coefficients in the Taylor expansions at λ of ψ_i and E_i coincide. The latter assertion is obvious when the ith column of V_R is 0 or the ith column of V_R is analytic at λ. Suppose $d_j E_j$, the ith column of V_R, has a pole at λ. Then the ith column of W is $p_i d_i E_i + d_j E_j$ and the jth column of W is $p_j d_j E_j + d_\kappa E_\kappa$ for some κ and scalar polynomials p_i, p_j with $p_i(\lambda) p_j(\lambda) \neq 0$. So

$$p_j(p_i d_i E_i + d_j E_j) - (p_j d_j E_j + d_\kappa E_\kappa) = p_i p_j d_i E_i - d_\kappa E_\kappa.$$

Inducting, if necessary, on κ we can find an $m \times 1$ rational vector function $\tilde{\psi} = p d_i E_i - d_\kappa E_\kappa$ with $d_\kappa(\lambda) \neq 0$ such that $\tilde{\psi} = W\phi$ with ϕ a vector polynomial which does not vanish at λ. So $\psi = (pd_i)^{-1}\tilde{\psi}$ is a right pole function for W at λ of order l and the first l coefficients in the Taylor expansions at λ of ψ and E_i coincide. Thus, (C_π, A_π) is a right pole pair for W over σ as asserted.

In the remaining part of the proof we show that (A_ζ, B_ζ) is a left null pair for W over σ. Suppose that $\{\Xi_{m-\tilde{l}+1}, \Xi_{m-\tilde{l}+2}, ..., \Xi_m\}$ is a canonical set of left null functions for W_1 at $\lambda \in \sigma$ of orders $l_1, l_2, ..., l_{\tilde{l}}$. Then the last $\tilde{l}$ diagonal entries of D vanish at λ to the orders $l_1, l_2, ..., l_{\tilde{l}}$. By (iii), the last $\tilde{l}$ diagonal entries of diag $(\tilde{d}_1, \tilde{d}_2, ..., \tilde{d}_\eta)$ vanish at λ to the orders $l_1, l_2, ..., l_{\tilde{l}}$. By (ii) and Result 2 in [VK], the multiplicity of the zero at λ of W is equal to $\sum_{i=1}^{\tilde{l}} l_i$. Now if $\Xi_i W_1$ vanishes at λ to order l_j, then it follows from the construction that $\Xi_i W$ vanishes at λ to order l_j. Since obviously

$$dim \left(\begin{bmatrix} \Xi_{m-\tilde{l}+1} \\ \Xi_{m-\tilde{l}+2} \\ \vdots \\ \Xi_m \end{bmatrix} \tilde{W}(\lambda) \right) = \tilde{l},$$

it follows from Proposition 3.11 that $\{\Xi_{m-\tilde{l}+1}, \Xi_{m-\tilde{l}+2}, ..., \Xi_m\}$ is a canonical set of left null functions for W at λ of orders $l_1, l_2, ..., l_{\tilde{l}}$. Hence (A_ζ, B_ζ) is a left null pair for W over σ.

□

The following example illustrates the construction in Lemma 4.7.

Example 4.8 Let H be a rational matrix function with a Smith-McMillan factorization

$$E(z)D(z)F(z) = \begin{bmatrix} z^2-1 & z+2 & z^3 \\ z-2 & 1 & z^2 \\ 0 & 0 & 1 \end{bmatrix} \cdot$$

$$\cdot \begin{bmatrix} 1/\left(z(z-1)^2(z-2)\right) & & \\ & (z-3)/(z-2) & \\ & & z(z-3)^2(z-4)^2 \end{bmatrix} \begin{bmatrix} z^2 & z & 1 \\ z & 1 & 0 \\ 1 & 0 & 0 \end{bmatrix}, \quad (4.4)$$

and let $\sigma = \{z : \ |z| \leq 3\}$. We can read from the first two factors in the factorization (4.4) a right pole pair and a left null pair for H over σ:

$$(C_\pi, A_\pi) = \left(\begin{bmatrix} 0 & 2 & 3 & 4 & -1 \\ -1 & 1 & 0 & 1 & -2 \\ 0 & 0 & 0 & 0 & 0 \end{bmatrix}, \begin{bmatrix} 1 & 1 & & & \\ & 1 & & & \\ & & 2 & & \\ & & & 2 & \\ & & & & 0 \end{bmatrix} \right)$$

and

$$(A_\zeta, B_\zeta) = \left(\begin{bmatrix} 3 & 1 & & \\ & 3 & & \\ & & 3 & \\ & & & 0 \end{bmatrix}, \begin{bmatrix} 0 & 0 & 0 \\ 0 & 0 & 1 \\ 1 & 1 & 9 \\ 0 & 0 & 1 \end{bmatrix} \right).$$

We have $\eta = \nu = \mu = 2$. Also, $p_1(z) = z-3$ and $p_2(z) = z(z-3)$. Hence

$$Q(z) = \begin{bmatrix} z-3 & 0 \\ 1 & z(z-3) \\ 0 & 1 \end{bmatrix},$$

and

$$W(z) = E(z)D(z)Q(z)$$
$$= \begin{bmatrix} z^4 - 3z^2 + 2z & z^6 - 10z^5 + 32z^4 - 32z^3 \\ z^3 - 2z^2 + 2z - 2 & z^5 - 10z^4 + 32z^3 - 32z^2 \\ 0 & z^3 - 10z^2 + 32z - 32 \end{bmatrix} \cdot$$
$$\cdot \begin{bmatrix} (z-3)/\Big(z(z-1)^2(z-2)\Big) & \\ & z(z-3)^2/(z-2) \end{bmatrix}.$$

It can be verified that (C_π, A_π) and (A_ζ, B_ζ) are right pole and left null pairs for W over σ.

□

Let H be a "tall" rational matrix function with (C_π, A_π) as a right pole pair and (A_ζ, B_ζ) as a left null pair over σ, constructed as in Lemma 4.7, and let P be a matrix polynomial whose rows form a minimal polynomial basis for their span. The next lemma shows that if P satisfies the "consistency" conditions (iv) and (v) in Proposition 4.1, then we can construct a matrix function W such that (C_π, A_π) is a right pole pair and (A_ζ, B_ζ) is a left null pair for W over $\sigma(A_\pi) \cup \sigma(A_\zeta)$ and $PW = 0$.

Lemma 4.9 *Let $\sigma \subset \mathbf{C}$, and let $H \in \mathcal{R}^{m \times \eta}$ be a function with (C_π, A_π) as a right pole pair and (A_ζ, B_ζ) as a left null pair over σ. Suppose there is a point $\lambda \in \sigma$ such that the sum of the geometric multiplicity of a zero and the geometric multiplicity of a pole of H at λ equals η, and let P be a matrix polynomial such that the rows of P form a minimal polynomial basis for their span and P satisfies conditions (iv) and (v) in Proposition 4.1. Then we can construct a function $W \in \mathcal{R}^{m \times \eta}$ such that (C_π, A_π) is a right pole pair and (A_ζ, B_ζ) is a left null pair for W over $\sigma(A_\pi) \cup \sigma(A_\zeta)$, and $PW = 0$.*

Proof Find a Smith-McMillan factorization EDF of H and let $W_1 = ED$. Since the rank of the function D as a matrix over $\mathcal{R}$ is η, W_1 has full column rank. Call the columns of W_1 $w_1, w_2, ..., w_\eta$.

Let $\lambda_1, \lambda_2, ..., \lambda_r$ be the points of $\sigma(A_\pi) \cup \sigma(A_\zeta)$. Let κ be the largest geometric multiplicity of a zero of W_1 in σ and find functions $\phi_1, \phi_2, ..., \phi_\kappa$ as in Lemma 3.14. For each $i \in \{1, 2, ..., r\}$ find a subspace Λ_i of $\mathbf{C}^{1 \times m}$ as follows. Let the geometric multiplicities of the pole and zero of W_1 at λ_i be l and $\tilde{l}$, let

$$\Omega_1 = \text{span } \{[w_1]_{z=\lambda_i}, [w_2]_{z=\lambda_i}, ..., [w_l]_{z=\lambda_i}\}$$
$$\Omega_2 = \text{span } \{[w_{l+1}]_{z=\lambda_i}, [w_{l+2}]_{z=\lambda_i}, ..., [w_{\eta-\tilde{l}}]_{z=\lambda_i}\}$$

and let Ω be the subspace of $\mathbf{C}^{m \times 1}$ annihilated by $\phi_1(\lambda_i), \phi_2(\lambda_i), ..., \phi_{\tilde{l}}(\lambda_i)$. Let W^{ol} be the row space of the matrix polynomial P, and let $ker P(\lambda_i)$ be the subspace of $\mathbf{C}^{m \times 1}$ annihilated by $P(\lambda_i)$. Find a complement Ω_{pr} of $ker P(\lambda_i) \cap \Omega$ in Ω such that the projection of Ω_2 along

Ω_{pr} onto $ker P(\lambda_i)$ has dimension $\eta - \tilde{l} - l$ and intersects trivially with Ω_1. Let Λ_i be the subspace of $\mathbf{C}^{1\times m}$ which annihilates Ω_{pr}.

Using Corollary 3.15, find a subspace Ξ of $\mathcal{R}^{1\times m}$ such that $\Xi(\lambda_i) = \Lambda_i$ ($i = 1,2,...,r$) and the pair (A_ζ, B_ζ) is left-similar to the pair constructed from functions $\psi_1, \psi_2, ..., \psi_\kappa$ which are contained in Ξ. Construct W by projecting each column of W_1 along Ξ^{or} onto ker P. (Here ker P denotes the subspace of $\mathcal{R}^{m\times 1}$ annihilated by the polynomial P.)

We need to show that for any eigenvalue λ_i of A_ζ we can find a subspace Ω_{pr} in the way described above and that the corresponding subspace Λ_i satisfies the hypotheses of Corollary 3.15. Choose $\lambda_i \in \sigma(A_\pi) \cup \sigma(A_\zeta)$ and let $l, \tilde{l}, \Omega_1, \Omega_2$ and Ω be as above. Since the columns of W_1 are orthogonal at λ_i, $\Omega_1 \cap \Omega_2 = (0)$. Since $\phi_j W_1$ ($j = 1,2,...,\tilde{l}$) vanish at λ_i, $\Omega_1 + \Omega_2 \subset \Omega$. Since P satisfies condition (iv) in Proposition 4.1, $\Omega_1 \subset$ ker $P(\lambda_i)$. Plainly, there exists a complement Ω_{pr} of ker $P(\lambda_i) \cap \Omega$ in Ω such that the projection of Ω_2 onto ker $P(\lambda_i)$ along Ω_{pr} intersects trivially with Ω_1 and has the same dimension as Ω_2 provided that dim $(\Omega_1 + \Omega_2) \leq$ dim (ker $P(\lambda_i) \cap \Omega$). We show that the last inequality is valid. Let k be the number of rows of the matrix polynomial P. Then, since P satisfies condition (v) in Proposition 4.1, dim (ker $P(\lambda_i) \cap \Omega) = m - k - \tilde{l}$. Since dim $(\Omega_1 + \Omega_2) = \eta - \tilde{l}$, it suffices to show $\eta \leq m - k$. We show first that $\eta \leq n$. Let $\lambda \in \sigma$ be such that η diagonal entries of D have either a pole or a zero at λ. The i columns of $E(\lambda)$ which correspond to the diagonal entries of D with a pole at λ are annihilated by $P(\lambda)$ (see condition (iv) in Proposition 4.1) and by $\phi_1(\lambda), \phi_2(\lambda), ..., \phi_j(\lambda)$. Since, by condition (v) in Proposition 4.1, span $\{\phi_1(\lambda), \phi_2(\lambda), ..., \phi_j(\lambda)\}$ intersects trivially with the row space of $P(\lambda)$, $i + j + k \leq m$. So $\eta = i + j \leq m - k = n$ as asserted.

In order to show that the construction described above can be actually carried out, we need to verify that Ξ satisfies hypotheses of Corrolary 3.15 and that Ξ^{or} is an algebraic complement of ker P. Now, using the notation introduced above, Ξ satisfies hypotheses of Corollary 3.15 if, for every $\lambda_i \in \sigma(A_\zeta)$, the vectors $\phi_1(\lambda_i), \phi_2(\lambda_i), ..., \phi_{\tilde{l}}(\lambda_i) \in \Lambda_i$. But Ω is the subspace of $\mathbf{C}^{m\times 1}$ which is annihilated by $\phi_1(\lambda_i), \phi_2(\lambda_i), ..., \phi_{\tilde{l}}(\lambda_i)$, and $\Omega_{pr} \subset \Omega$. So $\phi_1(\lambda_i), \phi_2(\lambda_i), ..., \phi_{\tilde{l}}(\lambda_i)$ are contained in the annihilator of Ω, Λ_i. We verify now that Ξ^{or} and ker P are algebraic complements in $\mathcal{R}^{m\times 1}$. We also show that they are orthogonal on $\sigma(A_\pi) \cup \sigma(A_\zeta)$. If $\sigma(A_\pi) \cup \sigma(A_\pi)$ properly includes $\sigma(A_\zeta)$, let $\lambda_i \in \sigma(A_\pi) \backslash \sigma(A_\zeta)$. Then the corresponding subspace Ω is equal to $\mathbf{C}^{m\times 1}$, so $\mathbf{C}^{m\times 1} =$ ker $P(\lambda_i) \oplus \Omega_{pr}$. Hence, clearly, Ξ^{or} and ker P are algebraic complements. Suppose now $\sigma(A_\zeta) \neq 0$. In view of Theorem 2.8, it suffices to show that, for every $\lambda_i \in \sigma(A_\zeta)$,

$$\mathbf{C}^{1\times m} = \Lambda_i \oplus \textit{row space of } P(\lambda_i). \tag{4.5}$$

Equality (4.5) is equivalent to the direct sum decomposition $\mathbf{C}^{m\times 1} = \Omega_{pr} \oplus$ ker $P(\lambda_i)$. To see that the latter decomposition holds, choose vectors $c_{k+1}, c_{k+2}, ..., c_{m-\tilde{l}}$ such that the matrix

$$M = \begin{bmatrix} P(\lambda_i) \\ c_{k+1} \\ c_{k+2} \\ \vdots \\ c_{m-\tilde{l}} \\ \phi_1(\lambda_i) \\ \phi_2(\lambda_i) \\ \vdots \\ \phi_{\tilde{l}}(\lambda_i) \end{bmatrix}$$

is invertible. Then the first $m - \tilde{l}$ columns of M^{-1} form a basis for Ω and the last $\tilde{l}$ columns of M^{-1} form a basis for a subspace $\tilde{\Omega} \subset \ker P(\lambda_i)$. So

$$\begin{aligned} \mathbf{C}^{m\times 1} &= \Omega \oplus \tilde{\Omega} \\ &= \Omega_{pr} \oplus (\ker P(\lambda_i) \cap \Omega) \oplus \tilde{\Omega} \\ &= \Omega_{pr} \oplus \ker P(\lambda_i) \end{aligned}$$

as asserted. It follows that ker P and Ξ^{or} are orthogonal complements in $(\mathcal{R}^{m\times 1}, \sigma(A_\pi) \cup \sigma(A_\zeta))$. In particular, $\mathcal{R}^{m\times 1} = \Xi^{or} \oplus \ker P$. Thus, the construction of W described above is possible.

Clearly, the function W satisfies equality $PW = 0$. It remains to show that $(\mathbf{C}_\pi, A_\pi)$ is a right pole pair and (A_ζ, B_ζ) is a left null pair for W over $\sigma(A_\pi) \cup \sigma(A_\zeta)$. Since P satisfies condition (iv) in Proposition 4.1, in view of Proposition 4.6 the singular parts of H and W at each point λ of $\sigma(A_\pi) \cup \sigma(A_\zeta)$ coincide. Hence (C_π, A_π) is a right ole pair for W over $\sigma(A_\pi) \cup \sigma(A_\zeta)$.

Let $\lambda_i \in \sigma(A_\pi) \cup \sigma(A_\zeta)$. We show that the subspace V obtained by projecting

$$\text{span } \{[w_1]_{z=\lambda_i}, [w_2]_{z=\lambda_i}, ..., [w_\eta]_{z=\lambda_i}\}$$

onto ker $P(\lambda_i)$ along Ω_{pr} has dimension η. Then, by the Proposition 4.4, the columns of W are orthogonal on $\sigma(A_\pi) \cup \sigma(A_\zeta)$ and, by Corollary 4.5, the multiplicities of the zeros of W and W_1 at every point $\lambda \in \sigma$ are equal to the sum of multiplicities of a zero of each column of W_1 at λ. Now let V_1 and V_2 be the subspaces of V obtained by projecting span $\{[w_1]_{z=\lambda_i}, [w_2]_{z=\lambda_i}, ..., [w_{\eta-\tilde{l}}]_{z=\lambda_i}\}$ and span $\{[w_{\eta-\tilde{l}+1}]_{z=\lambda_i}, [w_{\eta-\tilde{l}+2}]_{z=\lambda_i}, ..., [w_\eta]_{z=\lambda_i}\}$, respectively, onto ker $P(\lambda_i)$ along Ω_{pr}. By the choice of Ω_{pr}, the subspace V_1 has dimension $\eta - \tilde{l}$. By Proposition 3.11, the matrix

$$\begin{bmatrix} \phi_1(\lambda_i) \\ \phi_2(\lambda_i) \\ \vdots \\ \phi_{\tilde{l}}(\lambda_i) \end{bmatrix} \left[[w_{\eta-\tilde{l}+1}]_{z=\lambda_i} \quad [w_{\eta-\tilde{l}+2}]_{z=\lambda_i} \quad \cdots \quad [w_\eta]_{z=\lambda_i} \right]$$

has rank $\tilde{l}$. Since $\phi_1(\lambda_i), \phi_2(\lambda_i), ..., \phi_{\tilde{l}}(\lambda_i)$ annihilate Ω_{pr}, the matrix

$$\begin{bmatrix} \phi_1(\lambda_i) \\ \phi_2(\lambda_i) \\ \vdots \\ \phi_{\tilde{l}}(\lambda_i) \end{bmatrix} \begin{bmatrix} [\tilde{w}_{\eta-\tilde{l}+1}]_{z=\lambda_i} & [\tilde{w}_{\eta-\tilde{l}+2}]_{z=\lambda_i} & \cdots & [\tilde{w}_{\eta}]_{z=\lambda_i} \end{bmatrix}$$

has rank $\tilde{l}$, where $\tilde{w}_i$ is the ith column of W. Hence dim $V_2 = \tilde{l}$. Since $\phi_k(\lambda_i)[w_j]_{z=\lambda_i}$ for $k = 1, 2, ..., \tilde{l}$ and $j = 1, 2, ..., \eta - \tilde{l}$,

$$V_1 \subset \ker \begin{bmatrix} \phi_1(\lambda_i) \\ \phi_2(\lambda_i) \\ \vdots \\ \phi_{\tilde{l}}(\lambda_i) \end{bmatrix}.$$

It follows that $V_1 \cap V_2 = (0)$ and dim V = dim V_1 + dim $V_2 = \eta$.

One can now easily see that if $\lambda_i \in \sigma(A_\pi) \cup \sigma(A_\zeta)$ is a zero of W_1 of order $\tilde{l}$, then $\{\phi_1, \phi_2, ..., \phi_{\tilde{l}}\}$ is a canonical set of left null functions for W at λ_i. Indeed, by Proposition 3.11 the sum of multiplicities of the zeros at λ of the functions $\phi_j W_1$ $(j = 1, 2, ..., \tilde{k})$ is equal to the multiplicity of the zero of W_1 at λ_i. Since $\phi_j W_1 = \phi_j W$ $(j = 1, 2, ..., k)$ and the multiplicities of the zeros of W_1 and W at λ_i coincide, functions $\phi_1, \phi_2, ..., \phi_{\tilde{l}}$ satisfy condition (i) in Proposition 3.11. Since the matrix (4.6) has full rank, $\phi_1, \phi_2, ..., \phi_{\tilde{l}}$ satisfy condition (ii) in Proposition 3.11 as well. Thus, $\{\phi_1, \phi_2, ..., \phi_{\tilde{l}}\}$ is a canonical set of left null functions for W at λ_i and it follows that (A_ζ, B_ζ) is a left null pair for W over $\sigma(A_\pi) \cup \sigma(A_\zeta)$.

□

We illustrate the construction in Lemma 4.9 with the following example.

Example 4.10 Let

$$H(z) = \begin{bmatrix} (z-1)^2/z & (z-1)^2(z^2-2) \\ (z-1)/z & (z-1)(z^2-2z+2) \\ (z-1)/z & z(z-1) \end{bmatrix}.$$

Then H has a Smith-McMillan factorization

$$E(z)D(z)F(z) = \begin{bmatrix} z-1 & z+1 & z^2+z \\ 1 & 1 & z \\ 1 & 0 & 1 \end{bmatrix} \begin{bmatrix} (z-1)/z & 0 \\ 0 & (z-1)^2(z-2) \\ 0 & 0 \end{bmatrix} \begin{bmatrix} 1 & z^2 \\ 0 & 1 \end{bmatrix},$$

and it can be easily verified that

$$(C_\pi, A_\pi) = \left(\begin{bmatrix} -1 \\ 1 \\ 1 \end{bmatrix}, \; [0] \right) \tag{4.7}$$

is a right pole pair for H over $\mathbf{C}$ and

$$(A_\zeta, B_\zeta) = \left(\begin{bmatrix} 1 & 1 & & \\ & 1 & & \\ & & 1 & \\ & & & 0 \end{bmatrix}, \begin{bmatrix} 0 & 0 & 0 \\ 0 & 0 & -1 \\ 0 & 1 & 0 \\ 0 & 1 & -1 \end{bmatrix} \right) \tag{4.8}$$

is a left null pair for H over $\mathbf{C}$. Let $P(z) = [z \quad z-1 \quad 1-z]$. Then (C_π, A_π), (A_ζ, B_ζ), and P satisfy conditions (iv) and (v) in Proposition 4.1, and we can use the construction in Lemma 4.9 to find a function $W \in \mathcal{R}^{3\times 2}$ with (C_π, A_π) as a right pole pair and (A_ζ, B_ζ) as a left null pair over the set $\{0, 1, 2\}$.

We put

$$\begin{aligned} W_1(z) &= E(z)D(z) \\ &= \begin{bmatrix} (z-1)^2/z & (z-1)^2(z+1)(z-2) \\ (z-1)/z & (z-1)^2(z-2) \\ (z-1)/z & 0 \end{bmatrix}. \end{aligned}$$

We may take $\{\phi_1, \phi_2\}$ with $\phi_1(z) = [0 \quad 1 \quad -1]$ and $\phi_2(z) = [0 \quad 1 \quad 0]$ as the "universal" canonical set of left null functions for H over $\mathbf{C}$ (cf. Lemma 3.14). For each $\lambda_i \in \sigma(A_\pi) \cup \sigma(A_\zeta)$, we find now the corresponding subspace Λ_i of $\mathbf{C}^{3\times 1}$.

$\underline{\lambda_1 = 0}$: The geometric multiplicity l of the pole of W at 0 equals 1 and the geometric multiplicity of the zero of W at 0 equals 0. Therefore

$$\begin{aligned} \Omega_1 &= \operatorname{span}\{[w_1]_{z=0}\} \\ &= \operatorname{span}\left\{ \begin{bmatrix} 1 \\ -1 \\ -1 \end{bmatrix} \right\}, \end{aligned}$$

and

$$\begin{aligned} \Omega_2 &= \operatorname{span}\{[w_2]_{z=0}\} \\ &= \operatorname{span}\left\{ \begin{bmatrix} 1 \\ 1 \\ 0 \end{bmatrix} \right\}. \end{aligned}$$

Also, $P(0) = [0 \quad -1 \quad 1]$ and

$$\ker P(0) = \operatorname{span}\left\{ \begin{bmatrix} 1 \\ 0 \\ 0 \end{bmatrix}, \begin{bmatrix} 0 \\ 1 \\ 1 \end{bmatrix} \right\}.$$

Since $\tilde{l} = 0$, $\Omega = \mathbf{C}^{3\times 1}$ and we can choose Ω_{pr} to be any complement of $\ker P(0)$ in $\mathbf{C}^{3\times 1}$. Equivalently, we can choose Λ_1 to be any complement of span $[0 \quad -1 \quad 1]$ in $\mathbf{C}^{3\times 1}$. We put

$$\Lambda_1 = \text{span } \{[1 \quad 0 \quad 0],\ [0 \quad 1 \quad 0]\}.$$

$\underline{\lambda_2 = 1}$: The geometric multiplicity l of the pole of $W_1(z)$ at $z = 1$ equals 0 and the geometric multiplicity $\tilde{l}$ of the zero of $W_1(z)$ at $z = 1$ equals 2. Consequently, $\Omega_1 = \Omega_2 = \emptyset$. Therefore we may take Λ_2 to be any complement of the row space of $P(1)$ in $\mathbf{C}^{3\times 1}$ which contains $\phi_1(1)$ and $\phi_2(1)$. Since $\mathbf{C}^{3\times 1} = \text{span}\ \{P(1), \phi_1(1), \phi_2(1)\}$, we may take

$$\begin{aligned}\Lambda_2 &= \text{span}\ \{\phi_1(1), \phi_2(1)\} \\ &= \text{span}\ \{[0 \quad 1 \quad -1],\ [0 \quad 1 \quad 0]\}.\end{aligned}$$

$\underline{\lambda_3 = 2}$: The geometric multiplicity l of the pole of $W_1(z)$ at $z = 2$ equals 0 and the geometric multiplicity $\tilde{l}$ of the zero of $W_1(z)$ at $z = 2$ equals 1. Consequently, we have $\Omega_1 = \emptyset$,

$$\Omega_2 = \text{span}\left\{\begin{bmatrix}1\\1\\1\end{bmatrix}\right\},$$

and

$$\begin{aligned}\Omega &= \ker\,[0 \quad 1 \quad -1] \\ &= \text{span}\left\{\begin{bmatrix}1\\0\\0\end{bmatrix}, \begin{bmatrix}0\\1\\1\end{bmatrix}\right\}.\end{aligned}$$

Since $P(2) = [2 \quad 1 \quad -1]$,

$$\ker\, P(2) = \text{span}\left\{\begin{bmatrix}1\\-2\\0\end{bmatrix}, \begin{bmatrix}1\\0\\2\end{bmatrix}\right\}.$$

So

$$ker P(2) \cap \Omega = \text{span}\left\{\begin{bmatrix}0\\1\\1\end{bmatrix}\right\}.$$

The unique complement of $\ker\, P(2) \cap \Omega$ in Ω is

$$\Omega_{pr} = \text{span}\left\{\begin{bmatrix}1\\0\\0\end{bmatrix}\right\}.$$

Consequently,

$$\Lambda_3 = \text{span}\ \{[0 \quad 1 \quad 0],\ [0 \quad 0 \quad 1]\}.$$

One can verify that the projection of Ω_2 onto $\ker\, P(2)$ along Ω_{pr} has dimension one.

We construct now the subspace Ξ of $\mathcal{R}^{3\times 1}$ as in Lemma 3.14. Since $\phi_1(\lambda_1) \notin \Lambda_1$, $\phi_1(\lambda_2) \in \Lambda_2$, and $\phi_1(\lambda_3) \in \Lambda_3$, we need to change the value of $\phi(z)$ at $z = 0$. Let the new value of $\phi_1(z)$ at $z = 0$ be $c_1 = [1 \quad 0 \quad 0]$. Since ϕ_1 is a null function for W_1 at $z = 1$ of order 2 and at $z = 2$ of order 1, we put $r_1(z) = (z-1)^2(z-2)$, and

$$\begin{aligned}\psi_1(z) &= \phi_1(z) + \frac{r_1(z)}{r_1(0)}\Big(c_1 - \phi_1(0)\Big) \\ &= \Big[-\tfrac{1}{2}(z-1)^2(z-2) \quad \tfrac{1}{2}z(z^2-4z+5) \quad -\tfrac{1}{2}z(z^2-4z+5)\Big].\end{aligned}$$

Since $\phi_2(\lambda_1) \in \Lambda_1$, $\phi_2(\lambda_2) \in \Lambda_2$, $\phi_2(\lambda_3) \in \Lambda_3$, and since $\phi_1(\lambda_i)$ and $\psi_1(\lambda_i)$ are linearly independent $(i = 1, 2, 3)$, we put

$$\psi_2 = \phi_2.$$

We have thus found functions ψ_1, ψ_2 such that $\Lambda_i = \text{span}\,\{\psi_1(\lambda_i),\ \psi_2(\lambda_i)\}$ $(i = 1, 2, 3)$, $\{\psi_1,\ \psi_2\}$ and $\{\psi_1\}$ are canonical sets of left null functions for W_1 at the points $z = 1$ and $z = 2$, and the pair (A_ζ, B_ζ) can be read from the Taylor coefficients of $\psi_1(z)$ and $\psi_2(z)$ at the points $z = 1$ and $z = 2$. Since $\mathcal{R}^3 = \text{span}\,\{\psi_1, \psi_2, P\}$, we put $\Xi = \text{span}\,\{\psi_1, \psi_2\}$.

Let

$$R(z) = \begin{bmatrix} \psi_1(z) \\ \psi_2(z) \\ P(z) \end{bmatrix}.$$

Since

$$W_1 = R^{-1} \begin{bmatrix} 1 & & \\ & 1 & \\ & & 0 \end{bmatrix} RW_1 + R^{-1} \begin{bmatrix} 0 & & \\ & 0 & \\ & & 1 \end{bmatrix} RW_1,$$

we put

$$\begin{aligned} W &= R^{-1} \begin{bmatrix} 1 & & \\ & 1 & \\ & & 0 \end{bmatrix} RW_1 \\ &= \begin{bmatrix} \dfrac{(z-2)(z-1)^5}{z(2z^4 - 9z^3 + 14z^2 - 7z + 2)} & \dfrac{(z-2)(z-1)^3(z^2 - 2z - 1)(z^2 - 2z + 2)}{2z^4 - 9z^3 + 14z^2 - 7z + 2} \\ \dfrac{(z-1)}{z} & (z-2)(z-1)^2 \\ \dfrac{(z-1)(z^5 - 3z^4 + 7z^2 - 5z + 2)}{z(2z^4 - 9z^3 + 14z^2 - 7z + 2)} & \dfrac{(z-2)^2(z-1)^4(z^2 + 2z - 1)}{2z^4 - 9z^3 + 14z^2 - 7z + 2} \end{bmatrix}. \end{aligned}$$

One can verify that the function W has the required properties. Indeed, obviously $PW = 0$. Also, from the Laurent expansions of W at $z = 1$:

$$W(z) = (z-1) \begin{bmatrix} 0 & 0 \\ 1 & 0 \\ 1 & 0 \end{bmatrix} + (z-1)^2 \begin{bmatrix} 0 & 0 \\ -1 & -1 \\ -1 & 0 \end{bmatrix} + (z-1)^3 \begin{bmatrix} 0 & 1 \\ 1 & 1 \\ 1 & 0 \end{bmatrix} + \dots,$$

at $z = 2$:

$$W(z) = \begin{bmatrix} 0 & 0 \\ \frac{1}{2} & 0 \\ \frac{1}{2} & 0 \end{bmatrix} + (z-2) \begin{bmatrix} \frac{1}{8} & -\frac{1}{2} \\ \frac{1}{4} & 1 \\ \frac{1}{2} & 0 \end{bmatrix} + (z-2)^2 \begin{bmatrix} \frac{13}{32} & -\frac{3}{8} \\ -\frac{1}{8} & 2 \\ \frac{9}{16} & \frac{7}{4} \end{bmatrix} + \dots,$$

and at $z = 0$:

$$W(z) = \frac{1}{z} \begin{bmatrix} 1 & 0 \\ -1 & 0 \\ -1 & 0 \end{bmatrix} + \begin{bmatrix} -2 & -2 \\ 1 & -2 \\ 0 & -2 \end{bmatrix} + z \begin{bmatrix} -\frac{3}{2} & -2 \\ 0 & 5 \\ 1 & 7 \end{bmatrix} + \dots,$$

it follows that (C_π, A_π) as in (4.7) is a right pole pair and (A_ζ, B_ζ) as in (4.8) is a left null pair for W over the set $\{0, 1, 2\}$.

□

The computations involved in Example 4.10 have been carried out using the symbolic manipulator MACSYMA.

Suppose next that a rational matrix function H, with a left annihilator H^{ol}, has (C_π, A_π) as a right pole pair and (A_ζ, B_ζ) as a left null pair over a subset σ of **C**. We show that if Λ is a subspace of H^{ol}, then we can construct a rational matrix function W with (C_π, A_π) as a right pole pair and (A_ζ, B_ζ) as a left null pair over σ and with the left annihilator Λ. In some sense this a converse of Lemma 4.7.

Lemma 4.11 *Let H be a subset of* **C**, *and let (C_π, A_π) and (A_ζ, B_ζ) be right pole and left null pairs for H over σ. Then, for each subspace Λ of H^{ol}, we can construct a function W such that (C_π, A_π) is a right pole pair and (A_ζ, B_ζ) is a left null pair for W over σ and Λ is the left annihilator of W.*

Proof We assume without loss of generality that H has full column rank. Find a basis $\{v_1, v_2, ..., v_\kappa\}$ for H^{ol} such that the function

$$\begin{bmatrix} v_1(z) \\ v_2(z) \\ \vdots \\ v_\kappa(z) \end{bmatrix}$$

has neither zeros nor poles in σ (e.g. a minimal polynomial basis for H^{ol} has such properties). Using Lemma 3.14, find a "universal" canonical set $\{\phi_1, \phi_2, ..., \phi_l\}$, such that the pair (A_ζ, B_ζ) is left-similar to a pair constructed from Taylor coefficients of ϕ_i's at the zeros of H in σ. Proceed as in Lemma 3.16 to find functions $\psi_1, \psi_2, ..., \psi_l$ such that the function

$$\begin{bmatrix} v_1 \\ \vdots \\ v_\kappa \\ \psi_1 \\ \vdots \\ \psi_l \end{bmatrix}$$

has neither zeros nor poles in σ.

Find a basis $\{w_1, w_2, ..., w_k\}$ for λ such that the function

$$\begin{bmatrix} w_1 \\ \vdots \\ w_k \end{bmatrix}$$

has neither zeros nor poles in σ. Select from among v_i's functions $v_{i_1}, v_{i_2}, ..., v_{i_{\kappa-k}}$ such that $\{v_{i_1}, v_{i_2}, ..., v_{i_{\kappa-k}}, w_1, w_2, ..., w_k\}$ is a basis for H^{ol}. Proceeding as in Lemma 3.16, find functions $\tilde{v}_{i_1}, \tilde{v}_{i_2}, ..., \tilde{v}_{i_{\kappa-k}}$ such that the rational matrix function

$$\begin{bmatrix} w_1 \\ \vdots \\ w_k \\ \tilde{v}_{i_1} \\ \vdots \\ \tilde{v}_{i_{\kappa-k}} \end{bmatrix}$$

has neither zeros nor poles in σ. Then, in view of Proposition 2.3, the function

$$R(z) = \begin{bmatrix} v_{i_1}(z) \\ \vdots \\ v_{i_{\kappa-k}}(z) \\ w_1(z) \\ \vdots \\ w_k(z) \\ \psi_1(z) \\ \vdots \\ \psi_l(z) \end{bmatrix}$$

is analytic and has no zeros in σ. Let EDF be a Smith-McMillan factorization of R, let $\tilde{R}$ be a regular rational matrix function obtained by appending bottom rows of F at the bottom of R, and let Q be a rational matrix function built from the first $\kappa - k$ columns of the function $\tilde{R}^{-1}$. Put

$$W = [H \quad Q].$$

It is easy to see that the function $\tilde{R}$ has neither zeros nor poles in σ. Hence Q has neither zeros nor poles in σ. Consequently, (C_π, A_π) is a right pole pair for W over σ. Also, the column spaces of H and Q are orthogonal on σ and $\psi_i Q = 0$ for $i = 1, 2, ..., l$. Hence, by Proposition 3.10, (A_ζ, B_ζ) is a left null pair for W over σ.

□

Finally, we note that we can remove a zero or a pole of a rational matrix function H at a point $\lambda \in \mathbf{C}$ without affecting the right pole or left zero structure of H in $\mathbf{C}\backslash\{\lambda\}$. Indeed, we can remove a pole of H at λ by simply multiplying H by a scalar polynomial with all roots at λ. If H has a zero at λ of multiplicity k, we can multiply H on the right by a unimodular matrix polynomial so that one of the columns of the resulting function vanishes at λ. Then we can divide this column by $(z - \lambda)$ (cf. Algorithm for Step 2 in [F] or Algorithm 4.1 in [BR]). Since such an operation decreases the multiplicity of the zero of

H at λ by one, a finite number of such operations removes the zero of H at λ. Clearly, such operations do not affect the right pole and left zero structure of H in $\mathbf{C}\backslash\{\lambda\}$.

We can now combine all the construction steps discussed above into a single algorithm for constructing a rational matrix function with a given left annihilator Λ, and with (C_π, A_π) and (A_ζ, B_ζ) as a right pole pair and a left null pair over $\sigma \subset \mathbf{C}_\infty$, from a regular rational matrix function with (C_π, A_π) and (A_ζ, B_ζ) as a right pole pair and a left null pair over σ.

Theorem 4.12 *Let H be a regular $m \times m$ rational matrix function with right pole and left null pairs over $\sigma \subset \mathbf{C}$ equal to (C_π, A_π) and (A_ζ, B_ζ), respectively. If $(A_\kappa, B_\kappa, D_\kappa)$ is a triple of matrices which satisfy conditions (iii)-(v) in Proposition 4.1, then we can construct a left invertible rational matrix function W with right pole and left null pairs over σ equal to (C_π, A_π) and (A_ζ, B_ζ) and a left kernel triple equal to $(A_\kappa, B_\kappa, D_\kappa)$.*

Proof We find first a Smith-McMillan factorization EDF of H and put $W_1 = ED$. Following the procedure in Lemma 4.7 we construct an $m \times \eta$ rational matrix function W_2 with (C_π, A_π) as a right pole pair and (A_ζ, B_ζ) as a left null pair over σ, where η is the largest sum of the geometric multiplicity of a pole and the geometric multiplicity of a zero of H at a single point of σ.

We identify the function W_2 obtained above with the function W_1 occuring in the proof of Lemma 4.9 and take the bottom μ rows of E^{-1} as the universal canonical set of left null functions for W_2 in σ, where μ is the largest geometric multiplicity of a zero of H in σ. Since the associated matrix polynomial of the triple $(A_\kappa, B_\kappa, D_\kappa)$ satisfies hypotheses of Lemma 4.9, we can apply the construction in Lemma 4.9 to obtain a function W_3 such that $PW_3 = 0$ and (C_π, A_π) and (A_ζ, B_ζ) are right pole and left null pairs for W_3 over $\sigma(A_\pi) \cup \sigma(A_\zeta)$. We also obtain functions $\psi_1, \psi_2, ..., \psi_\mu$ such that span $\{\psi_1, \psi_2, ..., \psi_\mu\}$ is orthogonal to the row space of the matrix polynomial P on $\sigma(A_\pi) \cup \sigma(A_\zeta)$. We apply Lemma 4.11 to W_3, with Λ equal to the row space of P (using functions $\psi_1, \psi_2, ..., \psi_\mu$ obtained in the previous step) to find a function W_4 with a left kernel triple $(A_\kappa, B_\kappa, D_\kappa)$, and with (C_π, A_π) as a right pole pair and (A_ζ, B_ζ) as a left null pair over $\sigma(A_\pi) \cup \sigma(A_\zeta)$. After removing the poles and zeros of W_4 in $\sigma\backslash\Big(\sigma(A_\pi) \cup \sigma(A_\zeta)\Big)$ as in the discussion preceding the theorem, we obtain a function W with the required properties.

□

REFERENCES

[BCR] J. A. Ball, N. Cohen and L. Rodman, Zero Data and Interpolation Problems for

Rectangular Matrix Polynomials, submitted to *Linear and Multilinear Algebra.*

[BGK] H. Bart, I. Gohberg and M. A. Kaashoek, *Minimal Factorization of Matrix and Operator Fuctions*, OT 1, Birkhäuser Verlag, Basel Boston Stuttgart, 1979.

[BGR1] J. A. Ball, I. Gohberg and L. Rodman, Realization and Interpolation of Rational Matrix Functions, in *Topics in Interpolation Theory of Rational Matrix Functions* (ed. I. Gohberg), OT 33, Birkhäuser Verlag, Basel Boston Berlin, pp. 1-72, 1988.

[BGR2] J. A. Ball, I. Gohberg and L. Rodman, Two-sided Lagrange-Sylvester Interpolation Problems for Rational Matrix Functions, *Proceeding Symposia in Pure Mathematics, Amer. Math. Soc.*, to appear.

[BGR3] J. A. Ball, I. Gohberg and L. Rodman, *Interpolation for Rational Matrix Functions,* monograph in preparation.

[BGR4] J. A. Ball, I. Gohberg and L. Rodman, Minimal Factorization of Meromorphic Matrix Functions in Terms of Local Data, in *Integral Equations and Operator Theory*, vol. 10, pp. 309-348, 1987.

[BGR5] J. A. Ball, I. Gohberg and L. Rodman, Two-Sided Nudelman Interpolation Problem for Rational Matrix Functions, in *Analysis and Partial Differential Equations* (ed. C. Sadosky), Marcel Dekker, New York, pp. 371-416, 1990.

[BGR6] J. A. Ball, I. Gohberg and L. Rodman, Sensitivity Minimization and Bitangetial Nevanlinna-Pick Interpolation in Contour Integral Form, in *Proc. IMA Workshop on Signal Processing*, Springer Verlag, to appear.

[BR] J. A. Ball and M. Rakowski, Minimal McMillan Degree Rational Matrix Functions with Prescribed Zero-Pole Structure, in *Linear Algebra and its Applications,* to appear.

[C] N. Cohen, On Spectral Analysis and Factorization of Rational Matrix Functions, PhD Dissertation, Weizmann Institute of Science, Rehovot, Israel, August 1984.

[CPW] G. Conte, A. M. Perdon and B. F. Wyman, Zero/Pole Structure of Linear Transfer Functions, in *Proceedings of the 24th IEEE Conference on Decision and Control*, Ft. Lauderdale, Florida, pp. 529-530, December 1985.

[F] G. D. Forney, Jr., Minimal bases of rational vector spaces, with applications to multivariable linear systems, in *SIAM Journal of Control*, vol. 13, no. 3, pp. 493-520, May 1975.

[GK1] I. Gohberg and M. A. Kaashoek, Regular Rational Matrix Functions with Prescribed Pole and Zero Structure, in *Topics in Interpolation Theory of Rational Matrix Functions* (ed. I. Gohberg), OT 33, Birkhäuser Verlag, Basel Boston Berlin, pp. 109-122, 1988.

[GK2] I. Gohberg and M. A. Kaashoek, An Inverse Spectral Problem for Rational Matrix Functions and Minimal Divisibility, in *Integral Equations and Operator Theory,* vol. 10, pp. 437-465, 1987.

[GKLR] I. Gohberg, M. A. Kaashoek, L. Lerer and L. Rodman, Minimal divisors of rational matrix functions with prescribed zero and pole structure, in *Topics in Operator Theory Systems and Networks* (ed. H. Dym and I. Gohberg), OT 12, Birkhäuser Verlag, Basel Boston Stuttgart, pp. 241-275, 1984.

[GLR] I. Gohberg, P. Lancaster and L. Rodman, *Matrix Polynomials*, Academic Press, New York London Paris, 1982.

[K] T. Kailath, *Linear Systems*, Prentice-Hall, Inc., Englewood Cliffs, N.J., 1980.

[M1] A. F. Monna, Sur les espaces normés non-archimédiens, I and II in *Proc. Ken. Ned. Akad. v. Wetensch.*, pp. 475-489, 1956, III and IV Ibid, pp. 459-476, 1957.

[M2] A. F. Monna, *Analyse non-archimédienne,* Springer-Verlag, Berlin Heidelberg New York, 1970.

[Ra] M. Rakowski, Zero-Pole Interpolation of Nonregular Rational Matrix Functions, PhD Dissertation, Virginia Tech, Blacksburg, December 1989.

[Ros] H. H. Rosenbrock, *State-space and Multivariable Theory*, John Wiley & Sons, Inc., New York, 1970.

[Ro] L. Rodman, Spectral Theory of Analytic Matrix Functions, PhD Dissertation, Tel-Aviv University, Tel-Aviv, Israel, June 1978.

[VDK] G. Verghese, P. Van Dooren and T. Kailath, Properties of the system matrix of a generalized state-sapce system, in *Int. J. Control*, vol. 30, no. 2, pp. 235-243, 1979.

[VK] G. Verghese and T. Kailath, Rational Matrix Structure, in Proceedings of the 18th IEEE Conference on Decision and Control, Fort Lauderdale, FL, vol. 2, pp. 1008-1012, December 12-14 1979. John Wiley & Sons, Inc., New York, 1970.

[WSCP] B. F. Wyman, M. K. Sain, G. Conte and A. M. Perdon, On the Zeros and Poles of a Transfer Function, in *Linear Algebra and its Applications*, 1989.

Department of Mathematics
Virginia Tech
Blacksburg VA 24061
USA

Operator Theory
Advances and Applications, Vol. 47

Structured Interpolation Theory

Hari Bercovici Ciprian Foias Allen Tannenbaum

This paper is dedicated to the memory of Mark Grigorovich Krein, one of the great mathematicians of our century.

Abstract

In this paper, we extend our previous work on the spectral commutant lifting theorem [4] to the case of structured singular values which appear in certain problems in control theory. We moreover give a new characterization of the structured singular value, and formulate and prove a structured version of the matrix Nevanlinna-Pick theorem.

1 Introduction

This paper is concerned with a new twist to the subject of interpolation in the disc by analytic functions in which we consider interpolation theory not with analytic matrices of bounded norm as in the classical Nevanlinna-Pick framework, but with bounded structured singular value. (See Section 2 for the precise definitions.) We will also describe a new characterization of the structured singular value from which we will derive an elementary proof to a result of Doyle [6], which relates the structured singular value to an invariant derived from certain scalings. See [6] and Section 3 below. The structured singular value was originated by Doyle [6] and Safonov [14] in connection with certain engineering control problems. Our results are based on our previous paper [4] in which we formulated and proved a spectral version of the commutant lifting theorem.

Recall the classical Nevanlinna-Pick theory [12], [13]. We are given $2n$ points

$$z_1, \ldots, z_n, w_1, \ldots, w_n \in \mathbf{D}$$

($\mathbf{D}$ denotes the open unit disc), and we would like to find necessary and sufficent conditions for the existence of an analytic function $f : \mathbf{D} \to \overline{\mathbf{D}}$ such that $f(z_j) = w_j$ for $1 \leq j \leq n$. As is well known a necessary and sufficient condition can be written for the existence of f in terms of the positivity of a certain Hermitian matrix (called the *Nevanlinna-Pick* matrix).

*This work was supported in part by grants from the Research Fund of Indiana University, National Science Foundation DMS-8858149, DMS-8521683, DMS-8802596, ECS-8704047, DMS-8811084, and by the Air Force Office of Scientific Research AFOSR-90-0024.

The standard approach to such interpolation problems is complex analytic. An important fact found by Donald Sarason [16] is that many results on interpolation by bounded analytic functions may be deduced in an operator theoretic manner, and using operator theory one can even extend interpolation theory to cover interpolation with points with infinite multiplicity. Sarason's theorem was strengthened by the Sz.-Nagy-Foias commutant lifting theorem [18], [19] which allows one to interpolate on the disc by bounded analytic *operator-valued* functions.

As alluded to above, the problem we will be considering in this paper had an applied motivation, namely in robust system design, i.e. the design of feedback control systems in the presence of parameter uncertainty. (See [6], [8], [9], [11], [14], [15], [20].) However, we believe that the mathematical problem which arises from this also has a purely intrinsic operator-theoretic interest.

In order to describe the results of the present paper, we consider the classical Nevanlinna-Pick theory in the matrix case. As above, let $z_j \in \mathbf{D}$ be distinct, $1 \leq j \leq n$, and let $F_1, \ldots, F_n$ be $N \times N$ matrices. Then we are interested in finding necessary and sufficient conditions for the existence of an analytic (in the disc) $N \times N$ matrix-valued function $F(z)$ with $F(z_j) = F_j$ $(1 \leq j \leq n)$, and such that $\|F\| \leq 1$. The existence of F can again be reduced to the determination of the positivity of a certain Nevanlinna-Pick matrix. (This fact can be deduced, e.g., from the commutant lifting theorem [18], [19].) However, in our case we will not be interested in bounding the norm, but instead in the *structured singular value* of the interpolating functions. This "structured Nevanlinna-Pick theorem" will be deduced as a consequence of a general commutant lifting result which generalizes the authors' previous spectral commutant lifting theorem [7].

This work was partially performed while the second-named author was visiting the Technion on a Lady Davis Fellowship. The author thanks the Technion and its Institute for Advanced Studies in Mathematics for their warm hospitality and support. Also we would like to thank Leonid Lerer and Alexander Marcus for useful discussions concerning the proof of Lemma 5 in Section 4, and Harry Dym for pointing out a number of typographical errors.

2 Classical Commutant Lifting Theorem

In what follows, $\mathcal{H}$ will denote a complex separable Hilbert space. By "operator" we shall always mean "bounded linear operator," unless explicitly stated otherwise. We let $\mathcal{L}(\mathcal{H})$ denote the set of operators on $\mathcal{H}$.

In this section, we shall give some background on the commutant lifting theorem [2], [16], [18], [19] and its connection to interpolation theory. We will follow specifically the treatment given in [4]. We have included the section to make the paper as self-contained as possible, but all the material which follows is standard. We want to write down in particular the Sarason set-up of the relationship of commutant lifting to Nevanlinna-Pick.

We begin with a result due to Sz.-Nagy [17]. Let $T : \mathcal{H} \to \mathcal{H}$ be a contraction, i.e. an operator such that $\|T\| \leq 1$. Then in [17] it is proven that there exists an isometry U on a Hilbert space $\mathcal{K}$ such that

$$\mathcal{K} = \bigvee_{n=0}^{\infty} U^n \mathcal{H}$$

and $P_{\mathcal{H}} U = T P_{\mathcal{H}}$, where $P_{\mathcal{H}} : \mathcal{K} \to \mathcal{H}$ denotes orthogonal projection. U is called the *minimal isometric dilation of* T. The *commutant lifting theorem* (proved in a special case in [16], and in complete generality in [18]) may be formulated as follows:

Theorem 1 (Commutant Lifting Theorem) *Let $\mathcal{H}$ and $\mathcal{H}'$ denote (complex separable) Hilbert spaces with $T : \mathcal{H} \to \mathcal{H}$, $T' : \mathcal{H}' \to \mathcal{H}'$ contractions. Let $A : \mathcal{H} \to \mathcal{H}'$ be an intertwining contraction for T and T', i.e. $AT = T'A$. Let $U : \mathcal{K} \to \mathcal{K}$ and $U' : \mathcal{K}' \to \mathcal{K}'$ denote the minimal isometric dilations of T and T' respectively. Then there exists a contraction $\hat{A} : \mathcal{K} \to \mathcal{K}'$ such that $U'\hat{A} = \hat{A}U$ and $P_{\mathcal{H}'}\hat{A} = AP_{\mathcal{H}}$ where $P_{\mathcal{H}'} : \mathcal{K}' \to \mathcal{H}'$ and $P_{\mathcal{H}} : \mathcal{K} \to \mathcal{H}$ denote orthogonal projections.*

Remarks 1.

(i) $\hat{A}$ is called an *intertwining dilation* of A.

(ii) For the commutant lifting theorem we do not need the minimal isometric dilation, but any isometric dilation of T satisfying $T^* = U^*|\mathcal{H}$. Moreover, in the proof we may take without loss of generality $\mathcal{H} = \mathcal{H}'$ and $T = T'$. In this case, $\hat{A}$ is called a *commuting dilation* of A.

The commutant lifting theorem allows one to derive most of the classical interpolation results involving bounded analtyic functions in a unified manner even when the functions are operator-valued. In order to give the reader an idea how this is done we will sketch the classical scalar Nevanlinna-Pick problem as in [16].

Recall that the problem of Nevanlinna-Pick concerns finding necessary and sufficient conditions for the existence of an analytic $f : \mathbf{D} \to \mathbf{D}$, such that $f(z_j) = w_j$, $j = 1, \ldots, n$. We assume as before that the z_j are distinct. We now put the Nevanlinna-Pick problem into the commutant lifting framework.

Indeed, we let

$$m(z) = \prod_{i=1}^{n} \frac{z - z_i}{1 - \overline{z_i} z}.$$

Set $H(m) := H^2 \ominus mH^2$. (All of the Hardy spaces in this paper will be defined in the unit disc $\mathbf{D}$ in the standard way.) One can then show that $H(m)$ is an n dimensional vector space with basis $f_1, \ldots, f_n$ where

$$f_j(z) := \left(\prod_{k \neq j} \frac{z - z_k}{1 - \overline{z_k} z} \right) \frac{1}{1 - \overline{z_j} z} \quad \text{for } j = 1, \ldots, n.$$

Now let $S : H^2 \to H^2$ denote the canonical unilateral right shift defined as multiplication by z. For $P_{H(m)} : H^2 \to H(m)$ orthogonal projection, set $S(m) := P_{H(m)} S | H(m)$ (the *compressed shift*). Again, one can compute [16] that

$$S(m) f_j = z_j f_j$$

for $1 \leq j \leq n$.

We can now define a linear operator $A : H(m) \to H(m)$ such that

$$Af_j = w_j f_j$$

for $j = 1, \ldots, n$. Clearly A commutes with T. We are now ready to sketch the proof of the Nevanlinna-Pick theorem:

Theorem 2 (Nevanlinna-Pick) *Notation as above. Then there exists an analytic $f : \mathbf{D} \to \overline{\mathbf{D}}$ such that $f(z_j) = w_j$ for $j = 1, \ldots, n$ if and only if the Nevanlinna-Pick matrix*

$$\left[\frac{1 - \overline{w_i} w_j}{1 - \overline{z_i} z_j}\right]_{i,j=1,\ldots,n} \geq 0.$$

Proof. Note that the non-negative definiteness of the Nevanlinna-Pick matrix is equivalent to the inequality $I - A^*A \geq 0$ or, equivalently to $\|A\| \leq 1$. By the commutant lifting theorem, this condition is equivalent to the existence of an operator $\hat{A}$, commuting with S, such that $\|\hat{A}\| \leq 1$ and

$$A = P_{H(m)} \hat{A} | H(m).$$

It is well-known that every operator $\hat{A}$ commuting with S is given by

$$\hat{A}u = fu, \quad u \in H^2,$$

for some $f \in H^\infty$, and $\|\hat{A}\| = \|f\|_\infty$. Thus we see that the non-negative definiteness of the Nevanlinna-Pick matrix is equivalent to the existence of a function $f \in H^\infty$, $\|f\|_\infty \leq 1$, such that

$$Au = P_{H(m)} fu = f(S(m))u, \quad u \in H(m).$$

This last equality simply means that $w_j = f(z_j)$, $j = 1, 2, \ldots, n$. The theorem follows from these observations. □

We will discuss the matrix structured singular value version of this result below.

3 Remarks on the Structured Singular Value

In this section, we introduce some notation and discuss some properties of the strutured singular value. We moreover state a result which we believe is new in this connection. Instead of working over diagonal sets of matrices as in [6], we work over an arbitrary finite dimensional C^*-algebra. See also Helton [11].

Let $\mathcal{E}$ be a finite dimensional complex Hilbert space, and $\Delta \subset \mathcal{L}(\mathcal{E})$, a C^*-algebra. For $A \in \mathcal{L}(\mathcal{E})$, $A \neq 0$, we define the *structured singular value*

$$\mu_\Delta(A) := [\inf\{\|X\| : X \in \Delta, \ -1 \in \sigma(AX)\}]^{-1}.$$

Moreover, we set

$$\hat{\mu}_\Delta(A) := \inf\{\|DAD^{-1}\| : D \in \Delta', D \text{ invertible}\}$$

where Δ' is the commutator of Δ. Note that for $\Delta = \mathcal{L}(\mathcal{E})$, $\mu_\Delta(A) = \|A\|$, while for $\Delta = \mathbf{C}I_{\mathcal{E}}$, $\mu_\Delta = \|A\|_{sp}$. Moreover, $\mu_\Delta(A) = \mu_\Delta(A^*)$.

We now summarize some of the elementary properties of μ_Δ and $\widehat{\mu}_\Delta$ based on [6]. However, we work here over an arbitrary finite dimensional C^*-algebra. Some of the properties which we mention are even valid in a more general setting. See Helton [11].

Let μ_1 denote either μ_Δ or $\widehat{\mu}_\Delta$. Then μ_1 enjoys the following properties:

(i) $\mu_1(\lambda A) = |\lambda|\mu_1(A)$.

(ii) $\|A\|_{sp} \le \mu_1(A) \le \|A\|$.

(iii) $\mu_1(AX) \le \|X\|\mu_1(A)$.

We also have the following useful observations [6] and [11]:

Lemma 1

$$\mu_\Delta(A) = \sup\{\|AX\|_{sp} : X \in \Delta,\ \ \|X\| \le 1\}.$$

Proof. For $X \in \Delta$, $-1 \in \sigma(AX) \Rightarrow -\|X\|^{-1} \in \sigma(AX/\|X\|) \Rightarrow \|AX/\|X\|\|_{sp} \ge 1/\|X\| \Rightarrow 1/\|X\| \le \sup\{\|AX\|_{sp} : X \in \Delta,\ \ \|X\| \le 1\}$. But

$$\begin{aligned}\mu_\Delta(A) &= \sup\{1/\|X\| : -1 \in \sigma(AX),\ \ X \in \Delta\} \\ &\le \sup\{\|AX\|_{sp} : X \in \Delta,\ \|X\| \le 1\}.\end{aligned}$$

Conversely,

$$\begin{aligned}\|AX\|_{sp} &\le \mu_\Delta(AX) \\ &\le \mu_\Delta(A)\|X\| \\ &\le \mu_\Delta(A)\end{aligned}$$

for every $X \in \Delta$, $\|X\| \le 1$. □

Lemma 2 *μ_Δ is continuous.*

Proof. We prove the result for μ_Δ. We first show that μ_Δ is upper semicontinuous. Let $X_o \in \Delta$ be such that

$$\|X_o\| < \frac{1}{\mu_\Delta(A)}.$$

Then $I + AX_o$ is invertible. Now let $\epsilon > 0$ be sufficiently small so that

$$\|X_o\| \le \frac{(1-\epsilon)}{\mu_\Delta(A)}.$$

Moreover, we define

$$\nu := \max\{\|(I + AX_o)^{-1}\| : \|X_o\| \le \frac{(1-\epsilon)}{\mu_\Delta(A)}\}.$$

For any A_1, we have

$$\begin{aligned} I + A_1 X_o &= I + AX_o + (A_1 - A)X_o \\ &= (I + AX_o)^{-1}[I + (I + AX_o)^{-1}(A_1 - A)X_o]. \end{aligned}$$

Thus for

$$\|A_1 - A\| < \frac{(1-\epsilon)}{\nu \mu_\Delta(A)},$$

we have $I + A_1 X_o$ is invertible, and

$$\mu_\Delta(A_1) \le \frac{\mu_\Delta(A)}{1-\epsilon}.$$

The lower semicontinuity is obvious, since from Lemma 1 we have that $\mu_\Delta(A)$ is the supremum of continuous functions. This completes the proof of the lemma. □

Lemma 3

$$\mu_\Delta(A) \le \widehat{\mu}_\Delta(A).$$

Proof. For $D \in \Delta'$, we have that

$$\mu_\Delta(A) = \mu_\Delta(DAD^{-1}) \le \|DAD^{-1}\|.$$

□

For certain diagonal algebras of matrices, it is argued in [6], [14], and [15] that the structured singular value μ_Δ is the natural object of study in robust control. Unfortunately the structured singular value is difficult to compute, so in practice it is $\widehat{\mu}_\Delta$ which is actually used for the control problems. It is therefore of interest to know when these two objects are equal.

In [6], Doyle has shown that in fact $\mu_\Delta = \widehat{\mu}_\Delta$ when the relevant diagonal algebra has three or fewer blocks. In Section 4, we give a very different proof of this fact based on the following result which we believe has independent interest. More precisely set $\mathcal{H} := \mathcal{L}(\mathcal{E})$, and define (the left multiplication operator) $L_A \in \mathcal{L}(\mathcal{E})$ by $L_A := AX$. Notice that $\mathcal{H}$ is a Hilbert space with respect to

$$\langle T_1, T_2 \rangle := Tr(T_2^* T_1),$$

where Tr denotes the trace. (This induces the Hilbert-Schmidt norm on $\mathcal{H}$.) Define

$$\bar{\mu}_\Delta(A) := \mu_{\tilde{\Delta}}(L_A)$$

where

$$\tilde{\Delta} := \{L_X : X \in \Delta'\}'.$$

We now have:

Theorem 3

$$\widehat{\mu}_\Delta(A) := \bar{\mu}_\Delta(A).$$

Proof. The proof is based on the construction of a certain operator B which is given in [3]. We give here a different argument from that of [3].

First note that for $D, D^{-1} \in \Delta'$, we have that

$$\bar{\mu}_\Delta(A) = \mu_{\tilde{\Delta}}(L_D L_A L_{D^{-1}}) = \bar{\mu}_\Delta(DAD^{-1}).$$

Since $\bar{\mu}_\Delta$ is also continuous (see Lemma 2), it follows that on

$$\overline{\{DAD^{-1} : D, D^{-1} \in \Delta'\}}$$

$\bar{\mu}_\Delta$ is constant (and so is μ_Δ). Notice also

$$\bar{\mu}_\Delta(A) = \bar{\mu}_\Delta(DAD^{-1}) \leq \|DAD^{-1}\|$$

so

$$\bar{\mu}_\Delta(A) \leq \widehat{\mu}_\Delta(A).$$

Moreover, we have that $\sigma(AX) \subset \sigma(L_A L_X)$, hence $\|AX\|_{sp} \leq \|L_A L_X\|_{sp}$ and so

$$\mu_\Delta(A) \leq \bar{\mu}_\Delta(A).$$

Next let $\|D_j A D_j^{-1}\| \to \widehat{\mu}_\Delta(A)$ for some $D_j, D_j^{-1} \in \Delta'$. Without loss of generality, we may assume that $D_j A D_j^{-1} \to A_o$, $\|A_o\| = 1$. Obviously $\|DA_oD^{-1}\| \geq \|A_o\|$ for all $D, D^{-1} \in \Delta'$. Now for $X \in \Delta'$, $\|X\| < 1$, we have that

$$\|(I - X)A_o(I + X + X^2 + \cdots)\| \geq 1.$$

Hence for every $X \in \Delta'$, and for sufficiently small $\epsilon > 0$, there exists h_ϵ with $\|h_\epsilon\| = 1$, such that

$$\|(I - \epsilon X)A_o(I + \epsilon X + \epsilon^2 X^2 + \cdots)h_\epsilon\|^2 \geq 1,$$

that is,

$$\langle A_o^* A_o h_\epsilon, h_\epsilon\rangle + 2\epsilon \Re\langle A_o^*(A_o X - X A_o)h_\epsilon, h_\epsilon\rangle + O(\epsilon^2) \geq 1$$

(where $\Re$ denotes the real part of the given complex number) or equivalently,

$$2\epsilon \Re\langle A_o^*(A_o X - X A_o)h_\epsilon, h_\epsilon\rangle + O(\epsilon^2) \geq \langle (I - A_o^* A_o)h_\epsilon, h_\epsilon\rangle.$$

Dividing by ϵ and letting $\epsilon \to 0$, we obtain $h = h_X \in \mathcal{H}$ such that

$$\Re\langle A_o^*(A_o X - X A_o)h, h\rangle \geq 0, \quad h \in \ker(I - A_o^* A_o), \quad \|h\| = 1. \tag{1}$$

Set $T := P_{\ker(I - A_o^* A_o)} A_o^*(A_o X - X A_o) | \ker(I - A_o^* A_o)$ for some fixed $X \in \Delta'$. Then for every $\zeta \in \partial \mathbf{D}$ (the unit circle), there exists $k_\zeta = k_{\zeta, X}$ such that

$$k_\zeta \in \ker(I - A_o^* A_o), \quad \Re(\zeta\langle T k_\zeta, k_\zeta\rangle) \geq 0, \quad \|k_\zeta\| = 1. \tag{2}$$

Now from the Toeplitz-Hausdorff theorem, we have that

$$W(T) := \{\langle Tk, k\rangle : \|k\| = 1, \ \ k \in \ker(I - A_o^* A_o)\}$$

is closed and convex. Hence if $0 \notin W(T)$, then there would exist $\zeta \in \partial D$ such that

$$\Re(\zeta\langle Tk, k\rangle) < 0 \ \ \forall k \in \ker(I - A_o^* A_o), \ \ \|k\| = 1,$$

contradicting (2). Summarizing the argument so far, we have proven that if A_o, $\|A_o\| = 1$, is a minimal element in the closure of

$$\{DAD^{-1} : D, D^{-1} \in \Delta'\},$$

then for every $X \in \Delta'$, there exists

$$k = k_X \in \ker(I - A_o^* A_o), \ \ \|k\| = 1$$

such that

$$\langle XA_o k, A_o k\rangle - \langle Xk, k\rangle \ (= \langle A_o^*(A_o X - XA_o)k, k\rangle = \langle Tk, k\rangle) = 0. \tag{3}$$

Now (3) can be written as

$$\langle A_o^* X A_o - X, k \otimes k^*\rangle = 0, \ \ (\text{in } \mathcal{H} = \mathcal{L}(\mathcal{E})), \tag{4}$$

where $(k \otimes k^*)v := \langle v, k\rangle k$. Set

$$\mathcal{L} := \{A_o^* X A_o - X : X \in \Delta'\},$$

$$\mathcal{K} := \{B : 0 \leq B \leq I, \ \ Tr(B) = 1, \ \ B\mathcal{E} \subset \ker(I - A_o^* A_o)\}.$$

Now if $\mathcal{K} \cap \mathcal{L}^\perp = \emptyset$, then in $\mathcal{L}$ we have that $0 \notin P_{\mathcal{L}}\mathcal{K}$ where $P_{\mathcal{L}}$ denotes orthogonal projection of $\mathcal{H}$ onto $\mathcal{L}$. Therefore there exists a unique minimal element

$$0 \neq L_o \in \mathcal{L} \cap P_{\mathcal{L}}\mathcal{K},$$

that is,

$$\|L_o\| = \min\{\|L\| : L \in \mathcal{L} \cap P_{\mathcal{L}}\mathcal{K}\}.$$

Then for any $K \in \mathcal{K}$, we have

$$\Re\langle K, L_o\rangle = \Re\langle P_{\mathcal{L}}K, L_o\rangle \geq \|L_o\|^2 = Tr(L_o^* L_o).$$

This means in particular for $L_o = A_o^* X_o A_o - X_o$ (where $X_o \in \Delta'$) we cannot have (4) for any $\|k\| = 1$, $k \in \ker(I - A_o^* A_o)$, a contradiction. Hence there exists $B \in \mathcal{K}$ such that

$$Tr((A_o^* X A_o - X)B) = 0 \ \ \ \forall X \in \Delta'.$$

Equivalently,

$$Tr(B^{1/2} A_o^* X A_o B^{1/2}) = Tr(B^{1/2} X B^{1/2}) \ \forall X \in \Delta'.$$

Since Δ' is a C^*-algebra, we also have

$$Tr(B^{1/2}A_o^*X^*XA_oB^{1/2}) = Tr(B^{1/2}X^*XB^{1/2}) \ \forall X \in \Delta',$$

that is,

$$\|L_XA_oB^{1/2}\|^2 = \|L_XB^{1/2}\|^2.$$

Thus the operator $V : \mathcal{H}_1 \to \mathcal{H}_2$ where $\mathcal{H}_1 := \{L_XB^{1/2} : X \in \Delta'\}$ and $\mathcal{H}_2 := \{L_XA_oB^{1/2} : X \in \Delta'\}$ defined by

$$VL_XA_oB^{1/2} = L_XB^{1/2}, \ \ X \in \Delta'$$

is an isometry. Obviously

$$VL_X|\mathcal{H}_1 = L_XV, \ \ \forall X \in \Delta'.$$

Now $\mathcal{H}_1$ and $\mathcal{H}_2$ reduce

$$\tilde{\Delta}' := \{L_X : X \in \Delta'\}.$$

By dimension theory [5] (see in particular Proposition 6, page 261), there exists W unitary $W \in \tilde{\Delta}$ such that $W|\mathcal{H}_1 = V$. This implies that

$$(L_{A_o} - W)B^{1/2} = 0,$$

or equivalently,

$$(W^*L_{A_o} - I)B^{1/2} = 0.$$

Since $B^{1/2} \neq 0$ (recall $Tr(B) = 1$), it follows that

$$\bar{\mu}_\Delta(A_o) \geq \|L_{A_o}W^*\|_{sp} \geq 1.$$

We conclude that

$$\bar{\mu}_\Delta(A) = \bar{\mu}_\Delta(A_o) \geq 1 = \widehat{\mu}_\Delta(A)$$

which completes the proof of the theorem. □

Remarks 2.

(i) Note that Theorem 3 implies that $\widehat{\mu}_\Delta$ can be regarded as a structured singular value on a bigger space. We note that the theorem is true for an arbitrary finite dimensional C^*-algebra Δ.

(ii) From Theorem 3 and Lemma 2 we can immediately infer that $\widehat{\mu}_\Delta$ is continuous.

4 Block Diagonal Δ

In this section, we specialize our treatment of the structured singular value to block diagonal C^*-algebras in order to prove a theorem of Doyle [6]. We first set up some notation.

Let

$$\Delta = \left\{ \begin{bmatrix} X_1 & \dots & 0 \\ & \ddots & \\ 0 & \dots & X_m \end{bmatrix} : X_j \in \mathcal{L}(\mathbf{C}^{n_j}) \ (1 \leq j \leq m) \right\}. \tag{5}$$

Then

$$\Delta' = \left\{ \begin{bmatrix} \lambda_1 I_{n_1} & \dots & 0 \\ & \ddots & \\ 0 & \dots & \lambda_m I_{n_m} \end{bmatrix} : \lambda_j \in \mathbf{C} \ \ (1 \le j \le m) \right\}. \tag{6}$$

Set $n := n_1 + \ldots + n_m$. If $X \in \mathcal{L}(\mathbf{C}^n)$, we can express X as a block matrix

$$\begin{bmatrix} X_{11} & X_{12} & \dots & X_{1m} \\ X_{21} & X_{22} & \dots & X_{2m} \\ \vdots & \vdots & \vdots & \vdots \\ X_{m1} & X_{m2} & \dots & X_{mm} \end{bmatrix} \tag{7}$$

where $X_{jk} \in \mathcal{L}(\mathbf{C}^{n_k}, \mathbf{C}^{n_j})$, $1 \le j, k \le m$. Moreover, if we view $\mathcal{L}(\mathbf{C}^n)$ as the vector space $\mathbf{C}^{n^2}$ in $\mathcal{L}(\mathcal{L}(\mathbf{C}^n)) \cong \mathcal{L}(\mathbf{C}^{n^2})$, we can express X as

$$[X_{11}\ X_{12}\ \dots\ X_{1m}\ X_{21}\ \dots\ X_{2m}\ \dots X_{m1}\ X_{m2}\ \dots\ X_{mm}]^t \tag{8}$$

where t denotes transpose. Hence the operator L_X for $X \in \Delta'$ has the matrix form

$$\operatorname{diag}(\lambda_1 I_{n_1^2}, \lambda_1 I_{n_1 n_2}, \dots, \lambda_1 I_{n_1 n_m}, \lambda_2 I_{n_2 n_1}, \dots, \lambda_2 I_{n_2 n_m}, \dots, \lambda_m I_{n_m n_1}, \dots, \lambda_m I_{n_m^2})$$

$$= \operatorname{diag}(\lambda_1 I_{n_1 n}, \lambda_2 I_{n_2 n}, \dots, \lambda_m I_{n_m n}). \tag{9}$$

Therefore $X \in \tilde{\Delta}$ (i.e., commutes with all the above diagonal operators) if and only if

$$X = \operatorname{diag}(X_1, X_2, \dots, X_m) \tag{10}$$

with $X_j \in \mathcal{L}(\mathbf{C}^{n_j n})$, $1 \le j \le m$.

On the other hand if in the representation (7), we have that

$$A = [A_{jk}]_{j,k=1}^m \in \mathcal{L}(\mathbf{C}^n),$$

then in the representation (9), (10), the left multiplication operator L_A has the form

$$\begin{bmatrix} L_{A_{11}} & L_{A_{12}} & \dots & L_{A_{1m}} \\ L_{A_{21}} & L_{A_{22}} & \dots & L_{A_{2m}} \\ \vdots & \vdots & \vdots & \vdots \\ L_{A_{m1}} & L_{A_{m2}} & \dots & L_{A_{mm}} \end{bmatrix} \tag{11}$$

where $L_{A_{jk}}$ is the left multiplication operator given by A_{jk} from

$$\mathcal{L}(\mathbf{C}^{n_k n}) = \mathcal{L}(\mathbf{C}^{n_1}, \mathbf{C}^{n_k}) \oplus \mathcal{L}(\mathbf{C}^{n_2}, \mathbf{C}^{n_k}) \oplus \cdots \oplus \mathcal{L}(\mathbf{C}^{n_m}, \mathbf{C}^{n_k})$$

into

$$\mathcal{L}(\mathbf{C}^{n_j n}) = \mathcal{L}(\mathbf{C}^{n_1}, \mathbf{C}^{n_j}) \oplus \mathcal{L}(\mathbf{C}^{n_2}, \mathbf{C}^{n_j}) \oplus \cdots \oplus \mathcal{L}(\mathbf{C}^{n_m}, \mathbf{C}^{n_j}).$$

Next from Theorem 3, we have that

$$\widehat{\mu}_\Delta(A) = \mu_{\tilde{\Delta}}(L_A),$$

and from our proof of this result, we have that

$$\mu_{\tilde{\Delta}}(L_A) = \max\{|\lambda| : L_A - \lambda X \text{ is singular for some unitary } X \in \tilde{\Delta}\}.$$

Obviously, without loss of generality there exists

$$V = (V_1, V_2, \ldots, V_m) \in \mathcal{L}(\mathbf{C}^{n_1 n}) \oplus \cdots \oplus \mathcal{L}(\mathbf{C}^{n_m n}),$$

$V \neq 0$, and $X \in \tilde{\Delta}$ unitary such that

$$\sum_{k=1}^{m} A_{jk} V_k = \rho X_j V_j, \quad 1 \leq j \leq m \tag{12}$$

with $\rho = \bar{\mu}_\Delta(A)$. Note that the X_j's exist if and only if

$$\|\sum_{k=1}^{m} A_{jk} V_k\|^2 = \rho^2 \|V_j\|^2, \quad 1 \leq j \leq m. \tag{13}$$

(Note that $V_k \in \mathcal{L}(\mathbf{C}^{n_k n}) = \mathcal{L}(\mathbf{C}^n, \mathbf{C}^{n_k})$. The norms we are taking are the Hilbert-Schmidt norms.)

Recapping the above discussion, we have shown that

$$\begin{aligned} \bar{\mu}_\Delta(A) \;&=\; \max\{\rho \geq 0 : \exists V_j \in \mathcal{L}(\mathbf{C}^n, \mathbf{C}^{n_j}) \text{ s.t. } \|\sum_{k=1}^{m} A_{jk} V_k\|^2 = \rho^2 \|V_j\|^2 \\ &\quad (1 \leq j \leq m) \text{ and } V = (V_1, \ldots, V_m) \neq 0\}. \end{aligned} \tag{14}$$

Similarly, we have that

$$\begin{aligned} \bar{\mu}_\Delta(A) \;&=\; \max\{\rho \geq 0 : \exists x_j \in \mathcal{L}(\mathbf{C}^{n_j}) \text{ s.t. } \|\sum_{k=1}^{m} A_{jk} x_k\|^2 = \rho^2 \|x_j\|^2 \\ &\quad (1 \leq j \leq m) \text{ and } x = (x_1, \ldots, x_m) \neq 0\}. \end{aligned} \tag{15}$$

Now let $\rho_o, V_{1o}, \ldots, V_{mo}$ be a maximal solution of (14). If $\alpha_1, \ldots, \alpha_{2n^2}$ are the real and imaginary parts of the coordinates of $V = (V_1, \ldots, V_m)$, then the conditions (13) can be equivalently expressed as

$$\rho^2 f_j(\alpha) - g_j(\alpha) = 0, \quad 1 \leq j \leq m \tag{16}$$

where $\alpha := (\alpha_1, \ldots, \alpha_{2n^2})$. Let $\alpha^o := (\alpha_1^o, \ldots, \alpha_{2n^2}^o)$ denote the coordinates of $(V_{1o}, \ldots, V_{mo})$. Consider the Jacobian

$$J(\rho^2, \alpha) =$$

$$\begin{bmatrix} f_1(\alpha) & \rho^2 \frac{\partial f_1}{\partial \alpha_1} - \frac{\partial g_1}{\partial \alpha_1} & \rho^2 \frac{\partial f_1}{\partial \alpha_2} - \frac{\partial g_1}{\partial \alpha_2} & \cdots & \rho^2 \frac{\partial f_1}{\partial \alpha_{2n^2}} - \frac{\partial g_1}{\partial \alpha_{2n^2}} \\ f_2(\alpha) & \rho^2 \frac{\partial f_2}{\partial \alpha_1} - \frac{\partial g_2}{\partial \alpha_1} & \rho^2 \frac{\partial f_2}{\partial \alpha_2} - \frac{\partial g_2}{\partial \alpha_2} & \cdots & \rho^2 \frac{\partial f_2}{\partial \alpha_{2n^2}} - \frac{\partial g_2}{\partial \alpha_{2n^2}} \\ \vdots & \vdots & \vdots & \vdots & \vdots \\ f_m(\alpha) & \rho^2 \frac{\partial f_m}{\partial \alpha_1} - \frac{\partial g_m}{\partial \alpha_1} & \rho^2 \frac{\partial f_m}{\partial \alpha_2} - \frac{\partial g_m}{\partial \alpha_2} & \cdots & \rho^2 \frac{\partial f_m}{\partial \alpha_{2n^2}} - \frac{\partial g_m}{\partial \alpha_{2n^2}} \end{bmatrix}$$

Since ρ_o, α^o is (by hypothesis) maximal, any maximal minor of $J(\rho_o^2, \alpha^o)$ not containing the first column must have determinant 0. Thus a necessary condition for maximality is that the matrix

$$J_1(\rho^2, \alpha) = \begin{bmatrix} \rho^2 \frac{\partial f_1}{\partial \alpha_1} - \frac{\partial g_1}{\partial \alpha_1} & \rho^2 \frac{\partial f_1}{\partial \alpha_2} - \frac{\partial g_1}{\partial \alpha_2} & \cdots & \rho^2 \frac{\partial f_1}{\partial \alpha_{2n^2}} - \frac{\partial g_1}{\partial \alpha_{2n^2}} \\ \vdots & \vdots & \vdots & \vdots \\ \rho^2 \frac{\partial f_m}{\partial \alpha_1} - \frac{\partial g_m}{\partial \alpha_1} & \rho^2 \frac{\partial f_m}{\partial \alpha_2} - \frac{\partial g_m}{\partial \alpha_2} & \cdots & \rho^2 \frac{\partial f_m}{\partial \alpha_{2n^2}} - \frac{\partial g_m}{\partial \alpha_{2n^2}} \end{bmatrix}$$

be singular at (ρ_o, α^o). This means that

$$\dim \ker J_1(\rho_o^2, \alpha_o) > 2n^2 - m.$$

Thus if we consider the system

$$\begin{cases} \Re[\langle \sum_{k=1}^m A_{jk} V_{ko}, \sum_{k=1}^m A_{jk} W_k \rangle - \rho_o^2 \langle V_{jo}, W_j \rangle] = 0 \quad 1 \leq j \leq m \\ W_k \in \mathcal{L}(\mathbf{C}^n, \mathbf{C}^{n_k}) \quad 1 \leq k \leq n, \end{cases} \tag{17}$$

the real dimension of the solution space is $> 2n^2 - m$.

Set

$$M_{jk} := \sum_{l=1}^m (A_{jk}^* A_{jl} - \rho_o^2 \delta_{jk} \delta_{jl}) V_{lo}, \quad 1 \leq j, k \leq m.$$

Then we may express (17) as

$$\Re \sum_{k=1}^m \langle M_{jk}, W_k \rangle = 0 \quad 1 \leq j \leq m. \tag{18}$$

Equivalently introducing the operator

$$A_j := (A_{j1}, A_{j2}, \ldots, A_{jm}) \in \mathcal{L}(\mathbf{C}^n, \mathbf{C}^{n_1}) \oplus \mathcal{L}(\mathbf{C}^n, \mathbf{C}^{n_2}) \oplus \cdots \mathcal{L}(\mathbf{C}^n, \mathbf{C}^{n_m}),$$

and similarly

$$W := (W_1, \ldots, W_m),$$

we have

$$\Re \langle A_j, W \rangle = 0 \quad 1 \leq j \leq m. \tag{19}$$

Hence there exist real numbers $r_1, \ldots, r_m$ not all zero such that

$$\sum r_j A_j = 0.$$

In other words, we have proven the following:

Lemma 4 *Notation as above. Then if $\rho_o, V_{o1}, \ldots, V_{om}$ is a maximal solution of (14), then for every $k = 1, \ldots, m$ we have that*

$$\sum_{j,l=1}^m r_j (A_{jk}^* A_{jl} - \rho_o^2 \delta_{jk} \delta_{jl}) V_{lo} = 0. \tag{20}$$

Now multiplying (20) by V_{ko}^* and summing over k, we obtain

$$\sum_{j=1}^{m} r_j B_j = 0 \tag{21}$$

where

$$B_j := \sum_{k,l=1}^{m} V_{ko}^*(A_{jk}^* A_{jl} - \rho_o^2 \delta_{jk}\delta_{jl})V_{lo} \quad 1 \leq j \leq m. \tag{22}$$

Note that the condition

$$\|\sum A_{jl}V_{lo}\|^2 = \rho_o^2 \|V_{jo}\|^2$$

means

$$Tr\ B_j = 0 \quad 1 \leq j \leq m.$$

Clearly the B_j are self-adjoint. We state the following elementary fact:

Lemma 5 *Let $B_1, B_2 \in \mathcal{L}(\mathbf{C}^n)$. If $Tr\ B_1 = Tr\ B_2 = 0$, $B_1^* = B_1$, $B_2^* = B_2$, then there exists $\xi \neq 0$, such that $\langle B_j\xi, \xi\rangle = 0$, $j = 1, 2$.*

Proof. Set $B = B_1 + iB_2$ and observe that $Tr(B) = 0$. If $e_1, e_2, \ldots, e_n$ is an orthonormal basis of $\mathbf{C}^n$, we deduce that

$$0 = \frac{Tr(B)}{n} = \frac{1}{n}\sum_{j=1}^{n}\langle Be_j, e_j\rangle$$

belongs to the numerical range of B, and hence there exists a unit vector ξ satisfying $\langle B\xi, \xi\rangle = 0$. This vector satisfies the requirements of our lemma. □

Remark 3. Lemma 5 was initially proven using a more involved argument. We are grateful to Professor Alexander Marcus for suggesting an argument which led to the proof given above.

We can now prove the following result [6]:

Corollary 1 (Doyle) *For Δ block diagonal with three or fewer blocks,*

$$\mu_\Delta(A) = \widehat{\mu}_\Delta(A).$$

Proof. We apply Lemma 5 to the B_j's given in equation (22). First consider the case $m = 2$. Then if we restrict B_1 and B_2 to

$$(\ker V_{1o} \cap \ker V_{2o})^\perp,$$

we can use Lemma 5 to infer the existence of a vector $0 \neq \xi \in (\ker V_{1o} \cap \ker V_{2o})^\perp$ such that

$$\langle B_1\xi, \xi\rangle = 0 = \langle B_2\xi, \xi\rangle.$$

Therefore setting

$$x_j = V_{jo}\xi \quad (j = 1, 2)$$

we have

$$\|\sum_k A_{jk}x_k\|^2 - \rho_o^2\|x_j\|^2 = \langle B_j\xi, \xi\rangle = 0, \quad j = 1, 2, \tag{23}$$

and so $\rho_o \leq \mu_\Delta(A)$. Since $\rho_o \geq \mu_\Delta(A)$, we have the result for $m = 2$.

If $m = 3$, without loss of generality we can assume $r_3 \neq 0$. Choose ξ as above. Then

$$r_3\langle B_3\xi, \xi\rangle = -r_1\langle B_1\xi, \xi\rangle - r_2\langle B_2\xi, \xi\rangle = 0.$$

Since $r_3 \neq 0$, we see that $\langle B_3\xi, \xi\rangle = 0$ and thus setting $x_3 := V_{3o}\xi$, in addition to (23), we also have

$$\|\sum_k A_{3k}x_k\|^2 - \rho_o^2\|x_3\|^2 = \langle B_3\xi, \xi\rangle = 0$$

so also here $\rho_o \leq \mu_\Delta(A)$. This completes the proof of the corollary. □

5 Structured Singular Values and Dilations

In this section, we will give a structured analogue of the commutant lifting theorem [18]. This will be applied to the structured version of classical matricial Nevanlinna-Pick interpolation below. The techniques we employ are very closely related to those in [4].

Set

$$T := S(m) \otimes I_{\mathcal{E}}$$

where m is a finite Blaschke product, $S(m)$ is the compressed shift, and $\mathcal{E}$ is a finite dimensional complex Hilbert space. Fix $\Delta \subset \mathcal{L}(\mathcal{E})$, a C^*-algebra. Define

$$I_{H^2} \otimes \Delta := \{I_{H^2} \otimes X : X \in \Delta\}$$

$$I_{H^2} \otimes \Delta' := \{I_{H^2} \otimes X : X \in \Delta'\}.$$

Notice that $\mathcal{H} := H^2(\mathcal{E}) \ominus mH^2(\mathcal{E})$ reduces both $I_{H^2} \otimes \Delta$ and $I_{H^2} \otimes \Delta'$. Now define for $A \in \{T\}'$ (the commutant of T),

$$\rho_T^\Delta(A) := \inf\{\|XAX^{-1}\| : X \text{ invertible}, \ X \in \{T\}', \ X \in (I_{H^2} \otimes \Delta|\mathcal{E})'\}.$$

Let U be the isometric dilation of T on $H^2(\mathcal{E})$ (so that U is defined by multiplication by z), and let $P : H^2(\mathcal{E}) \to \mathcal{H}$ denote orthogonal projection. For $B \in \{U\}'$, define

$$\rho_U^\Delta := \inf\{\|YBY^{-1}\| : Y \text{ invertible}, \ Y \in \{U\}', \ Y \in (I_{H^2} \otimes \Delta)'\}.$$

Lemma 6 *Let $D(\zeta) \in \Delta'$ for $\zeta \in \partial\mathbf{D}$ almost everywhere. Suppose that there exists $\Theta(\zeta)$ outer such that*

$$\Theta(\zeta)^*\Theta(\zeta) = D(\zeta)^*D(\zeta), \quad \zeta \in \partial\mathbf{D} \text{ almost everywhere.} \tag{24}$$

Then there exists an (analytic) outer function $\Theta_1 : \mathbf{D} \to \Delta'$ such that

$$\Theta_1(\zeta)^*\Theta_1(\zeta) = D(\zeta)^*D(\zeta), \quad \zeta \in \partial\mathbf{D} \text{ almost everywhere.}$$

Proof. Let $X \in \Delta$. Then

$$\begin{aligned} X^*\Theta(\zeta)^*\Theta(\zeta)X &= D(\zeta)^*X^*XD(\zeta) \\ &\le \|X\|^2 D(\zeta)^*D(\zeta) \\ &= \|X\|^2\Theta(\zeta)^*\Theta(\zeta). \end{aligned}$$

Thus there exists $F_X(z)$ analytic such that

$$\|F_X(z)\| \le \|X\|,$$

and moreover such that

$$\Theta(z)X = F_X(z)\Theta(z).$$

Note that the mapping

$$X \mapsto F_X(z)$$

is linear, that

$$XY \mapsto F_X(z)F_Y(z)$$

and $I \mapsto I$. Moreover, if X is unitary, then using the fact that the outer function $\Theta(z)$ is unique up to isometry (in the above spectral factorization (24)), we have that

$$F_X(z) \equiv \hat{X}$$

where $\hat{X}$ is constant unitary.

Next let $R = R^* \in \Delta$. Then $X := (R+iI)(R-iI)^{-1}$ is unitary. Thus $(F_R+iI)(F_R-iI)^{-1}$ is constant unitary, and so $F_R(z)$ is constant self-adjoint. Next for X invertible, we see taking the polar decomposition that $F_X(z)$ will be constant, and finally by continuity we derive that for all $X \in \Delta$, we have that $F_X(z) \equiv \hat{X}$ is constant. Thus we have shown that the mapping

$$X \mapsto F_X$$

is a *-monomorphism of $\Delta \to \mathcal{L}(\mathcal{E})$. In particular,

$$F_X = \Theta(0)X\Theta(0)^{-1}$$

and

$$F_X^* = \Theta(0)X^*\Theta(0)^{-1} \;\; \forall X \in \Delta.$$

It follows that

$$\begin{aligned} \Theta(0)^*\Theta(0)X &= \Theta(0)^*F_X\Theta(0) \\ &= X\Theta(0)^*\Theta(0) \;\; \forall X \in \Delta. \end{aligned}$$

Hence if $\Theta(0) = WR$ is the polar decomposition of $\Theta(0)$, then $R \in \Delta'$, and obviously W is unitary. Thus

$$\begin{aligned} F_X &= \Theta(0)X\Theta(0)^{-1} \\ &= WRXR^{-1}W^{-1} \\ &= WXW^{-1} \;\; \forall X \in \Delta. \end{aligned}$$

Setting $\Theta_1 = W^{-1}\Theta$, we have the required conclusion. □

We now have:

Proposition 1 *Let $A(z) : \overline{\mathbf{D}} \to \mathcal{L}(\mathcal{E})$ be continuous, and analytic in D. Suppose that*

$$\hat{\mu}_{\Delta}(A(z)) < 1 \;\; \forall z \in \overline{\mathbf{D}}.$$

Then there exists an analytic map $\Theta(z) : \mathbf{D} \to \Delta'$ with $\Theta(z)^{-1}$ analytic, such that

$$\|\Theta(z)A(z)\Theta(z)^{-1}\| < 1 \;\; \forall z \in \mathbf{D}.$$

Proof. For every $\zeta \in \partial\mathbf{D}$, there exist $D_\zeta, D_\zeta^{-1} \in \Delta'$ such that

$$\|D_\zeta A(\zeta) D_\zeta^{-1}\| < 1.$$

Since A is continuous, there exist an interval (arc) centered at ζ such that

$$\|D_\zeta A(\zeta') D_\zeta^{-1}\| < 1 \;\; \forall \zeta' \in I_\zeta.$$

By compactness, there exist a finite number of arcs $I_1, I_2, \ldots, I_m$ such that

$$\begin{aligned} I_1 \cup I_2 \cup \ldots \cup I_m &= \partial\mathbf{D} \\ I_j \cap I_k &= \emptyset \text{ for } j \neq k, \end{aligned}$$

and $D_1, D_1^{-1}, \ldots, D_m, D_m^{-1} \in \Delta'$ such that

$$\|D_j A(\zeta) D_j^{-1}\| < 1, \quad \zeta \in I_j.$$

Define

$$N(\zeta) = (D_j^* D_j)^{1/2}, \quad \zeta \in I_j.$$

Then from [19], Chapter V, there exists an outer function $\Theta(z)$ such that

$$\Theta(\zeta)^* \Theta(\zeta) = N(\zeta)^2$$

almost everywhere on $\partial\mathbf{D}$. By Lemma 6, we can take $\Theta(z) \in \Delta'$ for every $z \in \mathbf{D}$. Hence

$$\Theta(\zeta) = V(\zeta) D_j \;\; \forall \zeta \in I_j, \;\; 1 \leq j \leq m$$

with $V(\zeta)$ unitary. Thus for $\zeta \in I_j$,

$$\|\Theta(\zeta)A(\zeta)\Theta(\zeta)^{-1}\| = \|D_j A(\zeta) D_j^{-1}\| < 1,$$

and so by the maximum principle,

$$\sup_{z \in \mathbf{D}} \|\Theta(z)A(z)\Theta(z)^{-1}\| = \sup_{\zeta \in \partial\mathbf{D}} \|\Theta(\zeta)A(\zeta)\Theta(\zeta)^{-1}\| < 1.$$

□

We can now state the following result:

Corollary 2 *Let $B \in \{U\}'$, and suppose B is continuous on $\overline{\mathbf{D}}$. Then*

$$\rho_U^\Delta(B) = \sup_{z \in \mathbf{D}} \hat{\mu}_\Delta(B(z)).$$

Proof. Immediate from Proposition 1. □

6 Structured Commutant Lifting Theorem

In this section, we will prove our main result namely a structured generalization of the spectral commutant lifting theorem [4]. This will be applied to the a structured version of classical matricial Nevanlinna-Pick interpolation below.

As in [4] we need to recall some general facts about Banach algebras. More precisely, let $\mathcal{A}$ be a complex Banach algebra and let $x \in \mathcal{A}$ be an invertible element whose spectrum does not separate 0 from ∞. Then as is well-known

$$x = \exp(y)$$

for some $y \in \mathcal{A}$. In particular, if $\mathcal{A}$ is finite dimensional, then $\sigma(x)$ is finite for every $x \in \mathcal{A}$, and thus every invertible element x is an exponential.

Lemma 7 *Let $\mathcal{E}$ be a finite dimensional Hilbert space, and let $T \in L(\mathcal{E})$. Then every invertible operator $X \in \{T\}' \cap (I_{H^2} \otimes \Delta|\mathcal{H})'$ can be written as $X = \exp(Y)$ for some $Y \in \{T\}' \cap (I_{H^2} \otimes \Delta|\mathcal{H})'$.*

Proof. Apply the preceding remarks to the finite dimensional algebra

$$\{T\}' \cap (I_{H^2} \otimes \Delta|\mathcal{H})'.$$

□

We are now ready to state the main result of this section:

Theorem 4 (Structured Commutant Lifting Theorem) *Notation as above. Then for $A \in \{T\}'$,*

$$\rho_T^{\Delta}(A) = \inf\{\rho_U^{\Delta}(B) : B \text{ is a commuting dilation of } A\}.$$

Proof. For $Y, Y^{-1} \in \{U\}' \cap \{I_{H^2} \otimes \Delta\}'$, set

$$X := PY|\mathcal{H},$$

where $P = P_{\mathcal{H}} : H^2(\mathcal{E}) \to \mathcal{H}$ denotes orthogonal projection. Then $X^{-1} = PY^{-1}|\mathcal{H}$, since $mH^2(\mathcal{E})$ is invariant to Y and Y^{-1}. Moreover, for $X_o \in \Delta$, we have

$$\begin{aligned} X(I_{H^2} \otimes X_o)|\mathcal{H} &= PY(I_{H^2} \otimes X_o)|\mathcal{H} \\ &= P(I_{H^2} \otimes X_o)Y|\mathcal{H} \\ &= (I_{H^2} \otimes X_o)PY|\mathcal{H} \\ &= [(I_{H^2} \otimes X_o)|\mathcal{H}]X \end{aligned}$$

by the reducing property of $\mathcal{H}$ with respect to $I_{H^2} \otimes \Delta$. Since $XT = TX$, and

$$\begin{aligned} \rho_T^{\Delta}(A) &\leq \|XAX^{-1}\| \\ &\leq \|PYBY^{-1}|\mathcal{H}\| \\ &\leq \|YBY^{-1}\| \end{aligned}$$

we deduce that

$$\rho_T^{\Delta}(A) \leq \inf\{\rho_U^{\Delta}(B) : B \text{ commuting dilation of } A\}.$$

Conversely, let $X, X^{-1} \in \{T\}' \cap (I_{H^2} \otimes \Delta|\mathcal{H})'$. Note that we can write by Lemma 8

$$X = \exp(L)$$

with $L \in \{T\}' \cap (I_{H^2} \otimes \Delta|\mathcal{H})'$. Let $\hat{L}$ denote any commuting dilation of L. Without loss of generality, we may assume that $\hat{L} \in (I_{H^2} \otimes \Delta)'$. Indeed, if $\hat{L}$ is the maximum entropy dilation this follows from [7]. (Another way of seeing this, is to let $\mathcal{U}_\Delta$ denote the group of unitary operators in Δ, and μ the corresponding Haar measure. Then if L_o is any commuting dilation, we set

$$\hat{L} := \int_{\mathcal{U}_\Delta} (I_{H^2} \otimes W) L_o (I_{H^2} \otimes W)^{-1} d\mu(W).$$

Clearly, $\hat{L}$ is a commuting dilation of L which belongs to $(I_{H^2} \otimes \Delta)'$.)

Then

$$Y = \exp(\hat{L}) \in \{U\}' \cap (I_{H^2} \otimes \Delta)'$$

is such that

$$PY = XP, \quad PY^{-1} = X^{-1}P.$$

Let $\hat{B}$ be a commuting dilation of XAX^{-1} satisfying $\|\hat{B}\| = \|XAX^{-1}\|$. (The existence of such a $\hat{B}$ is guaranteed by the classical commutant lifting theorem.) Set

$$B = Y^{-1}\hat{B}Y.$$

Then

$$\rho_U^{\Delta}(B) \leq \|YBY^{-1}\| = \|\hat{B}\| = \|XAX^{-1}\|.$$

This immediately implies the required result. □

Remark 4. From the Structured Commutant Lifting Theorem and Corollary 2 we have that

$$\rho_T^{\Delta}(A) = \inf_B \sup_{z \in \mathbf{D}} \hat{\mu}_\Delta(B(z))$$

where $B \in \{U\}'$ is a commuting dilation of A.

For T as above with minimal isometric dilation U a shift, we will say that a commuting dilation B of $A \in \{T\}'$ is *rational* if it is defined by multiplication by a rational (matrix-valued) function.

Lemma 8 *Notation as above. Then for every invertible*

$$X \in \{T\}' \cap (I_{H^2} \otimes \Delta|\mathcal{H})',$$

there exists an invertible commuting dilation $\hat{X}$ of X, $\hat{X} \in \{U\}' \cap (I_{H^2} \otimes \Delta)'$ such that $\hat{X}$ is rational.

Proof. The proof is similar to that of Lemma 2 in [4]. First take $X = e^L$, $L \in \{T\}' \cap (I_{H^2} \otimes \Delta|\mathcal{H})'$. Let $\hat{L}$ be the maximal entropy dilation of L so that $\hat{L}$ is rational, and moreover $\hat{L} \in \{U\}' \cap (I_{H^2} \otimes \Delta)'$ (see [7]). Then $\hat{X} := e^{\hat{L}}$ satisfies

$$\|\hat{X} - \sum_{n=0}^{p} \frac{1}{n!}\hat{L}^n\| =: \eta(p) \to 0 \ \text{ for } p \to \infty.$$

On the other hand for every $p \geq 0$,

$$\hat{X}_p := \sum_{n=0}^{p} \frac{1}{n!}\hat{L}^n$$

is rational, and hence so is $\hat{X}_p^{-1}$. Moreover, for p sufficiently large (namely, such that $\eta(p) < \|\hat{X}^{-1}\|^{-1}$), we have

$$\|\hat{X} - \hat{X}_p\| \leq \eta(p)$$

and

$$\|\hat{X}^{-1} - \hat{X}_p^{-1}\| \leq \|\hat{X}^{-1}\|^2 \frac{\eta(p)}{1 - \|\hat{X}^{-1}\|\eta(p)} < \|\hat{X}^{-1}\|.$$

So if

$$\eta(p) < 1/2\|\hat{X}^{-1}\|$$

we have that

$$Q := P_{\mathcal{H}}(I - \hat{X}_p^{-1}\hat{X})|\mathcal{H}$$

satisfies $\|Q\| < 1$. Now choose an exact maximum entropy dilation $\hat{Q}$ of Q (so that $\hat{Q}$ is rational and $\hat{Q} \in \{U\}' \cap (I_{H^2} \otimes \Delta)'$), and define

$$\bar{X} := \hat{X}_p(I - \hat{Q}).$$

Then $\bar{X}$ is invertible, and

$$P_{\mathcal{H}}\bar{X}|\mathcal{H} = P_{\mathcal{H}}\hat{X}_p P_{\mathcal{H}}\hat{X}_p^{-1}\bar{X}|\mathcal{H} = P_{\mathcal{H}}\hat{X}|\mathcal{H} = X,$$

giving the desired result. □

We now have the following corollary to Theorem 4:

Corollary 3 *Notation as above. Then*

$$\rho_T^{\Delta}(A) = \inf\{\rho_U^{\Delta}(B) : B \text{ is a rational commuting dilation of } A\}. \tag{25}$$

Proof. Denote by μ_o the infimum of the right hand side of equality (25) above. By Theorem 4, we have that $\rho_T^{\Delta}(A) \leq \mu_o$. Suppose now to the contrary that $\rho_T^{\Delta}(A) < \mu_o$. Then there exists $X \in \{T\}' \cap (I_{H^2} \otimes \Delta|\mathcal{H})'$, X invertible such that

$$\|XAX^{\ 1}\| < \mu_o.$$

Now by Lemma 8, we can find an invertible rational commuting dilation $\hat{X} \in \{U\}' \cap (I_{H^2} \otimes \Delta)'$ of X. Let B be the exact maximum entropy commuting dilation of XAX^{-1}. Then B is rational. If we now set

$$\hat{B} := \hat{X}^{-1}B\hat{X},$$

we have that $\hat{X}$ is rational, from which we can draw the required contradiction to complete the proof of the corollary. □

7 Structured Nevanlinna-Pick Theory

In this section, we apply the above theory to a structured version of the Nevanlinna-Pick interpolation problem. In order to do this, we first put the Nevanlinna-Pick theory into the commutant lifting framework [16], [19], [7]. First let us recall the general problem of Nevanlinna-Pick in the matrix case. See also the closely related discussion above in the scalar case.

Let $\mathcal{E}$ be a finite dimensional Hilbert space, let $z_1, \ldots, z_n \in \mathbf{D}$ be mutually distinct, and let $F_1, \ldots, F_n \in \mathcal{L}(\mathcal{E})$. Then we want necessary and sufficient conditions for the existence of an analytic function $F : \mathbf{D} \to \mathcal{L}(\mathcal{E})$ with $\|F\|_\infty \leq 1$ such that

$$F(z_j) = F_j \tag{26}$$

for $j = 1, \ldots, n$. Define

$$m(z) := \prod_{j=1}^{n} \frac{z - z_j}{1 - \overline{z}_j z},$$

$$f_j := \left(\prod_{k \neq j} \frac{z - z_k}{1 - \overline{z}_k z} \right) \frac{1}{1 - \overline{z}_j z},$$

and

$$\mathcal{H} := H^2(\mathcal{E}) \ominus m H^2(\mathcal{E}).$$

We set $T := S(m) \otimes I_{\mathcal{E}}$, and note that

$$\mathcal{H} = f_1 \otimes \mathcal{E} + f_2 \otimes \mathcal{E} + \cdots + f_n \otimes \mathcal{E}.$$

This sum is direct but not orthogonal.

For the given interpolation data above, we define $A : \mathcal{H} \to \mathcal{H}$ by linearity and by

$$A(f_j \otimes \xi) = f_j \otimes F_j \xi$$

for all $\xi \in \mathcal{E}$, $j = 1, 2, \ldots, n$. Note that

$$T(f_j \otimes \xi) = z_j f_j \otimes \xi$$

for all $\xi \in \mathcal{E}$, $j = 1, 2, \ldots, n$. Thus

$$A \in \{T\}'.$$

It is easy to show that F satisfies the interpolation conditions

$$F(z_j) = F_j, \quad \forall j = 1, \ldots, n \tag{27}$$

if and only if

$$P_{\mathcal{H}} M_F = A P_{\mathcal{H}},$$

where $M_F : H^2(\mathcal{E}) \to H^2(\mathcal{E})$ is given by $(M_F f)(z) := F(z) f(z)$ for $f \in H^2(\mathcal{E})$, $z \in \mathbf{D}$. Thus, the commutant lifting theorem implies that there exists an F satisfying (27) and $\|F\|_\infty \leq 1$, if and only if $\|A\| \leq 1$. We will now show that the structured Nevanlinna-Pick problem can be given a similar solution, based on the structured commutant lifting theorem.

More precisely, define

$$\mathcal{I} := \{F : \mathbf{D} \to \mathcal{L}(\mathcal{E}) : F \text{ is rational, bounded in } \mathbf{D},\ F(z_j) = F_j,\ 1 \le j \le n\ \}.$$

We are interested in finding

$$\mu(\mathcal{I}) := \inf\{\sup_{z\in D} \mu_\Delta(F(z)) : F \in \mathcal{I}\}$$

$$\hat{\mu}(\mathcal{I}) := \inf\{\sup_{z\in D} \hat{\mu}_\Delta(F(z)) : F \in \mathcal{I}\}.$$

We denote the operator A associated with the matrices $F_1, \dots, F_n$ by $A(F_1, \dots, F_n)$. Set

$$\rho_T^\Delta(A) = \inf\{\|A(D_1F_1D_1^{-1}, \dots, D_nF_nD_n^{-1})\| \ :\ D_j \in \Delta',\ \ 1 \le j \le n,\ \ D_j \text{ invertible}\}$$

and

$$\rho_{T,\Delta}(A) = \sup\{\rho_T(A(F_1X_1, \dots, F_nX_n)) : \|A(X_1, \dots X_n)\| \le 1,\ X_1, \dots, X_n \in \Delta\},$$

where for an operator $M \in \{T\}'$ we let

$$\rho_T(M) = \inf\{\|XMX^{-1}\| : X \text{ is invertible and } X \in \{T\}'\ \}.$$

(In [7], $\rho_T(M)$ is called the T*-spectral radius.*) It is easy to show that

$$\rho_T^\Delta(A) \ge \rho_{T,\Delta}(A).$$

Indeed, just note that

$$\begin{aligned}\rho_T(A(F_1X_1, \dots, F_nX_n)) &= \inf\{\|A(Y_1F_1X_1Y_1^{-1}, \dots, Y_nF_nX_nY_n^{-1})\|\} \\ &\le \|A(D_1F_1D_1^{-1}X_1, \dots, D_nF_nD_n^{-1}X_n)\| \\ &\le \|A(D_1F_1D_1^{-1}, \dots, D_nF_nD_n^{-1})\|\|A(X_1, \dots, X_n)\| \\ &\le \|A(D_1F_1D_1^{-1}, \dots, D_nF_nD_n^{-1})\|\end{aligned}$$

We can now state:

Theorem 5 (Structured Nevanlinna-Pick) *Notation as above. Then*

$$\hat{\mu}(\mathcal{I}) = \rho_T^\Delta(A).$$

Proof. Follows immediately from the Structured Commutant Lifting Theorem, Remark 3, and the above set-up for the Nevanlinna-Pick theorem. □

Theorem 6 *Notation as above. Then*

$$\mu(\mathcal{I}) \geq \rho_{T,\Delta}(A).$$

Proof. Let $F \in \mathcal{I}$ satisfy

$$\mu_\Delta(F(z)) \leq \mu(\mathcal{I}) + \epsilon, \quad \forall z \in \mathbf{D}.$$

By our above discussion (see the proof of the Structured Commutant Lifting Theorem 4), if

$$\|A(X_1, \ldots, X_n)\| \leq 1, \quad X_1, X_2, \ldots, X_n \in \Delta,$$

there exists $X(z) : \mathbf{D} \to \Delta$ with $\|X\| \leq 1$, $X(z_j) = X_j$, $1 \leq j \leq n$. Then

$$\|F(z)X(z)\|_{sp} \leq \mu_\Delta(F(z)) \leq \mu(\mathcal{I}) + \epsilon.$$

By the spectral Nevanlinna-Pick theorem [4],

$$\rho_T(A(F(z_1)X(z_1), \ldots, F(z_n)X(z_n)) \leq \mu(\mathcal{I}) + \epsilon,$$

that is,

$$\rho_T(A(F_1X_1, \ldots, F_nX_n)) \leq \mu(\mathcal{I}) + \epsilon, \quad \forall \epsilon > 0.$$

Thus

$$\rho_T(A(F_1X_1, \ldots, F_nX_n)) \leq \mu(\mathcal{I})$$

for all $\|A(X_1, \ldots, X_n)\| \leq 1$, $X_1, \ldots, X_n \in \Delta$. □

Remark 5. We do not know at this point if we have equality in Theorem 6. This is still an interesting research problem.

8 Example

In this section we want to work an explicit formula for the analytic D-scaling in the 2×2 matrix case. We will take

$$\Delta = \Delta' = \left\{ \begin{bmatrix} \lambda_1 & 0 \\ 0 & \lambda_2 \end{bmatrix} : \lambda_1, \lambda_2 \in \mathbf{C} \right\}.$$

Note that for $A \in \mathcal{L}(\mathbf{C}^2)$ that

$$\mu_\Delta(A) = \hat{\mu}_\Delta(A)$$

by virtue of Corollary 1.

Let $\|A\|_{HS}$ denote the Hilbert-Schmidt norm of A. Then looking at the characteristic polynomial of A^*A it is easy to compute that

$$\|A\| \leq 1 \Leftrightarrow \|A\|_{HS}^2 + \sqrt{\|A\|_{HS}^4 - 4|\det A|^2} \leq 2. \tag{28}$$

One can then show that $\|A\| \leq 1$ if and only if

$$\|A\|_{HS}^2 \leq \begin{cases} 2 \\ 1 + |\det A|^2 \end{cases} . \tag{29}$$

Now again it is easy to compute that for Δ as above, we have that $\widehat{\mu}_\Delta(A) = \mu_\Delta(A) \leq 1$ if and only if

$$\inf_{\lambda \neq 0} \left\| \begin{bmatrix} a_{11} & \lambda a_{12} \\ \lambda^{-1} a_{21} & a_{22} \end{bmatrix} \right\| \leq 1. \tag{30}$$

Thus if (30) holds, for every $\theta > 1$, there exists $\lambda_\theta \neq 0$ such that

$$\left\| \begin{bmatrix} \theta^{-1} a_{11} & \theta^{-1} \lambda_\theta a_{12} \\ \lambda_\theta^{-1} \theta^{-1} a_{21} & \theta^{-1} a_{22} \end{bmatrix} \right\| \leq 1 \tag{31}$$

and this is equivalent to

$$|a_{11}|^2 + |\lambda_\theta|^2 |a_{12}|^2 + \frac{1}{|\lambda_\theta|^2} |a_{21}|^2 + |a_{22}|^2 \leq \begin{cases} 2\theta^2 \\ \theta^2 (1 + \theta^{-4} |\det A|^2). \end{cases} \tag{32}$$

Equivalently we have that

$$\inf_{\lambda \neq 0} (|\lambda|^2 |a_{12}|^2 + \frac{1}{|\lambda|^2} |a_{21}|^2) = 2|a_{12} a_{21}|. \tag{33}$$

Therefore (30) $\Rightarrow$ (32) $\Rightarrow$

$$|a_{11}|^2 + 2|a_{12} a_{21}| + |a_{22}|^2 \leq \begin{cases} 2 \\ 1 + |\det A|^2. \end{cases} \tag{34}$$

Conversely if (34) holds, then

Case 1. $a_{12} a_{21} = 0$. Then (see (33) letting $\lambda \to 0$ if $a_{12} \neq 0$, or $\lambda \to \infty$ if $a_{21} \neq 0$, we deduce (30) from (34).

Case 2. $a_{12} a_{21} \neq 0$. Set then

$$\lambda = \left| \frac{a_{21}}{a_{12}} \right|^{1/2}$$

to obtain from (29) that

$$\left\| \begin{bmatrix} a_{11} & \lambda a_{12} \\ \lambda^{-1} a_{21} & a_{22} \end{bmatrix} \right\| \leq 1$$

by virtue of (34).

We want to construct for $A : \mathbf{D} \to \mathcal{L}(\mathbf{C}^2)$ rational such that

$$\max_{z \in D} \mu_\Delta(A(z)) \leq 1, \tag{35}$$

an invertible analytic matrix (in D)

$$D(z) = \begin{bmatrix} \lambda_1(z) & 0 \\ 0 & \lambda_2(z) \end{bmatrix}$$

such that

$$\|D(z)A(z)D(z)^{-1}\| \leq 1 \ \ \forall z \in \mathbf{D}.$$

We divide the possibilities into 3 cases:

Case 1. $a_{21}a_{12} \neq 0$ on $\partial\mathbf{D}$. Consider the outer function $\lambda(z)$ such that

$$\lambda(\zeta) = \left|\frac{a_{21}(\zeta)}{a_{12}(\zeta)}\right|^{1/2}$$

on ∂D. Then $\lambda(z)^{-1}$ is analytic in $\mathbf{D}$ and bounded. Set

$$D(z) := \begin{bmatrix} \lambda(z) & 0 \\ 0 & 1 \end{bmatrix}.$$

Then for for $\zeta \in \partial D$, from the above argument (assuming (35)) we have that

$$|a_{11}(\zeta)|^2 + |\lambda(\zeta)a_{12}(\zeta)|^2 + |\frac{a_{21}(\zeta)}{\lambda(\zeta)}|^2 + |a_{22}(\zeta)|^2 \equiv$$

$$|a_{11}(\zeta)|^2 + 2|a_{12}(\zeta)a_{21}(\zeta)| + |a_{22}(\zeta)|^2 \leq \left\{ \begin{array}{c} 2 \\ 1 + |\det A(\zeta)|^2, \end{array} \right.$$

that is,

$$\|D(\zeta)A(\zeta)D(\zeta)^{-1}\| \leq 1 \ \ \forall \zeta \in \partial\mathbf{D}.$$

By the maximum principle,

$$\|D(z)A(z)D(z)^{-1}\| \leq 1 \ \ \forall z \in \mathbf{D}.$$

Case 2. $a_{21}a_{12} \equiv 0$. Then at least one of the a_{21} or a_{12} is $\equiv 0$, and a constant $\lambda \to 0$ or $\lambda \to \infty$ will do the trick.

Case 3. $a_{21}a_{12}$ is not identically 0 but $a_{21}a_{12}$ has zeros on $\partial\mathbf{D}$. We have then by the continuity of μ_Δ that for $\epsilon > 0$, there exists $\delta_\epsilon > 0$ such that

$$\mu_\Delta(A(z)) \leq 1 + \epsilon$$

for $|z| \leq 1 + \delta_\epsilon$. Without loss of generality, we can assume that $(a_{21}a_{12})(\zeta) \neq 0$ for $|\zeta| = 1 + \delta_\epsilon$. Then the above argument shows that there exists $D_\epsilon(z)$ analytic and invertible on $(1 + \delta_\epsilon)\mathbf{D}$ such that

$$\|D_\epsilon(z)\frac{1}{1+\epsilon}A(z)D_\epsilon(z)^{-1}\| \leq 1$$

on $(1 + \delta_\epsilon)\mathbf{D}$, so

$$\|D_\epsilon(z)A(z)D_\epsilon(z)^{-1}\| \leq 1 + \epsilon$$

on **D**. Hence

$$\inf_{\epsilon>0} \sup_{\mathbf{D}} \|D_\epsilon(z)A(z)D_\epsilon(z)^{-1}\| \leq 1.$$

This completes our construction of the D-scalings.

REFERENCES

[1] V. M. Adamjan, D. Z. Arov, and M. G. Krein, "Infinite Hankel block matrices and related extension problems," *Amer. Math. Society Translations* **111** (1978), 133-156.

[2] H. Bercovici, *Operator Theory and Arithmetic in H^∞*, AMS Publications **26**, Providence, Rhode Island, 1988.

[3] H. Bercovici, C. Foias, and A. Tannenbaum, "On the optimal solutions in spectral commutant lifting theory," in preparation.

[4] H. Bercovici, C. Foias, and A. Tannenbaum, "A spectral commutant lifting theorem," to appear in *Trans. AMS.*

[5] J. Dixmier, *Von Neumann Algebras*, North Holland Publishing, New York, 1981.

[6] J. C. Doyle, "Analysis of feedback systems with structured uncertainties," *IEE Proc.* **129** (1982), 242-250.

[7] H. Dym and I. Gohberg, "A maximal entropy principle for contractive interpolants," *J. Functional Analysis* **65** (1986), 83-125.

[8] B. A. Francis, *A Course in H^∞ Control Theory*, Lecture Notes in Control and Information Sciences **88** Springer-Verlag, 1987.

[9] B. A. Francis and A. Tannenbaum, "Generalized interpolation in control theory," *Mathematical Intelligencer* **10** (1988), 48-53.

[10] J. Garnett, *Bounded Analytic Functions*, Academic Press, New York, 1981.

[11] J. W. Helton, *Optimization in operator theory, analytic function theory and electrical engineering*, CBMS Publications **68**, Providence, Rhode Island, 1985.

[12] R. Nevanlinna, "Über beschrankte Funftionen, die in gegebenen Punkten vorgeschriebene Werte annehmen, *Ann. Acad. Sci. Fenn.* **13** (1919), 27-43.

[13] G. Pick, "Über die Beschränkungen analytischer Funktionen, welche durch vorgegebenen Funktionswerte bewiskt sind, *Math. Ann.* **77** (1916), 7-23.

[14] M. G. Safonov, *Stability Robustness of Multivariable Feedback Systems*, MIT Press, Cambridge, Mass., 1980

[15] M. G. Safonov, "Optimal H^∞ synthesis of robust controllers for systems with structured uncertainty," Proc. of 25-th IEEE Conference on Decision and Control, December 1986, 1822-1825.

[16] D. Sarason, "Generalized interpolation in H^∞," *Trans. Amer. Math. Soc.* **127** (1967), 179-203.

[17] B. Sz.-Nagy, "Sur les contractions de l'espace de Hilbert," *Acta Sci. Math.* **15** (1953), 87-92.

[18] B. Sz.-Nagy and C. Foias, "Dilation des commutants," *C. R. Acad. Sci Paris, Série A* **266** (1968), 493-495.

[19] B. Sz.-Nagy and C. Foias, *Harmonic Analysis of Operators on Hilbert Space*, North Holland, 1970.

[20] A. Tannenbaum, *Invariance and System Theory: Algebraic and Geometric Aspects*, Lecture Notes in Mathematics **845**, Springer-Verlag, 1981.

Hari Bercovici and Ciprian Foias
Department of Mathematics,
Indiana University,
Bloomington, Indiana 47405

Allen Tannenbaum
Department of Electrical Engineering,
University of Minnesota,
Minneapolis, Minnesota 55455
and
Technion, Israel Institute of Technology,
Haifa, Israel

Operator Theory
Advances and Applications, Vol. 47

Extension Theorems for Contraction Operators on Kreĭn Spaces

MICHAEL A. DRITSCHEL AND JAMES ROVNYAK*

To the memory of Mark Grigor'evič Kreĭn

Abstract notions of Julia and defect operators are used as a foundation for a theory of matrix extension and commutant lifting problems for contraction operators on Kreĭn spaces. The account includes a self-contained treatment of key propositions from the theory of Potapov, Ginsburg, Kreĭn, and Shmul'yan on the behavior of a contraction operator on negative subspaces. This theory is extended by an analysis of the behavior of the adjoint of a contraction operator on negative subspaces. Together, these results provide the technical input for the main extension theorems.

CONTENTS

* The second author was supported by the National Science Foundation.

Introduction

Kreĭn spaces are indefinite generalizations of Hilbert spaces which are important in both abstract operator theory and its applications. We are concerned with everywhere defined and continuous linear operators on a Kreĭn space $\mathcal{H}$ to a Kreĭn space $\mathcal{K}$. The set of all such operators is denoted $\mathbf{B}(\mathcal{H},\mathcal{K})$. An operator $T \in \mathbf{B}(\mathcal{H},\mathcal{K})$ is a contraction if

$$\langle Tf, Tf\rangle_{\mathcal{K}} \leq \langle f, f\rangle_{\mathcal{H}}$$

for all vectors f in $\mathcal{H}$. If both T and T^* are contractions, then T is said to be a bicontraction.

Our purpose is to show the possibility of proving extension theorems for contraction operators on Kreĭn spaces which are strikingly similar to those of the Hilbert space case. By an extension of an operator $T \in \mathbf{B}(\mathcal{H},\mathcal{K})$ we mean, for example, a row extension

$$\begin{pmatrix} T & E \end{pmatrix} \in \mathbf{B}(\mathcal{H}\oplus\mathcal{E},\mathcal{K}),$$

column extension

$$\begin{pmatrix} T \\ F \end{pmatrix} \in \mathbf{B}(\mathcal{H},\mathcal{K}\oplus\mathcal{F}),$$

or a two-by-two matrix extension

$$\begin{pmatrix} T & E \\ F & G \end{pmatrix} \in \mathbf{B}(\mathcal{H}\oplus\mathcal{E},\mathcal{K}\oplus\mathcal{F}),$$

where $\mathcal{E}$ and $\mathcal{F}$ are Kreĭn spaces. Many theorems have the hypothesis that $\mathcal{F}$ is a Hilbert space. In applications, this hypothesis is often assured whenever T is a contraction, and so the hypothesis is not as restrictive as might appear. Applications include commutant lifting theorems, in which a commutator relation is extended to minimal isometric and minimal unitary dilations.

We assume familiarity with operator theory on Hilbert spaces, but we do not presume that the reader is necessarily at ease in the indefinite environment of Kreĭn spaces. Our aim has been to give a self-contained treatment for such readers, and experts may view parts of the paper as expository. Background material is summarized in §1.1. While it is essential to master concepts of regular subspaces, projections, negative subspaces, isometries, and direct sums, if a few things are taken for granted or gleaned from standard monographs, not much more is needed as preparation. Kreĭn space extensions of the Hilbert space notions of defect and Julia operators are introduced in §1.2. Julia operators are a kind of unitary two-by-two matrix extension of a given operator and are needed to formulate certain of the extension theorems in Chapters 2 and 3.

The backbone of the study of contraction operators on Kreĭn spaces is a theory, due to Ginsburg, Kreĭn, and Shmul'yan, which analyzes how a contraction operator acts with respect to negative subspaces. One should also mention Potapov, whose treatment of the finite dimensional case helped to motivate developments. The discussion in §1.3 includes the Potapov-Ginsburg transform and its interpretation as a scattering operator, characterizations of bicontractions, and mappings of operator spheres.

The main new results are those of Chapters 2 and 3. These characterize contractive and bicontractive extensions of given contractive and bicontractive operators. Theorems on row, column, and two-by-two matrix extensions, as well as commutant lifting theorems, are proved for Kreĭn spaces in forms essentially identical to the Hilbert space case. Similar results were previously obtained by Alpay, de Branges, Constantinescu and Ghéondea, and Dritschel, using other hypotheses and somewhat different methods. Our approach is based on an analysis of the adjoint of a contraction relative to negative subspaces in §2.1.

The principal methods and main results of Chapters 2 and 3 appear in the first author's doctoral dissertation (University of Virginia, May, 1989). The present account is an expanded version of the dissertation which includes simplifications and additional results.

Chapter 1: Operator Theory on Kreĭn spaces

1.1 Definitions and Preliminaries

Kreĭn spaces are generalizations of Hilbert spaces and are the fundamental underlying objects in our study. Our philosophy on notation is to reserve the simplest notation for Kreĭn spaces and operators which act on them. This is generally standard Hilbert space notation.

A. Kreĭn Spaces and Continuous Operators

A Kreĭn space $\mathcal{H}$ is a scalar product space which is isomorphic to the direct sum of a Hilbert space and the anti-space of a Hilbert space. By a *scalar product space* we mean a complex vector space $\mathcal{H}$ together with a scalar product $\langle \cdot, \cdot \rangle$ which obeys the same axioms of linearity and symmetry as for Hilbert spaces and is nondegenerate in the sense that the only vector f in $\mathcal{H}$ such that $\langle f, g \rangle = 0$ for all g in $\mathcal{H}$ is $f = 0$. The anti-space of a scalar product space $(\mathcal{H}, \langle \cdot, \cdot \rangle)$ is $(\mathcal{H}, -\langle \cdot, \cdot \rangle)$. Notions of isomorphism, subspace, orthogonality, direct sum, and linear operator are defined as in linear algebra. Orthogonality is indicated by $\perp$, direct sum by $\dot{+}$, and orthogonal direct sum by $\oplus$. Occasionally, a subscript $\mathcal{H}$ is used to show dependence on the underlying space.

A *fundamental decomposition* of a Kreĭn space $\mathcal{H}$ is a direct sum representation $\mathcal{H} = \mathcal{H}_+ \oplus \mathcal{H}_-$ of $\mathcal{H}$, where $\mathcal{H}_+$ and $\mathcal{H}_-$ are subspaces of $\mathcal{H}$ such that $\mathcal{H}_+$ is a Hilbert space and $\mathcal{H}_-$ the anti-space of a Hilbert space in the scalar product of $\mathcal{H}$. In general, fundamental decompositions are not unique. The choice of a fundamental decomposition $\mathcal{H} = \mathcal{H}_+ \oplus \mathcal{H}_-$ induces a Hilbert space inner product, norm, and strong topology on $\mathcal{H}$. Namely, the Hilbert space inner product of $f_+ + f_-$ and $g_+ + g_-$ ($f_+, g_+ \in \mathcal{H}_+$ and $f_-, g_- \in \mathcal{H}_-$) is $\langle f_+, f_+ \rangle - \langle f_-, f_- \rangle$. The strong topology of this Hilbert space is independent of the choice of fundamental decomposition and is also called the *Mackey topology* of $\mathcal{H}$. It is used to define convergence and continuity in the usual way. The norm of the Hilbert space depends on the choice of fundamental decomposition, but two such norms are equivalent.

If $\mathcal{H}$ is a Kreĭn space, the dimensions of $\mathcal{H}_+$ and $\mathcal{H}_-$ in any fundamental decomposition $\mathcal{H} = \mathcal{H}_+ \oplus \mathcal{H}_-$ are independent of the choice of fundamental decomposition. These dimensions are called the *positive* and *negative indices* and of $\mathcal{H}$. A *Pontryagin space* is a Kreĭn space with finite negative index. Two Kreĭn spaces are isomorphic if and only if they have the same positive and negative indices.

Given a Kreĭn space $\mathcal{H}$ with fundamental decomposition $\mathcal{H} = \mathcal{H}_+ \oplus \mathcal{H}_-$, we define operators J, P_+, P_- on $\mathcal{H}$ by

$$Jf = f_1 - f_2, \qquad P_+ f = f_1, \qquad P_- f = f_2,$$

whenever

$$f = f_1 + f_2, \qquad f_1 \in \mathcal{H}_+,\ f_2 \in \mathcal{H}_-.$$

We call J the *signature operator* or *fundamental symmetry*, and P_+ and P_- the *associated projections* for the given fundamental decomposition. The signature operator J serves to identify the fundamental decomposition. We write $\mathcal{H}_J$ for $\mathcal{H}$ viewed as a Hilbert space relative to the given fundamental decomposition. Thus,

$$\langle f, g\rangle_{\mathcal{H}_J} = \langle Jf, g\rangle_{\mathcal{H}}$$

for any vectors f and g in $\mathcal{H}$. The *absolute value* of $\mathcal{H}_-$, $|\mathcal{H}_-|$, is a Hilbert space defined as the anti-space of $\mathcal{H}_-$. In this notation, $\mathcal{H}_J = \mathcal{H}_+ \oplus |\mathcal{H}_-|$.

If $\mathcal{H}$ is a Kreĭn space and J_1 and J_2 are signature operators for two fundamental decompositions of $\mathcal{H}$, then $J_2 = U^{-1}J_1U$ where U is an isomorphism of $\mathcal{H}$ onto itself.

If $\mathcal{H}$ and $\mathcal{K}$ are Kreĭn spaces, $\mathbf{B}(\mathcal{H})$ and $\mathbf{B}(\mathcal{H},\mathcal{K})$ denote the sets of everywhere defined continuous operators on $\mathcal{H}$ to itself and on $\mathcal{H}$ to $\mathcal{K}$, respectively. Every $A \in \mathbf{B}(\mathcal{H},\mathcal{K})$ has a unique *adjoint* $A^* \in \mathbf{B}(\mathcal{K},\mathcal{H})$ satisfying

$$\langle Af, g\rangle_{\mathcal{K}} = \langle f, A^*g\rangle_{\mathcal{H}}, \qquad f \in \mathcal{H}, g \in \mathcal{K}.$$

The identity operator is written 1. By viewing $\mathcal{H}$ and $\mathcal{K}$ as Hilbert spaces relative to some fundamental decompositions, we may induce norm, weak operator, and strong operator topologies on $\mathbf{B}(\mathcal{H},\mathcal{K})$ which are independent of the choice of fundamental decompositions. Any two operator norms obtained in this way are equivalent.

Kreĭn space adjoints and Hilbert space adjoints must be distinguished. Let $\mathcal{H}$ and $\mathcal{K}$ be Kreĭn spaces with fundamental symmetries $J_{\mathcal{H}}$ and $J_{\mathcal{K}}$. The Kreĭn space adjoint of an operator $A \in \mathbf{B}(\mathcal{H},\mathcal{K})$ is an operator $A^* \in \mathbf{B}(\mathcal{K},\mathcal{H})$. The Hilbert space adjoint of $A \in \mathbf{B}(\mathcal{H}_{J_{\mathcal{H}}}, \mathcal{K}_{J_{\mathcal{K}}})$ is an operator $A^\times \in \mathbf{B}(\mathcal{K}_{J_{\mathcal{K}}}, \mathcal{H}_{J_{\mathcal{H}}})$ related to A^* by

$$A^* = J_{\mathcal{H}} A^\times J_{\mathcal{K}}.$$

In general, we reserve $*$ for Kreĭn space adjoints and $\times$ for Hilbert space adjoints.

Let $\mathcal{H}$ and $\mathcal{K}$ be Kreĭn spaces. As in the Hilbert space case, we say that

(i) $A \in \mathbf{B}(\mathcal{H})$ is *selfadjoint* if $A^* = A$,
(ii) $A \in \mathbf{B}(\mathcal{H})$ is a *projection* if A is selfadjoint and $A^2 = A$,
(iii) $A \in \mathbf{B}(\mathcal{H},\mathcal{K})$ is *isometric* if $A^*A = 1$, and
(iv) $A \in \mathbf{B}(\mathcal{H},\mathcal{K})$ is *unitary* if both A and A^* are isometric.

A partial ordering of selfadjoint operators is defined in the usual way: if $A, B \in \mathbf{B}(\mathcal{H})$ are selfadjoint, $A \geq 0$ means that $\langle Af, f\rangle_{\mathcal{H}} \geq 0$ for all f in $\mathcal{H}$, and $A \geq B$ means that $A - B \geq 0$.

Note that the associated projections $P_\pm$ for a fundamental decomposition $\mathcal{H} = \mathcal{H}_+ \oplus \mathcal{H}_-$ are projections in the sense of the preceding definition.

B. Subspaces and Projections

A *subspace* of a Kreĭn space $\mathcal{H}$ is a nonempty linear set $\mathcal{M}$ in $\mathcal{H}$ (not necessarily closed). The Kreĭn space orthogonal complement $\mathcal{M}^{\perp}$ of $\mathcal{M}$ coincides with the Hilbert space orthogonal complement of $J\mathcal{M}$ in $\mathcal{H}_J$ for any signature operator J for $\mathcal{H}$. For any subspaces $\mathcal{M}$ and $\mathcal{N}$ of $\mathcal{H}$,

$$\mathcal{M}^{\perp\perp} = \overline{\mathcal{M}} \qquad \text{and} \qquad (\mathcal{M} + \mathcal{N})^{\perp} = \mathcal{M}^{\perp} \cap \mathcal{N}^{\perp},$$

where $\overline{\mathcal{M}}$ is the closure of $\mathcal{M}$. If $\mathcal{M}$ and $\mathcal{N}$ are closed, then also

$$(\mathcal{M} \cap \mathcal{N})^{\perp} = \overline{(\mathcal{M}^{\perp} + \mathcal{N}^{\perp})}.$$

A subspace $\mathcal{M}$ of $\mathcal{H}$ is dense in $\mathcal{H}$ if and only if $\mathcal{M}^{\perp} = \{0\}$.

In contrast with the Hilbert space case, the relation $\mathcal{M} + \mathcal{M}^{\perp} = \mathcal{H}$ may fail for a closed subspace $\mathcal{M}$ of a Kreĭn space $\mathcal{H}$. Moreover, a closed subspace $\mathcal{M}$ of a Kreĭn space $\mathcal{H}$ need not itself be a Kreĭn space in the scalar product of $\mathcal{H}$. These pathologies are excluded in an important class of subspaces. By a *regular subspace* of a Kreĭn space $\mathcal{H}$ we mean a closed subspace $\mathcal{M}$ of $\mathcal{H}$ which is a Kreĭn space in the scalar product of $\mathcal{H}$. An analogue of the projection theorem for Hilbert spaces holds for regular subspaces of a Kreĭn space.

THEOREM 1.1.1. *If $\mathcal{M}$ is a closed subspace of a Kreĭn space $\mathcal{H}$, the following assertions are equivalent:*

(i) $\mathcal{M}$ is regular;
(ii) $\mathcal{H} = \mathcal{M} \oplus \mathcal{M}^{\perp}$;
(iii) $\mathcal{M}$ is the range of a projection operator P.

In this case, if $\mathcal{M}$ is viewed as a Kreĭn space in the scalar product of $\mathcal{H}$, the strong topology of $\mathcal{M}$ coincides with the restriction of the strong topology of $\mathcal{H}$ to $\mathcal{M}$, and the inclusion of $\mathcal{M}$ in $\mathcal{H}$ is continuous.

We occasionally write $\mathrm{Pr}^{\mathcal{H}}_{\mathcal{M}}$ or simply $\mathrm{Pr}_{\mathcal{M}}$ for the projection operator on a Kreĭn space $\mathcal{H}$ whose range is the regular subspace $\mathcal{M}$. If P is a projection on $\mathcal{H}$ with range $\mathcal{M}$, then $1 - P$ is a projection with range $\mathcal{M}^{\perp}$. Therefore $\mathcal{M}^{\perp}$ is regular whenever $\mathcal{M}$ is regular. The class of regular subspaces of a Kreĭn space $\mathcal{H}$ is not in general closed under intersection and union.

A subspace $\mathcal{M}$ of a Kreĭn space $\mathcal{H}$ is

(i) *negative* if $\langle f, f\rangle_{\mathcal{H}} \leq 0$ for all f in $\mathcal{M}$,
(ii) *maximal negative* if $\mathcal{M}$ is negative and not a proper subset of another negative subspace,

(iii) *uniformly negative* if for some (and hence any) fundamental symmetry J on $\mathcal{H}$, there is a $\delta_J > 0$ such that

$$\langle f, f \rangle_{\mathcal{H}} \leq -\delta_J \|f\|_{\mathcal{H}_J}^2$$

for all f in $\mathcal{M}$, and

(iv) *maximal uniformly negative* if $\mathcal{M}$ is uniformly negative and not a proper subset of another uniformly negative subspace.

An equivalent form of (iv) is that $\mathcal{M}$ is

(iv′) *maximal uniformly negative* if $\mathcal{M}$ is maximal negative and uniformly negative.

By reversing the sense of inequalities of scalar products, we obtain parallel definitions for a subspace to be *positive*, *maximal positive*, *uniformly positive*, *and maximal uniformly positive*. A subspace which is either positive or negative is said to be *definite*. Maximal positive and maximal negative subspaces are closed.

Properties of definite subspaces are derived from a graph representation. Let $\mathcal{M}$ be a negative subspace of a Kreĭn space $\mathcal{H}$. Fix a fundamental decomposition $\mathcal{H} = \mathcal{H}_+ \oplus \mathcal{H}_-$ of $\mathcal{H}$. If $h = g + f$, $g \in \mathcal{H}_+$, $f \in \mathcal{H}_-$, we also write

$$h = \begin{pmatrix} g \\ f \end{pmatrix}.$$

In this representation, no nonzero element of $\mathcal{M}$ has the form $\binom{g}{0}$, and so $\mathcal{M}$ is the graph

$$\mathcal{G}(K) = \left\{ \begin{pmatrix} Kf \\ f \end{pmatrix} : f \in \operatorname{dom} K \right\}$$

of a Hilbert space contraction operator K, with domain $\operatorname{dom} K \subset |\mathcal{H}_-|$ and range $\operatorname{ran} K \subset \mathcal{H}_+$, which is called the *angle operator* for $\mathcal{M}$. Every Hilbert space contraction operator K with $\operatorname{dom} K \subset |\mathcal{H}_-|$ and $\operatorname{ran} K \subset \mathcal{H}_+$ is the angle operator of some negative subspace $\mathcal{M}$ of $\mathcal{H}$. The following properties are more or less immediate:

(i) $\mathcal{M}$ is closed if and only if $\operatorname{dom} K$ is closed in $|\mathcal{H}_-|$;

(ii) $\mathcal{M}$ is maximal negative if and only if $\operatorname{dom} K = |\mathcal{H}_-|$;

(iii) $\mathcal{M}$ is uniformly negative if and only if $\|K\| < 1$;

(iv) $\mathcal{M}$ is maximal uniformly negative if and only if both $\operatorname{dom} K = |\mathcal{H}_-|$ and $\|K\| < 1$.

Positive subspaces have a similar graph representation, and parallel results hold.

THEOREM 1.1.2. *Let $\mathcal{H}$ be a Kreĭn space.*

(i) A closed subspace $\mathcal{M}$ of $\mathcal{H}$ is maximal negative if and only if $\mathcal{M}^{\perp}$ is maximal positive in $\mathcal{H}$.

(ii) A closed subspace $\mathcal{M}$ of $\mathcal{H}$ is maximal uniformly negative if and only if $\mathcal{M}^{\perp}$ is maximal uniformly positive in $\mathcal{H}$.

(iii) Every negative subspace of $\mathcal{H}$ is contained in a maximal negative subspace of $\mathcal{H}$.

(iv) Every uniformly negative subspace of $\mathcal{H}$ is contained in a maximal uniformly negative subspace of $\mathcal{H}$.

Moreover, each of these statements remains true if the words "positive" and "negative" are interchanged.

The condition for a negative subspace $\mathcal{M}$ of a Kreĭn space $\mathcal{H}$ to be maximal negative is frequently used in this form: For some and hence any fundamental decomposition $\mathcal{H} = \mathcal{H}_+ \oplus \mathcal{H}_-$ of $\mathcal{H}$, $\mathrm{Pr}_{\mathcal{H}_-}\mathcal{M} = \mathcal{H}_-$.

A closed positive subspace of a Kreĭn space $\mathcal{H}$ is a Hilbert space if and only if it is uniformly positive. An example is $\mathcal{H}_+$ in any fundamental decomposition $\mathcal{H} = \mathcal{H}_+ \oplus \mathcal{H}_-$ of $\mathcal{H}$. This subspace is also maximal uniformly positive. Conversely, if $\mathcal{H}_+$ is a maximal uniformly positive subspace of a Kreĭn space $\mathcal{H}$ and $\mathcal{H}_- = \mathcal{H}_+^{\perp}$, then $\mathcal{H} = \mathcal{H}_+ \oplus \mathcal{H}_-$ is a fundamental decomposition of $\mathcal{H}$.

C. Isometries and Partial Isometries

The definition of a partial isometry in the Kreĭn space setting is similar to that for Hilbert spaces. Let $A \in \mathbf{B}(\mathcal{H}, \mathcal{K})$, where $\mathcal{H}$ and $\mathcal{K}$ are Kreĭn spaces. We call A a *partial isometry* if there exist regular subspaces $\mathcal{M}$ of $\mathcal{H}$ and $\mathcal{N}$ of $\mathcal{K}$ such that A maps $\mathcal{M}$ isometrically onto $\mathcal{N}$ and $\ker A = \mathcal{M}^{\perp}$. We call $\mathcal{M}$ the *initial space* and $\mathcal{N}$ the *final space* of A in this situation. The following properties are immediate.

(i) For any f and g in $\mathcal{H}$, the identity

$$\langle Af, Ag\rangle_{\mathcal{K}} = \langle f, g\rangle_{\mathcal{H}}$$

holds if either f or g is in $\mathcal{M}$.

(ii) The adjoint A^* of the operator A is a partial isometry with initial space $\mathcal{N}$ and final space $\mathcal{M}$.

(iii) If P is the projection of $\mathcal{H}$ on $\mathcal{M}$ and Q is the projection of $\mathcal{K}$ on $\mathcal{N}$, then $A^*A = P$, $\ker A = \ker P$, and $AA^* = Q$, $\ker A^* = \ker Q$.

It is not evident from the definition of a partial isometry that an isometry is a partial isometry. This is true and a consequence of the following nonspatial characterization of partial isometries.

THEOREM 1.1.3. *Let $A \in \mathbf{B}(\mathcal{H},\mathcal{K})$, where $\mathcal{H}$ and $\mathcal{K}$ are Kreĭn spaces. Then A is a partial isometry if and only if $AA^*A = A$.*

Proof. Suppose that A is a partial isometry with initial space $\mathcal{M}$ and final space $\mathcal{N}$. If f is in $\mathcal{M}^\perp = \ker A$, then

$$AA^*Af = 0 = Af.$$

If f is in $\mathcal{M}$, then $A^*Af = f$ by what was noted above, and hence again $AA^*Af = Af$. By linearity, $AA^*A = A$.

Conversely, assume that $AA^*A = A$, and set $P = A^*A$. Then P is selfadjoint and idempotent:

$$P^2 - P = A^*(AA^*A - A) = 0.$$

Hence P is a projection operator on $\mathcal{H}$. In a similar way, $A^*AA^* = A^*$ and $Q = AA^*$ is a projection operator on $\mathcal{K}$. By Theorem 1.1.1, $\mathcal{M} = P\mathcal{H}$ and $\mathcal{N} = Q\mathcal{K}$ are regular subspaces of $\mathcal{H}$ and $\mathcal{K}$. Now

$$A\mathcal{M} = AA^*A\mathcal{M} \subset AA^*\mathcal{K} = Q\mathcal{K} = \mathcal{N}$$

and

$$A^*\mathcal{N} = A^*AA^*\mathcal{N} \subset A^*A\mathcal{H} = P\mathcal{H} = \mathcal{M}.$$

Hence $A\mathcal{M} = \mathcal{N}$ and $A|\mathcal{M}$ maps $\mathcal{M}$ isometrically onto $\mathcal{N}$. It remains to show that $\ker A = \mathcal{M}^\perp$. Let $f \in \mathcal{H}$. If $Af = 0$, then $Pf = A^*Af = 0$, and so $f \in \mathcal{M}^\perp$. Conversely, if $f \in \mathcal{M}^\perp$, then $A^*Af = Pf = 0$ and $Af = AA^*Af = 0$. Thus A is a partial isometry. ■

COROLLARY 1.1.4. *Let $\mathcal{H}$ and $\mathcal{K}$ be Kreĭn spaces. An isometry $A \in \mathbf{B}(\mathcal{H},\mathcal{K})$ is a partial isometry with initial space $\mathcal{H}$. In particular, the range of an isometry $A \in \mathbf{B}(\mathcal{H},\mathcal{K})$ is a regular subspace of $\mathcal{K}$.*

Proof. If $A \in \mathbf{B}(\mathcal{H},\mathcal{K})$ is an isometry, then $A^*A = 1$ and so $AA^*A = A$. Therefore A is a partial isometry by Theorem 1.1.3. The initial space of A is $(A^*A)\mathcal{H} = \mathcal{H}$. ■

THEOREM 1.1.5. *Let $\mathcal{H}$ and $\mathcal{K}$ be Kreĭn spaces, and let $A \in \mathbf{B}(\mathcal{H},\mathcal{K})$ be an isometry.*

(i) If $\mathcal{M}$ is a closed uniformly positive or closed uniformly negative subspace of $\mathcal{H}$, $A\mathcal{M}$ is of the same type in $\mathcal{K}$.
(ii) If $\mathcal{H} = \mathcal{H}_+ \oplus \mathcal{H}_-$ is a fundamental decomposition of $\mathcal{H}$, then $A\mathcal{H} = A\mathcal{H}_+ \oplus A\mathcal{H}_-$ is a fundamental decomposition of $A\mathcal{H}$.

THEOREM 1.1.6. *Let $A \in \mathbf{B}(\mathcal{H}, \mathcal{K})$, where $\mathcal{H}$ and $\mathcal{K}$ are Kreĭn spaces. The following assertions are equivalent:*

(i) A is a partial isometry;
(ii) A^*A is a projection operator and $\ker A^*A = \ker A$;
(iii) AA^* is a projection operator and $\ker AA^* = \ker A^*$.

Proofs of these results are not difficult and omitted. An example clarifies the condition on kernels in parts (ii) and (iii) of Theorem 1.1.6. Let $\mathcal{H}$ and $\mathcal{K}$ each be the space of pairs $\binom{a}{b}$ of complex numbers with

$$\left\langle \begin{pmatrix} a \\ b \end{pmatrix}, \begin{pmatrix} a \\ b \end{pmatrix} \right\rangle = |a|^2 - |b|^2.$$

If $A = \begin{pmatrix} 1 & 1 \\ 1 & 1 \end{pmatrix}$, then $A^* = \begin{pmatrix} 1 & -1 \\ -1 & 1 \end{pmatrix}$, and $A^*A = 0$ is trivially a projection. But A is not a partial isometry because $\ker A^*A$ and $\ker A$ do not coincide.

In contrast with the Hilbert space case, densely defined isometries, that is, linear mappings which preserve scalar products, do not necessarily have continuous extensions to everywhere defined isometries. There even exist everywhere defined isometries on Kreĭn spaces which are not continuous (Bognár **[12]**, p. 125). A simple condition for continuity, given in the next result, is sometimes useful.

THEOREM 1.1.7. *Let C be a densely defined isometry from a Hilbert space $\mathcal{H}$ to a Kreĭn space $\mathcal{K}$. Then C has an extension to an operator $\hat{C} \in \mathbf{B}(\mathcal{H}, \mathcal{K})$ if and only if* $\operatorname{ran} C$ *is uniformly positive.*

Proof. Necessity follows from Theorem 1.1.5. Conversely, if $\operatorname{ran} C$ is uniformly positive, then $\overline{\operatorname{ran} C}$ is a Hilbert space in the scalar product of $\mathcal{K}$. We can extend C by continuity to a Hilbert space isometry $\hat{C}$ from $\mathcal{H}$ to $\overline{\operatorname{ran} C}$ viewed as a Hilbert space in the scalar product of $\mathcal{K}$. Since the inclusion of $\overline{\operatorname{ran} C}$ in $\mathcal{K}$ is continuous by Theorem 1.1.1, $\hat{C}$ is continuous as an operator on $\mathcal{H}$ to $\mathcal{K}$. ■

D. Orthogonal Direct Sums

Orthogonal direct sums of Kreĭn spaces are convenient for matrix representations of operators. If a Kreĭn space $\mathcal{H}$ is represented as an orthogonal direct sum $\mathcal{H} = \mathcal{H}_1 \oplus \mathcal{H}_2$, then for any f in $\mathcal{H}_1$ and g in $\mathcal{H}_2$, the expressions $h = \binom{f}{g}$ and $h = f + g$ are used interchangeably. If a second Kreĭn space $\mathcal{K}$ is represented as $\mathcal{K} = \mathcal{K}_1 \oplus \mathcal{K}_2$, then in the usual way any operator $A \in \mathbf{B}(\mathcal{H}, \mathcal{K})$ has a representation in matrix form

$$A = \begin{pmatrix} A_{11} & A_{12} \\ A_{21} & A_{22} \end{pmatrix}.$$

This representation extends to orthogonal direct sums with any finite number of summands, and the usual rules of matrix calculus are valid. In fact, we can extend the matrix calculus to orthogonal direct sums with countably many summands. We shall only indicate the appropriate definitions of external and internal orthogonal direct sums of sequences, leaving the rest to the reader.

Let $\mathcal{H}_1$, $\mathcal{H}_2$, ... be given Kreĭn spaces with fundamental decompositions $\mathcal{H}_1^+ \oplus \mathcal{H}_1^-$, $\mathcal{H}_2^+ \oplus \mathcal{H}_2^-$, ... Define

$$\mathcal{H} = \mathcal{H}_1 \oplus \mathcal{H}_2 \oplus \cdots$$

to be the space of sequences $f = \{f_1, f_2, \ldots\}$ such that $f_n \in \mathcal{H}_n$ for all n and

$$\sum_{n=1}^{\infty} \|f_n\|^2 < \infty,$$

where norms are computed relative to the given fundamental decompositions. If $f = \{f_1, f_2, \ldots\}$ and $g = \{g_1, g_2, \ldots\}$ are in $\mathcal{H}$, set

$$\langle f, g \rangle_{\mathcal{H}} = \sum_{n=1}^{\infty} \langle f_n, g_n \rangle_{\mathcal{H}_n}.$$

Then $\mathcal{H}$ is a Kreĭn space. We call $\mathcal{H}$ an *external orthogonal direct sum* or simply *direct sum* of $\mathcal{H}_1$, $\mathcal{H}_2$, ... For each n, $\mathcal{H}_n$ has a natural embedding in $\mathcal{H}$ as a regular subspace. If $P_n^{\pm}$ is the projection of $\mathcal{H}$ onto $\mathcal{H}_n^{\pm}$ for every n, then

$$\sup_n \left\| \sum_{k=1}^{n} P_k^+ \right\| < \infty \qquad \text{and} \qquad \sup_n \left\| \sum_{k=1}^{n} P_k^- \right\| < \infty,$$

where norms are computed relative to any fundamental decomposition of $\mathcal{H}$. It is sufficient to prove this for a definite choice of fundamental decomposition of $\mathcal{H}$. Choosing $\mathcal{H} = \mathcal{H}^+ \oplus \mathcal{H}^-$, where $\mathcal{H}^{\pm} = \mathcal{H}_1^{\pm} \oplus \mathcal{H}_2^{\pm} \oplus \cdots$, we have, in fact,

$$\left\| \sum_{k=1}^{n} P_k^{\pm} \right\| \le 1$$

for all n. It should be noted that the definition of $\mathcal{H}$ depends on the choice of fundamental decompositions for the summands. However, any two spaces obtained in this way with different choices of fundamental decompositions for the summands are naturally isomorphic.

There are several possibilities for the definition of an internal orthogonal direct sum. We use a strong hypothesis which results in a notion which is isomorphic with the previous definition of an external orthogonal direct sum.

THEOREM 1.1.8. *Let $\mathcal{H}$ be a Kreĭn space, and let $\mathcal{M}_1$, $\mathcal{M}_2$, ... be given pairwise orthogonal regular subspaces of $\mathcal{H}$ with fundamental decompositions $\mathcal{M}_1^+ \oplus \mathcal{M}_1^-$, $\mathcal{M}_2^+ \oplus \mathcal{M}_2^-$, ... Assume that the projections $P_1^\pm$, $P_2^\pm$, ... on the subspaces $\mathcal{M}_1^\pm$, $\mathcal{M}_2^\pm$, ... satisfy*

$$\sup_n \left\| \sum_{k=1}^n P_k^+ \right\| < \infty \qquad \text{and} \qquad \sup_n \left\| \sum_{k=1}^n P_k^- \right\| < \infty,$$

where norms are computed relative to any fundamental decomposition of $\mathcal{H}$. Then:

(i) The closed span $\mathcal{M}$ of $\mathcal{M}_1$, $\mathcal{M}_2$, ... is a regular subspace of $\mathcal{H}$. The closed span $\mathcal{M}^\pm$ of $\mathcal{M}_1^\pm$, $\mathcal{M}_2^\pm$, ... is uniformly positive with the plus signs and uniformly negative with the minus signs, and $\mathcal{M} = \mathcal{M}^+ \oplus \mathcal{M}^-$.

(ii) Every $f \in \mathcal{M}$ has a norm convergent representation $f = f_1 + f_2 + \cdots$, where $f_n \in \mathcal{M}_n$ for all n and $\sum_{n=1}^\infty \|f_n\|^2 < \infty$ with norms computed relative to any fundamental decomposition of $\mathcal{H}$. The representation is unique. Conversely, every square summable sequence of elements of $\mathcal{M}_1$, $\mathcal{M}_2$, ... determines an element of $\mathcal{M}$ in this way.

(iii) If P, $P^\pm$, $P_1^\pm$, $P_2^\pm$, ... are the projections of $\mathcal{H}$ on the subspaces $\mathcal{M}$, $\mathcal{M}^\pm$, $\mathcal{M}_1^\pm$, $\mathcal{M}_2^\pm$, ..., then $P = \sum_{n=1}^\infty P_n$ and $P^\pm = \sum_{n=1}^\infty P_n^\pm$ with convergence in the strong operator topology.

In this situation, we write $\mathcal{M} = \mathcal{M}_1 \oplus \mathcal{M}_2 \oplus \cdots$ and call $\mathcal{M}$ the *internal orthogonal direct sum* or *direct sum* of $\mathcal{M}_1$, $\mathcal{M}_2$, ... An internal orthogonal direct sum is defined whenever there exist projections satisfying the hypotheses of the theorem, but the direct sum itself is independent of the choice of projections. The role of the hypothesis on projections is to insure that the difference between internal and external orthogonal direct sums is essentially a notational one. Every internal orthogonal direct sum is naturally isomorphic to an external orthogonal direct sum, and conversely. In practice, the distinction between internal and external orthogonal direct sums is typically ignored.

LEMMA 1.1.9. *Let $\mathcal{N}_1$, $\mathcal{N}_2$, ... be regular subspaces of a Kreĭn space $\mathcal{H}$ such that $\mathcal{N}_1 \supset \mathcal{N}_2 \supset \cdots$. Assume that the projections Q_1, Q_2, ... of $\mathcal{H}$ onto the subspaces $\mathcal{N}_1$, $\mathcal{N}_2$, ... satisfy $\sup_n \|Q_n\| < \infty$, where norms are computed relative to any fundamental decomposition of $\mathcal{H}$. Then*

$$\mathcal{N} = \bigcap_1^\infty \mathcal{N}_n \qquad \text{and} \qquad \mathcal{M} = \bigvee_1^\infty \mathcal{N}_n^\perp$$

are regular subspaces of $\mathcal{H}$ with $\mathcal{N} = \mathcal{M}^\perp$. If Q is the projection of $\mathcal{H}$ onto $\mathcal{N}$, then $Q = \lim_{n\to\infty} Q_n$ with convergence in the strong operator topology.

Proof of Lemma 1.1.9. For any $f \in \mathcal{H}$, $Q_1 f$, $Q_2 f$, ... is a bounded sequence, and hence $g = \lim_{k\to\infty} Q_{n_k} f$ exists weakly for some subsequence. Elementary Hilbert space considerations show that $g \in \mathcal{N}$. Write $h = f - g$, so that

$$h = \lim_{k\to\infty} (1 - Q_{n_k}) f$$

weakly. Since

$$(1 - Q_{n_k}) f \in \mathcal{N}_{n_k}^\perp \subset \mathcal{N}^\perp$$

for every k, we have $h \in \mathcal{N}^\perp$. Thus $f = g + h$ with $g \in \mathcal{N}$ and $h \in \mathcal{N}^\perp$. Therefore $\mathcal{N} + \mathcal{N}^\perp = \mathcal{H}$. This implies that $\mathcal{N} \cap \mathcal{N}^\perp = \{0\}$, because any vector in $\mathcal{N} \cap \mathcal{N}^\perp$ is orthogonal to $\mathcal{N} + \mathcal{N}^\perp = \mathcal{H}$. By Theorem 1.1.1, $\mathcal{N}$ is a regular subspace of $\mathcal{H}$.

In the preceding construction, suppose that $f \in \mathcal{N}^\perp$. Then $g = 0$ because $\mathcal{H} = \mathcal{N} \oplus \mathcal{N}^\perp$. Therefore

$$f = h = \lim_{k\to\infty} (1 - Q_{n_k}) f$$

belongs to $\mathcal{M} = \vee_1^\infty \mathcal{N}_n^\perp$, and so $\mathcal{N}^\perp \subset \mathcal{M}$. The reverse inclusion is obvious, and so $\mathcal{N}^\perp = \mathcal{M}$. In particular, $\mathcal{M}$ is a regular subspace of $\mathcal{H}$. We clearly have

$$\lim_{n\to\infty} Q_n f = Qf$$

if f is in the span of $\mathcal{N}$, $\mathcal{N}_1^\perp$, $\mathcal{N}_2^\perp$, ... Since this span is dense in $\mathcal{H}$ and the projections Q_1, Q_2, ... are bounded in norm, the relation holds for all $f \in \mathcal{H}$ by a routine approximation. ■

Proof of Theorem 1.1.8. For each $n \geq 1$, let $\mathcal{N}_n$ be the orthogonal complement of $\mathcal{M}_1 \oplus \cdots \oplus \mathcal{M}_n$ in $\mathcal{H}$. The projection Q_n of $\mathcal{H}$ onto $\mathcal{N}_n$ is $Q_n = 1 - P_1 - \cdots - P_n$. By Lemma 1.1.9,

$$\mathcal{N} = \bigcap_1^\infty \mathcal{N}_n \qquad \text{and} \qquad \mathcal{M} = \bigvee_1^\infty \mathcal{N}_n^\perp$$

are regular subspaces of $\mathcal{H}$ with $\mathcal{N} = \mathcal{M}^{\perp}$. If Q is the projection of $\mathcal{H}$ onto $\mathcal{N}$, then $Q = \lim_{n\to\infty} Q_n$ in the strong operator topology. By the definition of the subspaces $\mathcal{N}_1, \mathcal{N}_2, \ldots$, $\mathcal{M}$ is the closed span of $\mathcal{M}_1, \mathcal{M}_2, \ldots$. Moreover, if P is the projection of $\mathcal{H}$ onto $\mathcal{M}$, then

$$P = 1 - Q = \lim_{n\to\infty}(1 - Q_n) = \lim_{n\to\infty}\sum_{k=1}^{n} P_k$$

with convergence in the strong operator topology.

We repeat the construction with $\mathcal{M}_1, \mathcal{M}_2, \ldots$ replaced by $\mathcal{M}_1^{\pm}, \mathcal{M}_2^{\pm}, \ldots$. Thus the closed span of $\mathcal{M}_1^{\pm}, \mathcal{M}_2^{\pm}, \ldots$ is a regular subspace $\mathcal{M}^{\pm}$ of $\mathcal{H}$. Since $\mathcal{M}^{\pm}$ is positive/negative and regular, $\mathcal{M}^{\pm}$ is uniformly positive/uniformly negative. If $P^{\pm}$ is the projection of $\mathcal{H}$ onto $\mathcal{M}^{\pm}$, then

$$P^{\pm} = \lim_{n\to\infty}\sum_{k=1}^{n} P_k^{\pm}$$

with convergence in the strong operator topology. Clearly $P = P^+ + P^-$ and $P^+P^- = P^-P^+ = 0$. Thus $\mathcal{M} = \mathcal{M}^+ \oplus \mathcal{M}^-$. The assertions (i) and (iii) are now proved.

The assertion (ii) is independent of the choice of fundamental decomposition of $\mathcal{H}$. It is convenient to choose a fundamental decomposition $\mathcal{H} = \mathcal{H}_+ \oplus \mathcal{H}_-$ of $\mathcal{H}$ such that $\mathcal{H}_{\pm} \supset \mathcal{M}^{\pm}$. Then the subspaces $\mathcal{M}_1^+, \mathcal{M}_2^+, \ldots$ and $\mathcal{M}_1^-, \mathcal{M}_2^-, \ldots$ are pairwise orthogonal in both the Kreĭn space sense and the Hilbert space sense. In this situation, (ii) is clear. ∎

1.2 Defect Operators and Julia Operators

As preparation for the study of operators on Kreĭn spaces, we consider two related problems. One is to represent a given selfadjoint operator H in the form $H = AA^*$, where A is an operator with zero kernel. The other is to embed any given operator T in a unitary matrix $U = \begin{pmatrix} T & * \\ * & * \end{pmatrix}$. These constructions are needed to circumvent certain Hilbert space notions which do not have adequate counterparts in the indefinite case.

Let $\mathcal{H}$ be a Kreĭn space, and let $H \in \mathbf{B}(\mathcal{H})$ be a selfadjoint operator. By $h_+(H)$ we mean the supremum of all $r = 1, 2, \ldots$ such that there exists a nonnegative and invertible matrix of the form

$$\left[\langle Hf_j, f_k\rangle_{\mathcal{H}}\right]_{j,k=1}^{r}, \qquad f_1, \ldots, f_r \in \mathcal{H}.$$

Set $h_+(H) = 0$ if no such r exists, and $h_-(H) = h_+(-H)$. We call $h_{\pm}(H)$ the *positive* and *negative hermitian indices* of H. Notice that $H \geq 0$ if and only if $h_-(H) = 0$, and $H \leq 0$ if and only if $h_+(H) = 0$.

THEOREM 1.2.1. *Let $\mathcal{A}$ and $\mathcal{H}$ be Kreĭn spaces, $\mathcal{A}$ separable, and let $A \in \mathbf{B}(\mathcal{A}, \mathcal{H})$. If A has zero kernel, then the positive and negative indices of $\mathcal{A}$ coincide with $h_+(AA^*)$ and $h_-(AA^*)$, respectively.*

Proof. Let $\mathcal{A} = \mathcal{A}_+ \oplus \mathcal{A}_-$ be a fundamental decomposition. It is sufficient to show that $h_+(AA^*) = \dim \mathcal{A}_+$, since we then obtain $h_-(AA^*) = \dim \mathcal{A}_-$ by suitably reversing signs.

The dimension of $\mathcal{A}_+$ is the supremum of all $r = 1, 2, \ldots$ such that there exists a nonnegative and invertible matrix of the form

$$\left[\langle g_j, g_k \rangle_{\mathcal{A}}\right]_{j,k=1}^r, \qquad g_1, \ldots, g_r \in \mathcal{A},$$

and zero if no such r exists. If $f_1, \ldots, f_r \in \mathcal{H}$,

$$\left[\langle AA^* f_j, f_k \rangle_{\mathcal{H}}\right]_{j,k=1}^r = \left[\langle A^* f_j, A^* f_k \rangle_{\mathcal{A}}\right]_{j,k=1}^r,$$

and so $h_+(AA^*) \leq \dim \mathcal{A}_+$.

To see that equality holds, consider vectors $g_1, \ldots, g_r$ in $\mathcal{A}$ such that

$$\left[\langle g_j, g_k \rangle_{\mathcal{A}}\right]_{j,k=1}^r$$

is nonnegative and invertible. Since $\ker A = \{0\}$, the range of A^* is dense in $\mathcal{A}$, and there exist vectors $f_{1n}, \ldots, f_{rn}$ in $\mathcal{H}$ such that

$$\lim_{n \to \infty} A^* f_{jn} = g_j, \qquad j = 1, \ldots, r.$$

Hence for all sufficiently large n, the matrix

$$\left[\langle A^* f_{jn}, A^* f_{kn} \rangle_{\mathcal{A}}\right]_{j,k=1}^r = \left[\langle AA^* f_{jn}, f_{kn} \rangle_{\mathcal{H}}\right]_{j,k=1}^r$$

is nonnegative and invertible, and so $h_+(AA^*) \geq \dim \mathcal{A}_+$. Thus equality holds. ∎

THEOREM 1.2.2. *Let $\mathcal{H}$ be a Kreĭn space, and let $H \in \mathbf{B}(\mathcal{H})$ be a selfadjoint operator. Then there is a Kreĭn space $\mathcal{A}$ and an operator $A \in \mathbf{B}(\mathcal{A}, \mathcal{H})$ with zero kernel such that $H = AA^*$.*

Proof. Let $J_{\mathcal{H}}$ be the signature operator for some fundamental decomposition $\mathcal{H} = \mathcal{H}_+ \oplus \mathcal{H}_-$. Then $HJ_{\mathcal{H}}$ is a selfadjoint operator on the Hilbert space $\mathcal{H}_+ \oplus |\mathcal{H}_-|$. Let its spectral decomposition be $HJ_{\mathcal{H}} = \int \lambda dE(\lambda)$, and set $R = \int |\lambda|^{1/2} dE(\lambda)$, $\mathcal{M}_+ = E((0, \infty))$, and $\mathcal{M}_- = E((-\infty, 0))$.

Let $\mathcal{A}$ be a Kreĭn space with fundamental decomposition $\mathcal{A} = \mathcal{A}_+ \oplus \mathcal{A}_-$ and signature operator $J_{\mathcal{A}}$ such that $\dim \mathcal{A}_\pm = \dim \mathcal{M}_\pm$. Choose an isometry

W on $\mathcal{A}_+ \oplus |\mathcal{A}_-|$ to $\mathcal{H}_+ \oplus |\mathcal{H}_-|$ such that $W\mathcal{A}_\pm = \mathcal{M}_\pm$. Define $A \in \mathbf{B}(\mathcal{A},\mathcal{H})$ by $A = RW$. Then $\ker A = \{0\}$, and $AA^* = RWW^*R^* = (RWJ_{\mathcal{A}}W^\times R)J_{\mathcal{H}} = (HJ_{\mathcal{H}})J_{\mathcal{H}} = H$. ■

DEFINITION 1.2.3. *Let $T \in \mathbf{B}(\mathcal{H},\mathcal{K})$, $\mathcal{H}$ and $\mathcal{K}$ Kreĭn spaces.*

(i) By a **defect operator** *for T we mean an operator $\tilde{D} \in \mathbf{B}(\tilde{\mathcal{D}},\mathcal{H})$, where $\tilde{\mathcal{D}}$ a Kreĭn space, such that $\tilde{D}$ has zero kernel and $1 - T^*T = \tilde{D}\tilde{D}^*$.*

(ii) By a **Julia operator** *for T we mean a unitary operator U having the form*

$$U = \begin{pmatrix} T & D \\ \tilde{D}^* & L \end{pmatrix} \in \mathbf{B}(\mathcal{H} \oplus D, \mathcal{K} \oplus \tilde{\mathcal{D}}),$$

where $\mathcal{D}$ and $\tilde{\mathcal{D}}$ are Kreĭn spaces, $D \in \mathbf{B}(\mathcal{D},\mathcal{K})$ and $\tilde{D} \in \mathbf{B}(\tilde{\mathcal{D}},\mathcal{H})$ have zero kernels, and $L \in \mathbf{B}(\mathcal{D},\tilde{\mathcal{D}})$.

In (i), the positive and negative indices of $\tilde{\mathcal{D}}$ are determined by T and given by $h_\pm(1-T^*T)$ when $\tilde{\mathcal{D}}$ is separable by Theorem 1.2.1. In (ii), $\tilde{D}$ is a defect operator for T and D is a defect operator for T^*. Hence in (ii), the positive and negative indices of $\mathcal{D}$ and $\tilde{\mathcal{D}}$ are determined by T and coincide when these spaces are separable with $h_\pm(1 - TT^*)$ and $h_\pm(1 - T^*T)$, respectively. It is also easy to see that if U is a Julia operator for T, then U^* is a Julia operator for T^*.

For an example, let $\mathcal{H}$ and $\mathcal{K}$ be Kreĭn spaces, and let $T \in \mathbf{B}(\mathcal{H},\mathcal{K})$ be a partial isometry with initial space $\mathcal{M}$ and final space $\mathcal{N}$. Choose $\mathcal{D} = \mathcal{N}^\perp$ and $\tilde{\mathcal{D}} = \mathcal{M}^\perp$ in the scalar products of $\mathcal{K}$ and $\mathcal{H}$, respectively. Define $D \in \mathbf{B}(\mathcal{D},\mathcal{K})$ and $\tilde{D} \in \mathbf{B}(\tilde{\mathcal{D}},\mathcal{H})$ to be the inclusion mappings, and let $L \in \mathbf{B}(\mathcal{D},\tilde{\mathcal{D}})$ be the zero operator. Then $\left(\begin{smallmatrix} T & D \\ \tilde{D}^* & L \end{smallmatrix}\right) \in \mathbf{B}(\mathcal{H} \oplus D, \mathcal{K} \oplus \tilde{\mathcal{D}})$ is a Julia operator for T.

Theorem 1.2.2 insures the existence of a defect operator for any given operator T. As we show next, it also implies the existence of a Julia operator for any given operator T.

THEOREM 1.2.4. *Let $\mathcal{H}$ and $\mathcal{K}$ be Kreĭn spaces, and let $T \in \mathbf{B}(\mathcal{H},\mathcal{K})$. If $\tilde{D} \in \mathbf{B}(\tilde{\mathcal{D}},\mathcal{H})$ is a defect operator for T, there exists a Julia operator of the form*

$$U = \begin{pmatrix} T & D \\ \tilde{D}^* & L \end{pmatrix} \in \mathbf{B}(\mathcal{H} \oplus D, \mathcal{K} \oplus \tilde{\mathcal{D}}).$$

Proof. We seek a Kreĭn space $\mathcal{D}$ and operators $D \in \mathbf{B}(\mathcal{D},\mathcal{K})$ and $L \in \mathbf{B}(\mathcal{D},\tilde{\mathcal{D}})$ such that D has zero kernel and $U = \left(\begin{smallmatrix} T & D \\ \tilde{D}^* & L \end{smallmatrix}\right)$ is unitary, that is,

$$\left.\begin{aligned} T^*T + \tilde{D}\tilde{D}^* &= 1, \\ T^*D + \tilde{D}L &= 0, \\ D^*D + L^*L &= 1, \end{aligned}\right\} \qquad (1.2.1a, b, c)$$

and

$$\left.\begin{aligned} TT^* + DD^* &= 1, \\ \tilde{D}^*T^* + LD^* &= 0, \\ \tilde{D}^*\tilde{D} + LL^* &= 1. \end{aligned}\right\} \qquad (1.2.2a, b, c)$$

The relation (1.2.1a) holds by the assumption that $\tilde{D} \in \mathbf{B}(\tilde{\mathcal{D}}, \mathcal{H})$ is a defect operator for T. It implies that

$$V = \begin{pmatrix} T \\ \tilde{D}^* \end{pmatrix} \in \mathbf{B}(\mathcal{H}, \mathcal{K} \oplus \tilde{\mathcal{D}})$$

is an isometry. Then $1 - VV^*$ is the projection onto $\ker V^*$. Factor $1 - VV^* = BB^*$, where $B \in \mathbf{B}(\mathcal{D}, \mathcal{K} \oplus \tilde{\mathcal{D}})$ for some Kreĭn space $\mathcal{D}$ and $\ker B = \{0\}$. Since BB^* is a projection and $\ker B^* = \ker BB^*$, B is a partial isometry by Theorem 1.1.6. In fact, it is an isometry with range $\ker V^*$. Thus $V^*B = 0$ and $B^*B = 1$.

If we write elements of $\mathcal{K} \oplus \tilde{\mathcal{D}}$ in column form, then

$$B = \begin{pmatrix} D \\ L \end{pmatrix},$$

where $D \in \mathbf{B}(\mathcal{D}, \mathcal{K})$ and $L \in \mathbf{B}(\mathcal{D}, \tilde{\mathcal{D}})$. The relations (1.2.1b,c) follow from $V^*B = 0$ and $B^*B = 1$. The relations (1.2.2a,b,c) follow from $1 - VV^* = BB^*$. It remains to verify that D has zero kernel. If $Df = 0$ for some vector f in $\mathcal{D}$, then $Lf = 0$ by (1.2.1b) because $\tilde{D}$ has zero kernel by assumption. Hence $Bf = 0$ and $f = 0$. ■

1.3 Contraction and Bicontraction Operators

Let $\mathcal{H}$ and $\mathcal{K}$ be Kreĭn spaces. An operator $T \in \mathbf{B}(\mathcal{H}, \mathcal{K})$ is said to be a *contraction* if

$$\langle Tf, Tf\rangle_{\mathcal{K}} \leq \langle f, f\rangle_{\mathcal{H}}, \qquad f \in \mathcal{H},$$

and a *bicontraction* if both T and T^* are contractions. An example of a contraction which is not a bicontraction is the embedding of a Hilbert space into its direct sum with the anti-space of a nonzero Hilbert space. The structural properties of contractions depend on the way in which such operators map negative subspaces.

Theorem 1.3.1. *Let $\mathcal{H}$ and $\mathcal{K}$ be Kreĭn spaces, and let $T \in \mathbf{B}(\mathcal{H}, \mathcal{K})$ be a contraction. Choose fundamental decompositions for $\mathcal{H}$ and $\mathcal{K}$, and let norms be computed with respect to the associated Hilbert spaces. Set*

$$\delta = \left\{\|T\| + \left[1 + \|T\|^2\right]^{1/2}\right\}^{-1}.$$

(i) For any $f \in \mathcal{H}$ with $\langle f, f\rangle_{\mathcal{H}} \leq 0$, $\|Tf\| \geq \delta\|f\|$.
(ii) The kernel of T is a closed uniformly positive subspace of $\mathcal{H}$.

Proof. Let J be the chosen fundamental symmetry on $\mathcal{H}$. Since T is a contraction, $1 - T^*T \geq 0$ in the partial ordering of selfadjoint operators on $\mathcal{H}$. Therefore $C = J(1 - T^*T)$ is nonnegative as an operator on $\mathcal{H}_J$. For any $f \in \mathcal{H}$,

$$\|f\| - \|T\|\|Tf\| \leq \|Cf\| \leq \|C\|^{1/2}\langle Cf, f\rangle_{\mathcal{H}_J}^{1/2} \leq \left[1 + \|T\|^2\right]^{1/2}\langle Cf, f\rangle_{\mathcal{H}_J}^{1/2}.$$

If $\langle f, f\rangle_{\mathcal{H}} \leq 0$, then also

$$\langle Cf, f\rangle_{\mathcal{H}_J} = \langle f, f\rangle_{\mathcal{H}} - \langle Tf, Tf\rangle_{\mathcal{K}} \leq -\langle Tf, Tf\rangle_{\mathcal{K}} \leq \|Tf\|^2.$$

Combining these inequalities, we obtain

$$\|f\| - \|T\|\|Tf\| \leq \left[1 + \|T\|^2\right]^{1/2}\|Tf\|,$$

which proves (i). If $g \in \ker T$, then $Cg = J(1 - T^*T)g = Jg$ and

$$\|g\|^2 = \|Cg\|^2 \leq \|C\|\langle Cg, g\rangle_{\mathcal{H}_J} = \|C\|\langle g, g\rangle_{\mathcal{H}}.$$

Thus (ii) follows. ■

COROLLARY 1.3.2. *Let $\mathcal{H}$ and $\mathcal{K}$ be Kreĭn spaces, and let $T \in \mathbf{B}(\mathcal{H}, \mathcal{K})$ be a contraction. Then*

(i) *T maps any closed negative subspace of $\mathcal{H}$ in a one-to-one way onto a closed negative subspace of $\mathcal{K}$, and*

(ii) *T maps any closed uniformly negative subspace of $\mathcal{H}$ in a one-to-one way onto a closed uniformly negative subspace of $\mathcal{K}$.*

Proof. Part (i) follows from Theorem 1.3.1. To prove (ii), we must show that if $\mathcal{M}$ is a uniformly negative subspace of $\mathcal{H}$, then $T\mathcal{M}$ is a uniformly negative subspace of $\mathcal{K}$.

Assume that fundamental symmetries are chosen for $\mathcal{H}$ and $\mathcal{K}$ and norms are computed with respect to the associated Hilbert spaces. If $\mathcal{M}$ is uniformly negative in $\mathcal{H}$, there is an $\eta > 0$ such that

$$\langle f, f \rangle_{\mathcal{H}} \leq -\eta \|f\|^2, \qquad f \in \mathcal{M}.$$

Hence for $f \in \mathcal{M}$,

$$\langle Tf, Tf \rangle_{\mathcal{K}} \leq \langle f, f \rangle_{\mathcal{H}} \leq -\eta \|f\|^2 \leq -\eta \|T\|^{-2} \|Tf\|^2,$$

and $T\mathcal{M}$ is uniformly negative in $\mathcal{K}$. This proves (ii). ∎

The main results on contraction operators are derived using a scattering formalism. The formalism uses two Kreĭn spaces $\mathcal{H}$ and $\mathcal{K}$ and fundamental decompositions $\mathcal{H} = \mathcal{H}_+ \oplus \mathcal{H}_-$ and $\mathcal{K} = \mathcal{K}_+ \oplus \mathcal{K}_-$. In addition, let $\mathcal{L}$ be the direct sum of the anti-space of $\mathcal{H}$ together with $\mathcal{K}$. Thus $\mathcal{L}$ is the space of pairs $\binom{f}{g}$ with $f \in \mathcal{H}$, $g \in \mathcal{K}$, and

$$\left\langle \begin{pmatrix} f \\ g \end{pmatrix}, \begin{pmatrix} f \\ g \end{pmatrix} \right\rangle_{\mathcal{L}} = -\langle f, f \rangle_{\mathcal{H}} + \langle g, g \rangle_{\mathcal{K}}.$$

We use the fundamental decomposition of $\mathcal{L}$ given by $\mathcal{L} = \mathcal{L}_+ \oplus \mathcal{L}_-$,

$$\mathcal{L}_+ = \mathcal{K}_+ \oplus \mathcal{H}_- \qquad \text{and} \qquad \mathcal{L}_- = \mathcal{H}_+ \oplus \mathcal{K}_-.$$

The elements of $\mathcal{L}$ are also represented as pairs $\binom{v}{u}$ with v in $\mathcal{L}_+$ and u in $\mathcal{L}_-$.

If $T \in \mathbf{B}(\mathcal{H}, \mathcal{K})$ is a contraction, its graph

$$\mathcal{G}(T) = \left\{ \begin{pmatrix} f \\ Tf \end{pmatrix} : f \in \mathcal{H} \right\}$$

is a closed negative subspace of $\mathcal{L}$. It is negative because for any $f \in \mathcal{H}$,

$$\left\langle \begin{pmatrix} f \\ Tf \end{pmatrix}, \begin{pmatrix} f \\ Tf \end{pmatrix} \right\rangle_{\mathcal{L}} = -\langle f, f \rangle_{\mathcal{H}} + \langle Tf, Tf \rangle_{\mathcal{K}} \leq 0.$$

It is closed because the domain of T is closed. By the *scattering operator* or *Potapov-Ginsburg transform* of T we mean the angle operator S for $\mathcal{G}(T)$ relative to the fundamental decomposition $\mathcal{L} = \mathcal{L}_+ \oplus \mathcal{L}_-$. Thus

$$\mathcal{G}(T) = \left\{ \begin{pmatrix} Su \\ u \end{pmatrix} : u \in \operatorname{dom} S \right\} \subset \mathcal{L}_+ \oplus \mathcal{L}_-.$$

By construction, the scattering operator S is a Hilbert space contraction with closed domain $\operatorname{dom} S \subset |\mathcal{L}_-|$ and $\operatorname{ran} S \subset \mathcal{L}_+$.

Write T in matrix form

$$T = \begin{pmatrix} T_{11} & T_{12} \\ T_{21} & T_{22} \end{pmatrix} \in \mathbf{B}(\mathcal{H}_+ \oplus \mathcal{H}_-, \mathcal{K}_+ \oplus \mathcal{K}_-). \tag{1.3.1}$$

Denote the signature operators and projections for the given fundamental decompositions of $\mathcal{H}$ and $\mathcal{K}$ by $J_{\mathcal{H}}$, $J_{\mathcal{K}}$ and

$$P_\pm : \mathcal{H} \to \mathcal{H}_\pm, \qquad Q_\pm : \mathcal{K} \to \mathcal{K}_\pm. \tag{1.3.2}$$

Define operators

$$Q_+T + P_- = \begin{pmatrix} T_{11} & T_{12} \\ 0 & 1 \end{pmatrix} \in \mathbf{B}(\mathcal{H}_+ \oplus \mathcal{H}_-, \mathcal{K}_+ \oplus \mathcal{H}_-), \tag{1.3.3}$$

and

$$P_+ + Q_-T = \begin{pmatrix} 1 & 0 \\ T_{21} & T_{22} \end{pmatrix} \in \mathbf{B}(\mathcal{H}_+ \oplus \mathcal{H}_-, \mathcal{H}_+ \oplus \mathcal{K}_-). \tag{1.3.4}$$

The preceding notation is assumed in Theorems 1.3.3–1.3.5.

THEOREM 1.3.3. *Let $T \in \mathbf{B}(\mathcal{H}, \mathcal{K})$ be a contraction with scattering operator S. The domain of S is equal to* $\operatorname{ran}(P_+ + Q_-T)$, *and*

$$S = (Q_+T + P_-)(P_+ + Q_-T)^{-1}|\operatorname{dom} S.$$

Proof. We show that $P_+ + Q_-T$ is one-to-one and has closed range. First note that T_{22} is contraction on $\mathcal{H}_-$ to $\mathcal{K}_-$. For if $f \in \mathcal{H}_-$, then

$$\langle T_{22}f, T_{22}f\rangle_{\mathcal{K}_-} = \langle Q_-Tf, Q_-Tf\rangle_{\mathcal{K}} \le \langle Tf, Tf\rangle_{\mathcal{K}} \le \langle f, f\rangle_{\mathcal{H}} = \langle f, f\rangle_{\mathcal{H}_-}.$$

By Theorem 1.3.1, T_{22} is one-to-one and has closed range. By (1.3.4), $P_+ + Q_-T$ is one-to-one and has closed range.

For any $f \in \mathcal{H}$,

$$\Pr_{\mathcal{L}_-}\begin{pmatrix} f \\ Tf \end{pmatrix} = \begin{pmatrix} P_+ f \\ Q_- T(P_+ + P_-)f \end{pmatrix} = \begin{pmatrix} 1 & 0 \\ T_{21} & T_{22} \end{pmatrix}\begin{pmatrix} P_+ f \\ P_- f \end{pmatrix} = (P_+ + Q_- T)f$$

and

$$\Pr_{\mathcal{L}_+}\begin{pmatrix} f \\ Tf \end{pmatrix} = \begin{pmatrix} Q_+ T(P_+ + P_-)f \\ P_- f \end{pmatrix} = \begin{pmatrix} T_{11} & T_{12} \\ 0 & 1 \end{pmatrix}\begin{pmatrix} P_+ f \\ P_- f \end{pmatrix} = (Q_+ T + P_-)f.$$

The computation of S is immediate from these identities and the definition of the angle operator. ■

The scattering operator S models a system in which a left input f_1 is partially transmitted in f_2, partially reflected in g_1, and partially absorbed. A right input g_2 is partially transmitted in g_1, partially reflected in f_2, and partially absorbed.

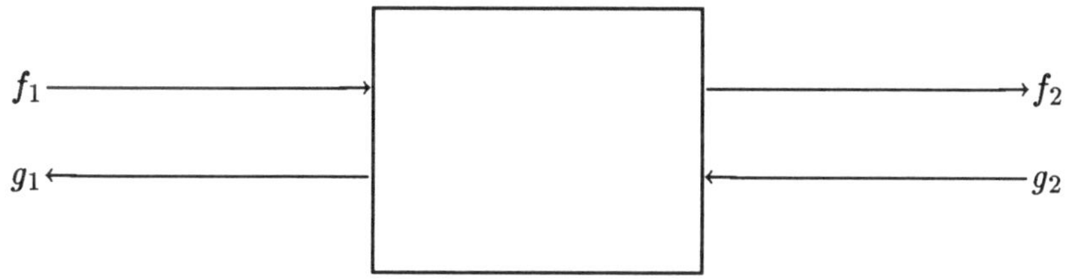

Here f_1, f_2, g_1, g_2 belong to $\mathcal{H}_+$, $\mathcal{K}_+$, $|\mathcal{H}_-|$, $|\mathcal{K}_-|$, respectively. The matrix entries in

$$S = \begin{pmatrix} T_\ell & R_r \\ R_\ell & T_r \end{pmatrix},$$

represent left and right transmission and reflection coefficients. The scattering operator

$$S : \begin{pmatrix} f_1 \\ g_2 \end{pmatrix} \to \begin{pmatrix} f_2 \\ g_1 \end{pmatrix}$$

associates net output to net input. The original operator

$$T : \begin{pmatrix} f_1 \\ g_1 \end{pmatrix} \to \begin{pmatrix} f_2 \\ g_2 \end{pmatrix}$$

is a *cascade scattering operator* for the system. *Passivity* of the system means that energy is not increased, or, what is the same thing, S is contractive in the Hilbert space sense:

$$\|f_2\|^2 + \|g_1\|^2 \leq \|f_1\|^2 + \|g_2\|^2.$$

This is equivalent to the inequality

$$\|f_2\|^2 - \|g_2\|^2 \le \|f_1\|^2 - \|g_1\|^2,$$

which holds because T is a contraction. When $\operatorname{dom} S \neq |\mathcal{L}_-|$, only certain inputs and outputs occur in the system.

The scattering formalism is particularly simple when $\operatorname{dom} S = |\mathcal{L}_-|$. In this case, from Theorem 1.3.3 and the identities (1.3.3) and (1.3.4), we see that T_{22} is invertible and

$$S = \begin{pmatrix} T_{11} - T_{12}T_{22}^{-1}T_{21} & T_{12}T_{22}^{-1} \\ -T_{22}^{-1}T_{21} & T_{22}^{-1} \end{pmatrix} \in \mathbf{B}(\mathcal{H}_+ \oplus |\mathcal{K}_-|, \mathcal{K}_+ \oplus |\mathcal{H}_-|). \tag{1.3.5}$$

Identities involving the scattering operator can therefore be reduced to straightforward matrix calculations.

THEOREM 1.3.4. *Let $T \in \mathbf{B}(\mathcal{H}, \mathcal{K})$ be a contraction with scattering operator S. The following assertions are equivalent:*

(i) T is a bicontraction;
(ii) the entry T_{22} in (1.3.1) is invertible;
(iii) $\operatorname{dom} S = |\mathcal{L}_-|$.

If T is a bicontraction, then

$$T = (Q_+S + Q_-)(P_+ + P_-S)^{-1}, \tag{1.3.6}$$

and the scattering operator for T^* is $S^\times$. Moreover, in this case,

$$T^* = (P_+S^\times + P_-)(Q_+ + Q_-S^\times)^{-1} \tag{1.3.7}$$

and

$$1 - T^*T = J_{\mathcal{H}}(P_+ + Q_-T)^\times(1 - S^\times S)(P_+ + Q_-T), \tag{1.3.8}$$

$$1 - TT^* = J_{\mathcal{K}}(Q_+ + P_-T^*)^\times(1 - SS^\times)(Q_+ + P_-T^*). \tag{1.3.9}$$

Recall that $*$ indicates Kreĭn space adjoint and $\times$ Hilbert space adjoint. Notation in (1.3.6)–(1.3.9) is similar to that used in (1.3.3) and (1.3.4). Explicitly, if

$$S = \begin{pmatrix} S_{11} & S_{12} \\ S_{21} & S_{22} \end{pmatrix},$$

then

$$Q_+S + Q_- = \begin{pmatrix} S_{11} & S_{12} \\ 0 & 1 \end{pmatrix} \in \mathbf{B}(\mathcal{H}_+ \oplus \mathcal{K}_-, \mathcal{K}_+ \oplus \mathcal{K}_-),$$

$$P_+ + P_-S = \begin{pmatrix} 1 & 0 \\ S_{21} & S_{22} \end{pmatrix} \in \mathbf{B}(\mathcal{H}_+ \oplus \mathcal{K}_-, \mathcal{H}_+ \oplus \mathcal{H}_-),$$

and,

$$P_+S^\times + P_- = \begin{pmatrix} S_{11}^\times & S_{21}^\times \\ 0 & 1 \end{pmatrix} \in \mathbf{B}(\mathcal{K}_+ \oplus \mathcal{H}_-, \mathcal{H}_+ \oplus \mathcal{H}_-),$$

$$Q_+ + Q_-S^\times = \begin{pmatrix} 1 & 0 \\ S_{12}^\times & S_{22}^\times \end{pmatrix} \in \mathbf{B}(\mathcal{K}_+ \oplus \mathcal{H}_-, \mathcal{K}_+ \oplus \mathcal{K}_-).$$

Proof of Theorem 1.3.4. (i) $\Rightarrow$ (ii) Assume that T is a bicontraction. As in the proof of Theorem 1.3.3, $T_{22} = Q_-TP_-|\mathcal{H}_-$ is one-to-one and has closed range as an operator on $\mathcal{H}_-$ to $\mathcal{K}_-$. Suppose that f is in $\mathcal{K}_-$ and $f \perp Q_-TP_-\mathcal{H}_-$. Then $f \perp T\mathcal{H}_-$ and $T^*f \perp \mathcal{H}_-$. So $T^*f \in \mathcal{H}_+$, and since T is bicontractive,

$$\langle T^*f, T^*f\rangle_{\mathcal{H}} \geq 0 \geq \langle f, f\rangle_{\mathcal{K}} \geq \langle T^*f, T^*f\rangle_{\mathcal{H}}.$$

Therefore equality holds throughout, and since $f \in \mathcal{K}_-$ we obtain $f = 0$. Hence $\operatorname{ran} T_{22} = \mathcal{K}_-$, and T_{22} is invertible.

(ii) $\Rightarrow$ (iii) If T_{22} is invertible, so is the operator (1.3.4). Then by Theorem 1.3.3, $\operatorname{dom} S = \operatorname{ran}(P_+ + Q_-T) = |\mathcal{L}_-|$.

(iii) $\Rightarrow$ (i). If $\operatorname{dom} S = |\mathcal{L}_-|$, the scattering operator S is given by (1.3.5). Multiplication of operator matrices shows that

$$S^\times = (P_+T^* + Q_-)(Q_+ + P_-T^*)^{-1}, \tag{1.3.10}$$

where

$$P_+T^* + Q_- = \begin{pmatrix} T_{11}^\times & -T_{21}^\times \\ 0 & 1 \end{pmatrix} \in \mathbf{B}(\mathcal{K}_+ \oplus |\mathcal{K}_-|, \mathcal{H}_+ \oplus |\mathcal{K}_-|),$$

$$Q_+ + P_-T^* = \begin{pmatrix} 1 & 0 \\ -T_{12}^\times & T_{22}^\times \end{pmatrix} \in \mathbf{B}(\mathcal{K}_+ \oplus |\mathcal{K}_-|, \mathcal{K}_+ \oplus |\mathcal{H}_-|).$$

Since $S^\times$ is a contraction, for all $g \in \mathcal{K}$,

$$\|(P_+T^* + Q_-)g\|^2_{\mathcal{H}_+ \oplus |\mathcal{K}_-|} \leq \|(Q_+ + P_-T^*)g\|^2_{\mathcal{K}_+ \oplus |\mathcal{H}_-|}.$$

Therefore

$$\|P_+T^*g\|^2 + \|Q_+g\|^2 \leq \|Q_+g\|^2 + \|P_-T^*g\|^2$$

and

$$\|P_+T^*g\|^2 - \|P_-T^*g\|^2 \leq \|Q_+g\|^2 - \|Q_-g\|^2,$$

where the norms are computed in $\mathcal{H}_+$, $|\mathcal{H}_-|$, $\mathcal{K}_+$, $|\mathcal{K}_-|$. It follows that T^* is contractive, and hence T is bicontractive.

Now assume that (i) - (iii) hold. Then (1.3.10) says that the scattering operator for T^* is $S^\times$, and (1.3.6) and (1.3.7) follow by multiplying matrices.

The identities (1.3.8) and (1.3.9) are equivalent because the scattering operator for T^* is $S^\times$. To prove (1.3.8), calculate using the identity for S in Theorem 1.3.3:

$$\begin{aligned}(P_+ + &Q_-T)^\times(1 - S^\times S)(P_+ + Q_-T)\\ &= (P_+ + Q_-T)^\times(P_+ + Q_-T) - (Q_+T + P_-)^\times(Q_+T + P_-)\\ &= \begin{pmatrix} 1 & T_{21}^\times \\ 0 & T_{22}^\times \end{pmatrix}\begin{pmatrix} 1 & 0 \\ T_{21} & T_{22} \end{pmatrix} - \begin{pmatrix} T_{11}^\times & 0 \\ T_{12}^\times & 1 \end{pmatrix}\begin{pmatrix} T_{11} & T_{12} \\ 0 & 1 \end{pmatrix}.\end{aligned}$$

By multiplication of matrices, this is the same as

$$\begin{pmatrix} 1 & 0 \\ 0 & -1 \end{pmatrix} - \begin{pmatrix} T_{11}^\times & T_{21}^\times \\ T_{12}^\times & T_{22}^\times \end{pmatrix}\begin{pmatrix} T_{11} & T_{12} \\ -T_{21} & -T_{22} \end{pmatrix} = J_{\mathcal{H}}(1 - T^*T),$$

which yields the result. ∎

The next result characterizes the Hilbert space contractions S which occur as scattering operators of bicontractions.

THEOREM 1.3.5. *Let $\mathcal{H}$ and $\mathcal{K}$ be Kreĭn spaces with fundamental decompositions $\mathcal{H} = \mathcal{H}_+ \oplus \mathcal{H}_-$ and $\mathcal{K} = \mathcal{K}_+ \oplus \mathcal{K}_-$. Let $S \in \mathbf{B}(\mathcal{H}_+ \oplus |\mathcal{K}_-|, \mathcal{K}_+ \oplus |\mathcal{H}_-|)$ be a contraction with matrix*

$$S = \begin{pmatrix} S_{11} & S_{12} \\ S_{21} & S_{22} \end{pmatrix}.$$

Then S is the scattering operator for a bicontraction $T \in \mathbf{B}(\mathcal{H}, \mathcal{K})$ if and only if S_{22} is invertible.

Proof. Necessity follows from Theorem 1.3.4 and (1.3.5). Conversely, suppose that S_{22} is invertible. Define $T \in \mathbf{B}(\mathcal{H}, \mathcal{K})$ by

$$T = \begin{pmatrix} S_{11} - S_{12}S_{22}^{-1}S_{21} & S_{12}S_{22}^{-1} \\ -S_{22}^{-1}S_{21} & S_{22}^{-1} \end{pmatrix}.$$

Multiplication of matrices verifies the identity

$$S = (Q_+T + P_-)(P_+ + Q_-T)^{-1}.$$

Since S is a contraction, for any $f \in \mathcal{H}$,

$$\|(Q_+T + P_-)f\|^2_{\mathcal{K}_+\oplus|\mathcal{H}_-|} \leq \|(P_+ + Q_-T)f\|^2_{\mathcal{H}_+\oplus|\mathcal{K}_-|},$$

equivalently,

$$\|Q_+Tf\|^2 + \|P_-f\|^2 \leq \|P_+f\|^2 + \|Q_-Tf\|^2,$$
$$\|Q_+Tf\|^2 - \|Q_-Tf\|^2 \leq \|P_+f\|^2 - \|P_-f\|^2,$$

which means that $T \in \mathbf{B}(\mathcal{H},\mathcal{K})$ is a contraction. Since $T_{22} = S_{22}^{-1}$ is invertible, T is a bicontraction by Theorem 1.3.4. By construction, S is the scattering operator for T. ■

Theorem 1.3.1 describes how a contraction maps negative subspaces. Bicontractions are characterized by how they map maximal negative subspaces.

THEOREM 1.3.6. *Let $\mathcal{H}$ and $\mathcal{K}$ be Kreĭn spaces. If $T \in \mathbf{B}(\mathcal{H},\mathcal{K})$ is a contraction, the following assertions are equivalent:*

(i) T is a bicontraction;

(ii) αT^* is a contraction for some positive number α;

(iii) T maps some maximal negative subspace of $\mathcal{H}$ onto a maximal negative subspace of $\mathcal{K}$;

(iv) T maps every maximal negative subspace of $\mathcal{H}$ onto a maximal negative subspace of $\mathcal{K}$.

In this case, T maps any maximal uniformly negative subspace of $\mathcal{H}$ in a one-to-one way onto a maximal uniformly negative subspace of $\mathcal{K}$.

Proof. Let $\mathcal{H} = \mathcal{H}_+ \oplus \mathcal{H}_-$ and $\mathcal{K} = \mathcal{K}_+ \oplus \mathcal{K}_-$ be fundamental decompositions of the spaces $\mathcal{H}$ and $\mathcal{K}$.

(i) ⇒ (ii) If (i) holds, (ii) holds with $\alpha = 1$.

(ii) ⇒ (iv) Assume (ii), and let $\mathcal{M}$ be a maximal negative subspace of $\mathcal{H}$. By Corollary 1.3.2, $T\mathcal{M}$ is a closed negative subspace of $\mathcal{K}$. By the graph representation of negative subspaces, $T\mathcal{M}$ is maximal negative if and only if there is no nonzero vector in $\mathcal{K}_-$ which is orthogonal to $T\mathcal{M}$. If f is in $\mathcal{K}_-$ and orthogonal to $T\mathcal{M}$, then for g in $\mathcal{M}$,

$$\langle \alpha T^*f, g\rangle_{\mathcal{H}} = \langle f, \alpha Tg\rangle_{\mathcal{K}} = 0.$$

It follows that αT^*f is in $\mathcal{M}^\perp$, which is maximal positive since $\mathcal{M}$ is maximal negative. Therefore,

$$\langle \alpha T^*f, \alpha T^*f\rangle_{\mathcal{H}} \geq 0 \geq \langle f, f\rangle_{\mathcal{K}}.$$

Since αT^* is contractive by assumption and f is in $\mathcal{K}_-$, $\langle f, f\rangle_{\mathcal{K}} = 0$ and $f = 0$. Hence $T\mathcal{M}$ is maximal negative.

(iv) $\Rightarrow$ (iii) This is trivial.

(iii) $\Rightarrow$ (i) Assume that there is a maximal negative subspace $\mathcal{M}$ of $\mathcal{H}$ such that $T\mathcal{M}$ is maximal negative in $\mathcal{K}$. We show that the entry T_{22} in (1.3.1) is invertible, and hence T is a bicontraction by Theorem 1.3.4.

Represent $\mathcal{M}$ as the graph $\mathcal{G}(K)$ of a contraction operator K on $|\mathcal{H}_-|$ to $\mathcal{H}_+$. Then sK is a contraction and its graph $\mathcal{G}(sK)$ is a maximal negative subspace of $\mathcal{H}$ for $0 \le s \le 1$. Since T is contractive, so is Q_-T where Q_- is the projection of $\mathcal{K}$ onto $\mathcal{K}_-$. Choose $\delta > 0$ for Q_-T as in Theorem 1.3.1. For all f in $\mathcal{H}_-$,

$$\|(T_{21}sK + T_{22})f\|^2_{|\mathcal{K}_-|} = \|Q_-T(sKf + f)\|^2_{|\mathcal{K}|} \ge \delta\|sKf + f\|^2_{|\mathcal{H}|} \ge \delta\|f\|^2_{|\mathcal{H}_-|}$$

uniformly for $0 \le s \le 1$. The range of $T_{21}K + T_{22}$ is all of $\mathcal{K}_-$ because T maps $\mathcal{M} = \mathcal{G}(K)$ onto a maximal negative subspace of $\mathcal{K}$ by assumption. Therefore $T_{21}K + T_{22}$ is an invertible operator on $\mathcal{K}_-$. By a Neumann series argument, $T_{21}sK + T_{22}$ is invertible for $0 \le s \le 1$. In particular, T_{22} is invertible.

We have proved the equivalence of the statements (i)–(iv). The last assertion of the theorem follows from Corollary 1.3.2. ■

THEOREM 1.3.7. *If $\mathcal{H}$ is a Pontryagin space, every contraction operator T from $\mathcal{H}$ into itself is bicontractive.*

Proof. In any fundamental decomposition $\mathcal{H} = \mathcal{H}_+ \oplus \mathcal{H}_-$, $\mathcal{H}_-$ has finite dimension. If P_- is the projection onto $\mathcal{H}_-$, then P_-T is a contraction because T is a contraction. By Corollary 1.3.2, $T_{22} = P_-TP_-|\mathcal{H}_-$ is a one-to-one mapping of $\mathcal{H}_-$ into itself. Hence T_{22} is invertible, and T is bicontractive by Theorem 1.3.6. ■

EXAMPLES 1.3.8. (i) Let $\mathcal{H}$ and $\mathcal{K}$ be Kreĭn spaces. An isometry $V \in \mathbf{B}(\mathcal{H}, \mathcal{K})$ is a bicontraction if and only if $\ker V^*$ is a uniformly positive subspace of $\mathcal{K}$. Necessity of the condition follows from Theorem 1.3.1. For the other direction, note that V is a partial isometry with initial space $\mathcal{H}$ and final space $V\mathcal{H}$ by Corollary 1.1.4. Since $\mathcal{K} = V\mathcal{H} \oplus \ker V^*$ with $\ker V^*$ uniformly positive, sufficiency of the condition follows from elementary properties of partial isometries.

(ii) Theorem 1.3.7 fails for Kreĭn spaces. Consider the Kreĭn space $\mathcal{H}$ of square summable sequences $a = (a_0, a_1, a_2, \dots)$ with

$$\langle a, a\rangle_{\mathcal{H}} = |a_0|^2 - |a_1|^2 - |a_2|^2 - \cdots.$$

The operator $T \in \mathbf{B}(\mathcal{H})$ defined by

$$T : (a_0, a_1, a_2, \dots) \to (0, a_0, a_1, \dots)$$

is contractive, but its adjoint

$$T^* : (a_0, a_1, a_2, \dots) \to (-a_1, a_2, a_3, \dots)$$

is not contractive. The range of T is a maximal negative subspace of $\mathcal{H}$ which is mapped by T onto a proper subspace of itself and hence onto a negative subspace which is not maximal.

(iii) In connection with Theorem 1.3.6, we note a counter-example to a related statement in Andô ([**4**], p. 31, Corollary 3.3.2). In our language, the statement is that an operator $T \in \mathbf{B}(\mathcal{H})$ is a scalar multiple of a bicontraction if it maps maximal negative subspaces onto maximal negative subspaces. A counterexample is $T = V^* \in \mathbf{B}(\mathcal{H})$, where $\mathcal{H}$ is the anti-space of a Hilbert space and V is an isometry whose range is not all of $\mathcal{H}$. The only maximal negative subspace of $\mathcal{H}$ is $\mathcal{H}$, and this mapped by T onto itself. No constant multiple of T is a contraction, since the kernel of T includes a vector f with $\langle f, f\rangle_{\mathcal{H}} < 0$.

Bicontractions are also characterized in terms of mappings of operator spheres.

THEOREM 1.3.9. *Let $\mathcal{H}$ and $\mathcal{K}$ be Kreĭn spaces, and let $T \in \mathbf{B}(\mathcal{H}, \mathcal{K})$ be a contraction with matrix (1.3.1) relative to some fundamental decompositions $\mathcal{H} = \mathcal{H}_+ \oplus \mathcal{H}_-$ and $\mathcal{K} = \mathcal{K}_+ \oplus \mathcal{K}_-$. Then T is a bicontraction if and only if whenever $X \in \mathbf{B}(|\mathcal{H}_-|, \mathcal{H}_+)$ is a contraction, then $T_{21}X + T_{22}$ is an invertible operator on $|\mathcal{H}_-|$ to $|\mathcal{K}_-|$ and*

$$Y = (T_{11}X + T_{12})(T_{21}X + T_{22})^{-1} \tag{1.3.11}$$

is a contraction in $\mathbf{B}(|\mathcal{K}_-|, \mathcal{K}_+)$.

Proof. If T is a bicontraction and $X \in \mathbf{B}(|\mathcal{H}_-|, \mathcal{H}_+)$ is a contraction, then

$$\mathcal{G}(X) = \left\{ \begin{pmatrix} Xf \\ f \end{pmatrix} : f \in \mathcal{H}_- \right\}$$

is a maximal negative subspace of $\mathcal{H}$ and

$$T\mathcal{G}(X) = \left\{ \begin{pmatrix} T_{11}Xf + T_{12}f \\ T_{21}Xf + T_{22}f \end{pmatrix} : f \in \mathcal{H}_- \right\}$$

is a maximal negative subspace of $\mathcal{K}$ by Theorem 1.3.6. Let Q_- be the projection of $\mathcal{K}$ onto $\mathcal{K}_-$. Then Q_-T is contractive, and $T_{21}X + T_{22}$ is one-to-one by Corollary 1.3.2. Since $Q_-T\mathcal{G}(X) = \mathcal{K}_-$, $T_{21}X + T_{22}$ is invertible. Therefore (1.3.11) defines an operator $Y \in \mathbf{B}(|\mathcal{K}_-|, \mathcal{K}_+)$. Since $T\mathcal{G}(X)$ is a negative subspace of $\mathcal{K}$,

$$\|(T_{11}X + T_{12})f\|^2_{\mathcal{K}_+} \leq \|(T_{21}X + T_{22})f\|^2_{|\mathcal{K}_-|}, \qquad f \in \mathcal{H}_-,$$

and Y is a contraction.

In the other direction, if the condition holds for $X = 0$, then T_{22} is invertible and T is a bicontraction by Theorem 1.3.4. ∎

In the next result, the domain and range spaces $\mathcal{H}$ and $\mathcal{K}$ coincide.

THEOREM 1.3.10. *Let $\mathcal{H}$ be a Kreĭn space with fundamental decomposition $\mathcal{H} = \mathcal{H}_+ \oplus \mathcal{H}_-$, and let $T \in \mathbf{B}(\mathcal{H})$ be a bicontraction. Define a mapping Φ of the set $\mathcal{C}$ of all contractions $X \in \mathbf{B}(|\mathcal{H}_-|, \mathcal{H}_+)$ into itself by*

$$\Phi(X) = (T_{11}X + T_{12})(T_{21}X + T_{22})^{-1}, \qquad X \in \mathcal{C}.$$

Then for $X \in \mathcal{C}$, we have $\Phi(X) = X$ if and only if the graph

$$\mathcal{M} = \left\{ \begin{pmatrix} Xf \\ f \end{pmatrix} : f \in \mathcal{H}_- \right\}$$

of X is invariant under T.

Proof. By its form, $\mathcal{M}$ is a maximal negative subspace of $\mathcal{H}$. If $\mathcal{M}$ is invariant under T, then $T\mathcal{M} = \mathcal{M}$ by Theorem 1.3.6. Since

$$T\mathcal{M} = \left\{ \begin{pmatrix} T_{11}Xf + T_{12}f \\ T_{21}Xf + T_{22}f \end{pmatrix} : f \in \mathcal{H}_- \right\},$$

we have $T_{11}X + T_{12} = X(T_{21}X + T_{22})$, that is, $\Phi(X) = X$. The other direction is obtained by reversing these steps. ■

Existence theorems for definite invariant subspaces follow with the aid of Theorem 1.3.10. A full treatment is beyond the scope of our discussion. But having come this far, we show the connection by giving one of the early results of the subject which is due to Kreĭn (Theorem 1.3.11). See the notes at the end of the chapter for literature references. Note that the set $\mathcal{C}$ in Theorem 1.3.10 is convex and compact in the weak operator topology of $\mathbf{B}(|\mathcal{H}_-|, \mathcal{H}_+)$. We also recall the SCHAUDER-TYCHONOFF THEOREM: *Every continuous mapping of a convex and compact subset of a locally convex linear topological vector space has a fixed point.* The Schauder-Tychonoff theorem is proved, for example, in Dunford and Schwarz ([**31**], p. 456). It is applicable in the situation of Theorem 1.3.10 whenever Φ is continuous. This is automatic in the case of a Pontryagin space.

THEOREM 1.3.11. *Let $\mathcal{H}$ be a Pontryagin space. If $T \in \mathbf{B}(\mathcal{H})$ is a contraction (and hence a bicontraction), there exists a maximal negative subspace $\mathcal{M}$ of $\mathcal{H}$ which is mapped by T onto itself.*

Proof. It is sufficient to show that the mapping Φ in Theorem 1.3.10 is continuous.

Let $\{X_\alpha\}_{\alpha\in\mathcal{D}}$ be a generalized sequence in $\mathcal{C}$ which converges in the weak operator topology to $X \in \mathcal{C}$. Then

$$\lim_\alpha (T_{11}X_\alpha + T_{12}) = T_{11}X + T_{12}$$

in the weak operator topology of $\mathbf{B}(|\mathcal{H}_-|, \mathcal{K}_+)$.

Let P_- be the projection of $\mathcal{H}$ onto $\mathcal{H}_-$. Choose $\delta > 0$ for the contraction $P_-T \in \mathbf{B}(\mathcal{H})$ as in Theorem 1.3.1. Then

$$\|T_{21}X_\alpha f + T_{22}f\| \geq \delta\|f\|, \qquad \alpha \in \mathcal{D}, f \in \mathcal{H}_-.$$

Therefore $\left\|(T_{21}X_\alpha + T_{22})^{-1}\right\| \leq 1/\delta$ for all $\alpha \in \mathcal{D}$. Now

$$\lim_\alpha (T_{21}X_\alpha + T_{22}) = T_{21}X + T_{22}$$

in the weak operator topology of $\mathbf{B}(|\mathcal{H}_-|)$ and hence also in the norm topology because $\mathcal{H}_-$ is finite dimensional. By Theorem 1.3.9, $T_{21}X + T_{22}$ is invertible, and

$$\lim_\alpha (T_{21}X_\alpha + T_{22})^{-1} = (T_{21}X + T_{22})^{-1}$$

in the operator norm of $\mathbf{B}(|\mathcal{H}_-|)$. Thus

$$\lim_\alpha \Phi(X_\alpha) = \Phi(X)$$

in the weak operator topology of $\mathbf{B}(|\mathcal{H}_-|, \mathcal{H}_+)$, and so Φ is continuous. ∎

1.4 Additional Results on Contractions and Bicontractions

The results here catalog properties of defect operators and Julia operators of contractions. We also give a useful theorem on the existence of a bicontractive extension of a densely defined contraction.

THEOREM 1.4.1. *Let $T \in \mathbf{B}(\mathcal{H}, \mathcal{K})$, $\mathcal{H}$ and $\mathcal{K}$ Kreĭn spaces, and let*

$$U = \begin{pmatrix} T & D \\ \tilde{D}^* & L \end{pmatrix} \in \mathbf{B}(\mathcal{H} \oplus \mathcal{D}, \mathcal{K} \oplus \tilde{\mathcal{D}})$$

be a Julia operator for T. The adjoint of the operator $(T \quad D) \in \mathbf{B}(\mathcal{H} \oplus \mathcal{D}, \mathcal{K})$ is an isometry. The following assertions are equivalent:

(i) T is a contraction;
(ii) $\tilde{\mathcal{D}}$ is a Hilbert space;
(iii) $(T \quad D)$ is a contraction;
(iv) $(T \quad D)$ is a bicontraction;
(v) the kernel of $(T \quad D)$ is uniformly positive in $\mathcal{H} \oplus \mathcal{D}$.

Proof. In the proof, we write $R = (T \quad D)$ and make repeated use of the relations (1.2.1) and (1.2.2). The operator R^* is an isometry because $RR^* = TT^* + DD^* = 1$ by the definition of a Julia operator.

(i) $\Rightarrow$ (ii) If T is a contraction, then $\tilde{D}\tilde{D}^* = 1 - T^*T \geq 0$. Since $\tilde{D}$ has zero kernel, the range of $\tilde{D}^*$ is dense in $\tilde{\mathcal{D}}$. The inequality $\tilde{D}\tilde{D}^* \geq 0$ then implies that $\langle g, g\rangle_{\tilde{\mathcal{D}}} \geq 0$ for all $g \in \mathcal{D}$, and therefore $\tilde{\mathcal{D}}$ is a Hilbert space.

(ii) $\Rightarrow$ (iii) If $\tilde{\mathcal{D}}$ is a Hilbert space, then

$$1 - R^*R = \begin{pmatrix} 1 - T^*T & -T^*D \\ -D^*T & 1 - D^*D \end{pmatrix} = \begin{pmatrix} \tilde{D}\tilde{D}^* & \tilde{D}L \\ L^*\tilde{D}^* & L^*L \end{pmatrix} = \begin{pmatrix} \tilde{D} \\ L^* \end{pmatrix} (\tilde{D}^* \quad L) \geq 0,$$

because $(\tilde{D}^* \quad L) \in \mathbf{B}(\mathcal{H} \oplus D, \tilde{\mathcal{D}})$. Thus R is a contraction.

(iii) $\Rightarrow$ (iv) Since R^* is an isometry, if R is a contraction it is automatically a bicontraction.

(iv) $\Rightarrow$ (v) This follows from Theorem 1.3.1.

(v) $\Rightarrow$ (i) Since R^* is isometric, this follows from Example 1.3.8 (i). ■

COROLLARY 1.4.2. *Let $T \in \mathbf{B}(\mathcal{H}, \mathcal{K})$, $\mathcal{H}$ and $\mathcal{K}$ Kreĭn spaces, and let $D \in \mathbf{B}(\mathcal{D}, \mathcal{K})$ be a defect operator for T^*. Then the operator $(T \quad D)$ has an isometric adjoint, and the following assertions are equivalent:*

(i) T is a contraction;
(ii) $(T \quad D)$ is a contraction;
(iii) $(T \quad D)$ is a bicontraction;
(iv) the kernel of $(T \quad D)$ is uniformly positive in $\mathcal{H} \oplus D$.

If these equivalent conditions are satisfied, then for any defect operator $\tilde{D} \in \mathbf{B}(\tilde{\mathcal{D}}, \mathcal{H})$ of T, $\tilde{\mathcal{D}}$ is a Hilbert space.

Proof. By Theorem 1.2.4, this is a consequence of Theorem 1.4.1. ■

COROLLARY 1.4.3. *Let $T \in \mathbf{B}(\mathcal{H}, \mathcal{K})$, $\mathcal{H}$ and $\mathcal{K}$ Kreĭn spaces, and let*

$$U = \begin{pmatrix} T & D \\ \tilde{D}^* & L \end{pmatrix} \in \mathbf{B}(\mathcal{H} \oplus D, \mathcal{K} \oplus \tilde{\mathcal{D}})$$

be a Julia operator for T. If T is a bicontraction, then $\mathcal{D}$ and $\tilde{\mathcal{D}}$ are Hilbert spaces, and both

$$(T \quad D) \in \mathbf{B}(\mathcal{H} \oplus D, \mathcal{K}) \qquad \text{and} \qquad \begin{pmatrix} T \\ \tilde{D}^* \end{pmatrix} \in \mathbf{B}(\mathcal{H}, \mathcal{K} \oplus \tilde{\mathcal{D}})$$

are bicontractions with isometric adjoints.

Proof. Since T^* has Julia operator

$$U^* = \begin{pmatrix} T^* & \tilde{D} \\ D^* & L^* \end{pmatrix} \in \mathbf{B}(\mathcal{K} \oplus \tilde{\mathcal{D}}, \mathcal{H} \oplus D),$$

the result follows by applying Theorem 3.1.1 to both T and T^*. ∎

In Hilbert spaces, densely defined contractions extend automatically by continuity to everywhere defined contractions. This is not true in Kreĭn spaces, but with an extra condition the existence of a continuous bicontractive extension is assured. The result can be used, for example, to construct continuous isometric operators from densely defined isometries.

THEOREM 1.4.4. *Let $\mathcal{H}$ and $\mathcal{K}$ be Kreĭn spaces, and let T_0 be a densely defined linear mapping from $\mathcal{H}$ to $\mathcal{K}$. Assume that*

$$\langle T_0 f, T_0 f\rangle_{\mathcal{K}} \leq \langle f, f\rangle_{\mathcal{H}}, \qquad f \in \operatorname{dom} T_0.$$

Assume also that $\operatorname{dom} T_0$ contains a maximal uniformly negative subspace $\mathcal{M}$ of $\mathcal{H}$ and that $T_0\mathcal{M}$ is a maximal uniformly negative subspace of $\mathcal{K}$. Then T_0 has an extension by continuity to a bicontractive operator $T \in \mathbf{B}(\mathcal{H}, \mathcal{K})$.

LEMMA 1.4.5. *Let $\mathcal{H}$ be a Kreĭn space with fundamental decomposition $\mathcal{H} = \mathcal{H}_+ \oplus \mathcal{H}_-$. Set $\alpha = \sqrt{2} - 1$. If $f \in \mathcal{H}_-$, $\|f\| \geq 1$, and $g \in \mathcal{H}$, $\|g\| < \alpha$, then*

$$\langle f + g, f + g\rangle < 0.$$

The norms in the lemma are computed with respect to the given fundamental decomposition of $\mathcal{H}$.

Proof of Lemma 1.4.5. Fix $f \in \mathcal{H}_-$, $\|f\| \geq 1$, and $g \in \mathcal{H}$, $\|g\| < \alpha$. Then

$$\begin{aligned}\langle f + g, f + g\rangle &= -\|f\|^2 + 2\operatorname{Re}\langle f, g\rangle + \langle g, g\rangle \\ &< -\|f\|^2 + 2\alpha\|f\| + \alpha^2.\end{aligned}$$

The function $\phi(x) = -x^2 + 2\alpha x + \alpha^2$ attains its maximum at $x = \alpha$, and therefore $\phi(x) \leq \phi(1)$ for $x \geq 1$. Thus $\langle f + g, f + g\rangle < -1 + 2\alpha + \alpha^2 = 0$. ∎

Proof of Theorem 1.4.4. Choose fundamental decompositions $\mathcal{H} = \mathcal{H}_+ \oplus \mathcal{H}_-$ and $\mathcal{K} = \mathcal{K}_+ \oplus \mathcal{K}_-$ such that

$$\mathcal{H}_+ = \mathcal{M}^{\perp}, \qquad \mathcal{H}_- = \mathcal{M},$$

and

$$\mathcal{K}_+ = (T_0\mathcal{M})^{\perp}, \qquad \mathcal{K}_- = T_0\mathcal{M}.$$

The conclusion is immediate if $\mathcal{H}$ is a Hilbert space, and we exclude this case in what follows. Let S be the restriction of T_0 to $\mathcal{H}_-$, viewed as a linear mapping of $\mathcal{H}_-$ onto $\mathcal{K}_-$. For $f \in \mathcal{H}_-$,

$$\|Sf\|^2 = -\langle T_0 f, T_0 f\rangle_{\mathcal{K}} \geq -\langle f, f\rangle_{\mathcal{H}} = \|f\|^2.$$

Therefore S has a continuous inverse, and so $S \in \mathbf{B}(\mathcal{H}_-, \mathcal{K}_-)$ by the open mapping theorem.

Set $\alpha = \sqrt{2} - 1$. We show that for $g \in \operatorname{dom} T_0$, $\|g\| < \alpha$,

$$\left\|\operatorname{Pr}_{\mathcal{K}_-} T_0 g\right\| < \|S\|.$$

Argue by contradiction, assuming that the inequality is not true. Choose $h \in \mathcal{H}_-$ such that $Sh = \operatorname{Pr}_{\mathcal{K}_-} T_0 g$. Then $\|S\| \le \left\|\operatorname{Pr}_{\mathcal{K}_-} T_0 g\right\| = \|Sh\| \le \|S\|\|h\|$, and so $\|h\| \ge 1$. By Lemma 1.4.5,

$$0 > \langle h - g, h - g\rangle_{\mathcal{H}} \ge \langle T_0(h-g), T_0(h-g)\rangle_{\mathcal{K}},$$

a contradiction, because $T_0(h-g)$ has zero projection in $\mathcal{K}_-$ and hence is in $\mathcal{K}_+$.

We show that for $g \in \operatorname{dom} T_0$, $\|g\| < \alpha$,

$$\left\|\operatorname{Pr}_{\mathcal{K}_+} T_0 g\right\| < 2\|S\|.$$

Since $\mathcal{H}$ is not a Hilbert space, we can choose $h \in \mathcal{H}_-$ with $\|h\| = 1$. By Lemma 1.4.5,

$$\begin{aligned} 0 &> \langle h - g, h - g\rangle_{\mathcal{H}} \ge \langle T_0(h-g), T_0(h-g)\rangle_{\mathcal{K}} \\ &= \left\|\operatorname{Pr}_{\mathcal{K}_+} T_0(h-g)\right\|^2 - \left\|\operatorname{Pr}_{\mathcal{K}_-} T_0(h-g)\right\|^2 \\ &= \left\|\operatorname{Pr}_{\mathcal{K}_+} T_0 g\right\|^2 - \left\|\operatorname{Pr}_{\mathcal{K}_-} T_0(h-g)\right\|^2. \end{aligned}$$

Therefore

$$\left\|\operatorname{Pr}_{\mathcal{K}_+} T_0 g\right\|^2 < \left\|\operatorname{Pr}_{\mathcal{K}_-} T_0(h-g)\right\|^2 \le \left[\left\|\operatorname{Pr}_{\mathcal{K}_-} T_0 h\right\| + \left\|\operatorname{Pr}_{\mathcal{K}_-} T_0 g\right\|\right]^2 < 4\|S\|^2,$$

which proves the assertion.

By what we have shown, for any $g \in \operatorname{dom} T_0$, $\|g\| < \alpha$,

$$\|T_0 g\|^2 = \left\|\operatorname{Pr}_{\mathcal{K}_+} T_0 g\right\|^2 + \left\|\operatorname{Pr}_{\mathcal{K}_-} T_0 g\right\|^2 < 5\|S\|^2.$$

It follows that T_0 has an extension to an operator $T \in \mathbf{B}(\mathcal{H}, \mathcal{K})$ with

$$\|T\| < \frac{\sqrt{5}}{\sqrt{2} - 1}\|S\|.$$

Clearly T is a contraction. Since T maps a maximal negative subspace of $\mathcal{H}$ onto a maximal negative subspace of $\mathcal{K}$, T is a bicontraction. ■

NOTES ON CHAPTER 1

Accounts of Kreĭn spaces and operators on them are given in Andô [4], Azizov and Iokhvidov [**10**], Bognár [**12**], and Kreĭn [**43**]. The book by Iokhvidov, Kreĭn, and Langer [**40**] is another useful source: it emphasizes Pontryagin spaces, but the results and methods often apply to general Kreĭn spaces. Kreĭn [**43**] is an authoritative source for the early history of the subject and applications that motivated its development. Azizov and Iokhvidov [**9**] survey literature in the years 1953–1978. For the finite dimensional case, see Gohberg, Lancaster, and Rodman [**38**] and Potapov [**55**].

Additional information on isometries and direct sums may be found in Bognár [**12**], Ghéondea [**36**], and McEnnis [**52,53**]. In particular, Theorem 1.1.8 on internal orthogonal direct sums is not as strong as possible. Ghéondea [**36**] gives necessary and sufficient conditions for the closed span of pairwise orthogonal regular subspaces to be regular. This condition is weaker than the uniform boundedness condition we give on the associated projections, which he shows to be equivalent to both convergence in the strong and the weak operator topologies of the sums given in 1.1.8(iii). McEnnis [**53**] gives necessary and sufficient conditions for the square summability of the representation of elements of an internal direct sum given in 1.1.8(ii).

Factorizations of selfadjoint operators in the form $H = AA^*$, discussed in §1.2, appear in many places, including Bognár [**12**]. The existence of a Julia operator in the Kreĭn space setting was first shown by Arsene, Constantinescu, and Ghéondea [**8**] by a different method (see Appendix B). This is an important result and plays a key role in what follows. The abstract definition of a Julia operator and existence proof in Theorem 1.2.4 seem to be new; uniqueness is discussed in Chapter 2 and Appendix B. For the Hilbert space case as well as references to the original papers of Julia and Halmos, see Sz.-Nagy and Foiaş [**66**].

The main results of §1.3 are due to Ginsburg [**37**] and Kreĭn and Shmul'yan [**45,47**]. In the finite dimensional case, some ideas go back to Potapov [**55**]. Additional results are given in Kreĭn and Shmul'yan [**46**].

Theorem 1.3.11 is the simplest of its kind. The method is due to Kreĭn [**42**], but the problem was known previously. The history of the problem is given in the survey of Azizov and Iokhvidov [**9**]. See their monograph [**10**] for a recent account. Additional information may be found in Iokhvidov, Kreĭn, and Langer [**40**]. Other sources are Kreĭn [**43**], Andô [4], and Bognár [**12**]. Best possible conditions for the existence of definite invariant subspaces are not known.

Literature citations for the Potapov-Ginsburg transform are given in Azizov and Iokhvidov [**9**]. Our treatment is adapted from Iokhvidov, Kreĭn, and Langer [**40**]. The transform is used in other situations, such as Arov [**5**], Dym [**32**], and Kreĭn and Langer [**44**]. The scattering interpretation appears in Alpay [**3**] and Dym [**32**].

Theorem 1.4.4 is due to Shmul'yan [**64**]. A similar result is proved by Yan [**67**].

Chapter 2: Matrix Extensions of Contraction Operators

2.1 The Adjoint of a Contraction

The theory of Ginsburg, Kreĭn, and Shmul'yan discussed in Chapter 1 is based on an analysis of how a contraction operator maps negative subspaces. In the case of bicontractions, the adjoint of a given contraction is again a contraction and subject to the same analysis. Contraction operators which are not bicontractions are also of interest. The main point of this section is that one can decompose the domain space of the adjoint of any Kreĭn space contraction in such a way as to discard a uniformly negative subspace, and then the restriction of the adjoint to what remains is a bicontraction. In this way we are able to apply the strong results on bicontractions to arbitrary contractions. In particular, we obtain a new characterization of bicontractions in terms of adjoint operators (Theorem 2.1.5).

To begin, we assume that bicontractive restrictions of the adjoint of a contraction operator exist, and we determine some consequences of this fact. Chiefly, there exist maximal uniformly negative subspaces of the range space with special properties. Then we shall reverse the process.

THEOREM 2.1.1. *Let $\mathcal{H}$ and $\mathcal{K}$ be Kreĭn spaces, and let $T \in \mathbf{B}(\mathcal{H},\mathcal{K})$ be a contraction operator. Let $\hat{\mathcal{K}}$ be a regular subspace of $\mathcal{K}$ such that $\hat{\mathcal{K}}^{\perp}$ is uniformly negative and $T^*|\hat{\mathcal{K}}$ is a bicontraction on $\hat{\mathcal{K}}$ to $\mathcal{H}$. Let $\mathcal{L}_-$ be a uniformly negative subspace of $\hat{\mathcal{K}}$ which is maximal in $\hat{\mathcal{K}}$, and let $\mathcal{K}_- = \mathcal{L}_- \oplus \hat{\mathcal{K}}^{\perp}$. Define*

$$\mathcal{L}_+ = \{v : v \in \mathcal{K}_- \text{ and } T^*v \perp T^*\mathcal{L}_-\}\,.$$

Then $\mathcal{K}_-$ is a maximal uniformly negative subspace of $\mathcal{K}$, and $\mathcal{K}_- = \mathcal{L}_- \dot{+} \mathcal{L}_+$.

Recall that $\dot{+}$ is the symbol for a direct sum which is not necessarily orthogonal.

Proof. Since $\hat{\mathcal{K}}^{\perp}$ is uniformly negative, $\mathcal{K}_- = \mathcal{L}_- \oplus \hat{\mathcal{K}}^{\perp}$ is closed and uniformly negative. Since $\mathcal{L}_-$ is a maximal uniformly negative subspace in $\hat{\mathcal{K}}$, the subspace defined by $\mathcal{K}_+ = \hat{\mathcal{K}} \ominus \mathcal{L}_-$ is closed and uniformly positive. Now

$$\mathcal{K} = \hat{\mathcal{K}} \oplus \hat{\mathcal{K}}^{\perp} = (\mathcal{K}_+ \oplus \mathcal{L}_-) \oplus \hat{\mathcal{K}}^{\perp} = \mathcal{K}_+ \oplus \mathcal{K}_-,$$

and so $\mathcal{K}_-$ is maximal uniformly negative in $\mathcal{K}$.

Since $T^*|\hat{\mathcal{K}}$ is a bicontraction and $\mathcal{L}_-$ is maximal uniformly negative in $\hat{\mathcal{K}}$, $T^*\mathcal{L}_-$ is a maximal uniformly negative subspace of $\mathcal{H}$ by Theorem 1.3.6. Choose a fundamental decomposition $\mathcal{H} = \mathcal{H}_+ \oplus \mathcal{H}_-$ by setting $\mathcal{H}_- = T^*\mathcal{L}_-$ and $\mathcal{H}_+ = (T^*\mathcal{L}_-)^\perp$.

Clearly $\mathcal{K}_- \supset \mathcal{L}_- + \mathcal{L}_+$. Let $k \in \mathcal{K}_-$. Write $T^*k = h_- + h_+$, $h_\pm \in \mathcal{H}_\pm$. Then $h_- = T^*u$ for some $u \in \mathcal{L}_-$, and so $h_+ = T^*v$, where $v = k - u \in \mathcal{K}_-$ and $T^*v \perp T^*\mathcal{L}_-$. Thus $k = u + v$ where $u \in \mathcal{L}_-$ and $v \in \mathcal{L}_+$. Therefore $\mathcal{K}_- = \mathcal{L}_- + \mathcal{L}_+$.

If $f \in \mathcal{L}_- \cap \mathcal{L}_+$, then $T^*f \in T^*\mathcal{L}_- \cap T^*\mathcal{L}_+ \subset \mathcal{H}_- \cap \mathcal{H}_+$ and $T^*f = 0$. Since $T^*|\mathcal{L}_-$ is one-to-one by Corollary 1.3.2, $f = 0$. Therefore the sum is direct. ■

Notice that in the situation of Theorem 2.1.1, $T^*|\mathcal{L}_-$ is one-to-one, its range is a maximal uniformly negative subspace of $\mathcal{H}$, and $T^*\mathcal{L}_+ \perp T^*\mathcal{L}_-$. We show that for any contraction $T \in \mathbf{B}(\mathcal{H}, \mathcal{K})$, $\mathcal{H}$ and $\mathcal{K}$ Kreĭn spaces, there exist subspaces of $\mathcal{K}$ having these properties.

THEOREM 2.1.2. *Let $\mathcal{H}$ and $\mathcal{K}$ be Kreĭn spaces, and let $T \in \mathbf{B}(\mathcal{H}, \mathcal{K})$ be a contraction. Let $D \in \mathbf{B}(\mathcal{D}, \mathcal{K})$ be a defect operator for T^*, and assume that $\mathcal{K}_-$, $\mathcal{D}_-$ are maximal uniformly negative subspaces of $\mathcal{K}$, $\mathcal{D}$. Then there exist closed subspaces $\mathcal{L}_-$ and $\mathcal{L}_+$ of $\mathcal{K}_-$ such that*

$$\mathcal{K}_- = \mathcal{L}_- \dot{+} \mathcal{L}_+,$$

and T^* maps $\mathcal{L}_-$ in a one-to-one way onto a maximal uniformly negative subspace of $\mathcal{H}$, $T^*\mathcal{L}_+ \perp T^*\mathcal{L}_-$, and $D^*\mathcal{L}_- \perp \mathcal{D}_-$.

LEMMA 2.1.3. *Let $T \in \mathbf{B}(\mathcal{H}, \mathcal{K})$ be a bicontraction, $\mathcal{H}$ and $\mathcal{K}$ Kreĭn spaces. Let $\hat{\mathcal{H}}$ be a regular subspace of $\mathcal{H}$ such that $\hat{\mathcal{H}}^\perp$ is uniformly positive. Then $\hat{T} = T|\hat{\mathcal{H}}$ is a bicontraction on $\hat{\mathcal{H}}$ to $\mathcal{K}$.*

Proof of Lemma 2.1.3. Since T is a contraction and $\hat{T}$ is a restriction of T, $\hat{T}$ is a contraction. For any $g \in \mathcal{K}$,

$$\begin{aligned}\left\langle \hat{T}^*g, \hat{T}^*g \right\rangle_{\mathcal{H}} &= \langle \mathrm{Pr}_{\hat{\mathcal{H}}} T^*g, \mathrm{Pr}_{\hat{\mathcal{H}}} T^*g \rangle_{\mathcal{H}} \\ &\le \langle \mathrm{Pr}_{\hat{\mathcal{H}}} T^*g, \mathrm{Pr}_{\hat{\mathcal{H}}} T^*g \rangle_{\mathcal{H}} + \langle \mathrm{Pr}_{\mathcal{H}\ominus\hat{\mathcal{H}}} T^*g, \mathrm{Pr}_{\mathcal{H}\ominus\hat{\mathcal{H}}} T^*g \rangle_{\mathcal{H}} \\ &= \langle T^*g, T^*g \rangle_{\mathcal{H}} \\ &\le \langle g, g \rangle_{\mathcal{K}}\end{aligned}$$

because $\mathcal{H} \ominus \hat{\mathcal{H}} = \hat{\mathcal{H}}^\perp$ is uniformly positive. ■

Proof of Theorem 2.1.2. Define $D_- = D|\mathcal{D}_- \in \mathbf{B}(\mathcal{D}_-, \mathcal{K})$. The operator

$$R = (T \quad D) \in \mathbf{B}(\mathcal{H} \oplus \mathcal{D}, \mathcal{K})$$

is a bicontraction by Corollary 1.4.2, and

$$R_- = (T \quad D_-) \in \mathbf{B}(\mathcal{H} \oplus \mathcal{D}_-, \mathcal{K})$$

is a bicontraction by Lemma 2.1.3. Hence $R_-^*\mathcal{K}_-$ is a maximal uniformly negative subspace of $\mathcal{H} \oplus \mathcal{D}_-$ by Theorem 1.3.6.

Let $\mathcal{H} = \mathcal{H}_+ \oplus \mathcal{H}_-$ be a fundamental decomposition, and write the graph representation of $R_-^*\mathcal{K}_-$ as

$$R_-^*\mathcal{K}_- = \left\{ \begin{pmatrix} Kf \\ f \end{pmatrix} : f \in \mathcal{H}_- \oplus \mathcal{D}_- \right\},$$

where $K \in \mathbf{B}(|\mathcal{H}_- \oplus \mathcal{D}_-|, \mathcal{H}_+)$ and $\|K\| < 1$. Then

$$\tilde{\mathcal{H}}_- = \left\{ \begin{pmatrix} Kf \\ f \end{pmatrix} : f \in \mathcal{H}_- \right\}$$

is a maximal uniformly negative subspace of $\mathcal{H}$. Set $\tilde{\mathcal{H}}_+ = \mathcal{H} \ominus \tilde{\mathcal{H}}_-$. Since $R_-^*|\mathcal{K}_-$ is one-to-one and $\tilde{\mathcal{H}}_- \subset R_-^*\mathcal{K}_-$, there is a subspace $\mathcal{L}_-$ of $\mathcal{K}_-$ which is mapped by R_-^* in a one-to-one way onto $\tilde{\mathcal{H}}_-$.

We show that $\mathcal{L}_-$ is closed. Let $f_1, f_2, \ldots \in \mathcal{L}_-$ and let $f_n \to f$ for some $f \in \mathcal{K}_-$. Then $R_-^* f_n \to R_-^* f$, and $R_-^* f \in \tilde{\mathcal{H}}_-$ because $R_-^* f_n \in \tilde{\mathcal{H}}_-$ for all n. By the definition of $\mathcal{L}_-$, we can write $R_-^* f = R_-^* g$ where $g \in \mathcal{L}_-$. Since $R_-^*|\mathcal{K}_-$ is one-to-one, $f = g \in \mathcal{L}_-$, and $\mathcal{L}_-$ is closed.

Next note that $T^*|\mathcal{L}_- = R_-^*|\mathcal{L}_-$, because

$$T^* f = \Pr_{\mathcal{H}} R_-^* f = R_-^* f$$

for any $f \in \mathcal{L}_-$ by the definition of $\mathcal{L}_-$. Since $R_-^*|\mathcal{L}_-$ is one-to-one and maps $\mathcal{L}_-$ onto $\tilde{\mathcal{H}}_-$, $T^*|\mathcal{L}_-$ is one-to-one and maps $\mathcal{L}_-$ onto $\tilde{\mathcal{H}}_-$. The preceding identity also shows that $D_-^* f = 0$ for any $f \in \mathcal{L}_-$, which implies that $D^*\mathcal{L}_- \perp \mathcal{D}_-$.

Define $\mathcal{L}_+ = \{v : v \in \mathcal{K}_- \text{ and } T^*v \perp T^*\mathcal{L}_-\}$. The subspace $T^*\mathcal{L}_- = \tilde{\mathcal{H}}_-$ is maximal uniformly negative, and exactly as in the proof of Theorem 2.1.1 we can show that $\mathcal{K}_- = \mathcal{L}_- \dot{+} \mathcal{L}_+$. By construction, $T^*\mathcal{L}_+ \perp T^*\mathcal{L}_-$. ■

The existence of bicontractive restrictions of the adjoint of any contraction operator follows.

THEOREM 2.1.4. *Let $\mathcal{H}$ and $\mathcal{K}$ be Kreĭn spaces, and let $T \in \mathbf{B}(\mathcal{H},\mathcal{K})$ be a contraction. There exists a regular subspace $\hat{\mathcal{K}}$ of $\mathcal{K}$ such that $\hat{\mathcal{K}}^{\perp}$ is uniformly negative and $T^*|\hat{\mathcal{K}}$ is a bicontraction. If $\hat{\mathcal{K}}'$ is a second such subspace with $\hat{\mathcal{K}}' \supset \hat{\mathcal{K}}$, then $\hat{\mathcal{K}}' = \hat{\mathcal{K}}$.*

Proof. Choose a defect operator $D \in \mathbf{B}(\mathcal{D},\mathcal{K})$ for T^*. Let $\mathcal{D}_-$ be a maximal uniformly negative subspace of $\mathcal{D}$, and define $D_- = D|\mathcal{D}_-$. By Corollary 1.4.2, $(T \quad D)$ is a bicontraction. In particular, D is a contraction and $D_-\mathcal{D}_- = D\mathcal{D}_-$ is a closed uniformly negative subspace of $\mathcal{K}$. Choose a maximal uniformly negative subspace $\mathcal{K}_-$ of $\mathcal{K}$ which contains $D_-\mathcal{D}_-$.

Choose closed subspaces $\mathcal{L}_-$ and $\mathcal{L}_+$ of $\mathcal{K}_-$ having the properties specified in Theorem 2.1.2. Define $\hat{\mathcal{K}} = \mathcal{L}_- \oplus \mathcal{K}_+$, where $\mathcal{K}_+ = \mathcal{K}_-^{\perp}$. We show that $\hat{\mathcal{K}}$ has the required properties. By construction, $\hat{\mathcal{K}}^{\perp}$ is uniformly negative.

For $f \in \hat{\mathcal{K}}$,

$$\begin{aligned}\langle f,f\rangle_{\hat{\mathcal{K}}} - \langle T^*f,T^*f\rangle_{\mathcal{H}} &= \langle (1-TT^*)f,f\rangle_{\mathcal{K}}\\ &= \langle D^*f,D^*f\rangle_{\mathcal{D}}\\ &\geq 0.\end{aligned}$$

The inequality is proved by contradiction in this way. If $\langle D^*f,D^*f\rangle_{\mathcal{K}} < 0$, then

$$D_-^*f = \mathrm{Pr}_{\mathcal{D}_-}D^*f \neq 0.$$

But $D_-^*\hat{\mathcal{K}} = D_-^*(\mathcal{L}_- \oplus \mathcal{K}_+) = \{0\}$ because $D^*\mathcal{L}_- \perp \mathcal{D}_-$ by the choice of $\mathcal{L}_-$ and $D_-\mathcal{D}_- \subset \mathcal{K}_-$ by the choice of $\mathcal{K}_-$. In view of this contradiction, we have shown that $T^*|\hat{\mathcal{K}}$ is a contraction from $\hat{\mathcal{K}}$ into $\mathcal{H}$.

The subspace $\mathcal{L}_-$ is maximal negative in $\hat{\mathcal{K}}$, and it is mapped by $T^*|\hat{\mathcal{K}}$ onto a maximal negative subspace of $\mathcal{H}$ by the choice of $\mathcal{L}_-$. Therefore $T^*|\hat{\mathcal{K}}$ is a bicontraction by Theorem 1.3.6. The last assertion is clear, since otherwise Theorem 1.3.6 would be violated. ■

The methods of this section yield a new characterization of bicontraction operators in the spirit of Theorem 1.3.6, but now with a condition on how adjoints map negative subspaces.

THEOREM 2.1.5. *Let $\mathcal{H}$ and $\mathcal{K}$ be Kreĭn spaces, and let $T \in \mathbf{B}(\mathcal{H},\mathcal{K})$ be a contraction. Then T is a bicontraction if and only if T^* maps some maximal negative subspace of $\mathcal{K}$ in a one-to-one way into a negative subspace of $\mathcal{H}$.*

Proof. Necessity follows from Theorem 1.3.6.

Conversely, assume that T^* maps the maximal negative subspace $\mathcal{N}$ of $\mathcal{K}$ into a negative subspace of $\mathcal{H}$. Let $D \in \mathbf{B}(\mathcal{D},\mathcal{K})$ be a defect operator for T^*. Let

$\mathcal{D} = \mathcal{D}_+ \oplus \mathcal{D}_-$, $\mathcal{H} = \mathcal{H}_+ \oplus \mathcal{H}_-$, and $\mathcal{K} = \mathcal{K}_+ \oplus \mathcal{K}_-$ be fundamental decompositions. The fundamental decomposition of $\mathcal{K}$ can be chosen so that $D\mathcal{D}_- \subset \mathcal{K}_-$, because $D\mathcal{D}_-$ is uniformly negative by an argument in the proof of Theorem 2.1.4. Set $D_- = D|\mathcal{D}_-$. Since the range of D_- is contained in $\mathcal{K}_-$, $D_-^*|\mathcal{K}_+ = 0$.

As in the proof of Theorem 2.1.2, the operator

$$R_-^* = (T \quad D_-) \in \mathbf{B}(\mathcal{H} \oplus \mathcal{D}_-, \mathcal{K})$$

is a bicontraction. By Theorem 1.3.6, R_-^* maps $\mathcal{N}$ in a one-to-one way onto a maximal negative subspace of $\mathcal{H} \oplus \mathcal{D}_-$. Therefore

$$R_-^*\mathcal{N} = \left\{ \begin{pmatrix} Kf \\ f \end{pmatrix} : f \in \mathcal{H}_- \oplus \mathcal{D}_- \right\}$$

for some contraction $K \in \mathbf{B}(|\mathcal{H}_- \oplus \mathcal{D}_-|, \mathcal{H}_+)$. Define

$$\mathcal{M} = \left\{ \begin{pmatrix} Kf \\ f \end{pmatrix} : f \in \mathcal{H}_- \right\}.$$

By its form, $\mathcal{M}$ is a maximal negative subspace of $\mathcal{H}$.

Since the range of $R_-^*|\mathcal{N}$ includes $\mathcal{M}$, there is a subspace $\mathcal{L}$ of $\mathcal{N}$ which is mapped by R_-^* onto $\mathcal{M}$. For $f \in \mathcal{L}$,

$$T^*f = \mathrm{Pr}_{\mathcal{H}}\, R_-^* f = R_-^* f,$$

because $R_-^* f \in \mathcal{M} \subset \mathcal{H}$. In other words, $T^*|\mathcal{L} = R_-^*|\mathcal{L}$. Now by assumption, T^* maps $\mathcal{N}$ in a one-to-one way into a negative subspace of $\mathcal{H}$. Since $\mathcal{L} \subset \mathcal{N}$ and $T^*\mathcal{L} = \mathcal{M}$ is maximal negative, we conclude that $\mathcal{L} = \mathcal{N}$. We therefore have $D_-^*|\mathcal{N} = 0$, because for $f \in \mathcal{N}$,

$$T^*f = R_-^*f = T^*f + D_-^*f$$

as elements of $\mathcal{H} \oplus \mathcal{D}_-$.

But $D_-^*|\mathcal{K}_+ = 0$, and so D_-^* annihilates $\mathcal{K}_+ + \mathcal{N}$. This is all of $\mathcal{K}$. In fact, $\mathcal{N}$ is maximal negative and has a representation as the set of vectors $Nu + u$, $u \in \mathcal{K}_-$, where $N \in \mathbf{B}(|\mathcal{K}_-|, \mathcal{K}_+)$ is a contraction. Therefore any u in $\mathcal{K}_-$ has a representation $u = -Nu + (Nu + u)$ with $-Nu \in \mathcal{K}_+$ and $Nu + u \in \mathcal{N}$. So $\mathcal{K}_- \subset \mathcal{K}_+ + \mathcal{N}$, and hence $\mathcal{K}_+ + \mathcal{N} = \mathcal{K}$. We have shown that $D_-^* = 0$.

Since $\mathrm{Pr}_{\mathcal{D}_-} D^* = D_-^* = 0$, the range of D^* is contained in $\mathcal{D}_+$. Therefore

$$\langle (1 - TT^*)f, f \rangle_{\mathcal{K}} = \langle D^*f, D^*f \rangle_{\mathcal{D}} \geq 0$$

for every f in $\mathcal{K}$, and T^* is a contraction. Since T is a contraction by assumption, T is a bicontraction. ■

2.2 Column Extensions

Let $\mathcal{H}$ and $\mathcal{K}$ be Kreĭn spaces, and let $T \in \mathbf{B}(\mathcal{H}, \mathcal{K})$. A *column extension* of T is an operator of the form

$$C = \begin{pmatrix} T \\ \tilde{E}^* \end{pmatrix} \in \mathbf{B}(\mathcal{H}, \mathcal{K} \oplus \tilde{\mathcal{E}}),$$

where $\tilde{\mathcal{E}}$ is a Kreĭn space and $\tilde{E} \in \mathbf{B}(\tilde{\mathcal{E}}, \mathcal{H})$. In this section we assume that $\tilde{\mathcal{E}}$ is a Hilbert space. We characterize the contractive column extensions of T when T is a contraction. We also characterize the bicontractive column extensions of T when T is a bicontraction.

The motivating example is obtained from a defect operator $\tilde{D} \in \mathbf{B}(\tilde{\mathcal{D}}, \mathcal{H})$ for T. If T is a contraction, $\tilde{\mathcal{D}}$ is a Hilbert space by Corollary 1.4.2. The operator

$$\begin{pmatrix} T \\ \tilde{D}^* \end{pmatrix} \in \mathbf{B}(\mathcal{H}, \mathcal{K} \oplus \tilde{\mathcal{D}})$$

is an isometry and hence a contractive column extension of T. If T is a bicontraction, then $\left(\begin{smallmatrix} T \\ \tilde{D}^* \end{smallmatrix}\right)$ is a bicontraction by Corollary 1.4.3.

Theorem 2.2.1. *Let $\mathcal{H}$ and $\mathcal{K}$ be Kreĭn spaces, and let $T \in \mathbf{B}(\mathcal{H}, \mathcal{K})$ be a given operator with defect operator $\tilde{D} \in \mathbf{B}(\tilde{\mathcal{D}}, \mathcal{H})$. Let*

$$C = \begin{pmatrix} T \\ \tilde{E}^* \end{pmatrix} \in \mathbf{B}(\mathcal{H}, \mathcal{K} \oplus \tilde{\mathcal{E}})$$

be a column extension of T with $\tilde{\mathcal{E}}$ a Hilbert space.

(i) *If T is a contraction, then C is a contraction if and only if $\tilde{E} = \tilde{D}\tilde{G}$ where $\tilde{G} \in \mathbf{B}(\tilde{\mathcal{E}}, \tilde{\mathcal{D}})$ is a contraction.*

(ii) *If T is a bicontraction, then C is a bicontraction if and only if $\tilde{E} = \tilde{D}\tilde{G}$ where $\tilde{G} \in \mathbf{B}(\tilde{\mathcal{E}}, \tilde{\mathcal{D}})$ is a contraction.*

Proof. (i) Assume that T is a contraction. Then $\tilde{\mathcal{D}}$ is a Hilbert space by Corollary 1.4.2. If C is a contraction, then

$$0 \le 1 - C^*C = 1 - T^*T - \tilde{E}\tilde{E}^* = \tilde{D}\tilde{D}^* - \tilde{E}\tilde{E}^*.$$

Hence

$$\|\tilde{E}^* f\|^2_{\tilde{\mathcal{E}}} \le \|\tilde{D}^* f\|^2_{\tilde{\mathcal{D}}}, \qquad f \in \mathcal{H}.$$

So $\tilde{E}^* = \tilde{G}^*\tilde{D}^*$ for a contraction $\tilde{G}^* \in \mathbf{B}(\tilde{\mathcal{D}}, \tilde{\mathcal{E}})$. Thus the condition is necessary. Sufficiency is verified by reversing these steps.

(ii) Assume that T is a bicontraction. Then T is a contraction, and the necessity of the condition follows from (i).

Conversely, let $\tilde{E} = \tilde{D}\tilde{G}$ where $\tilde{G} \in \mathbf{B}(\tilde{\mathcal{E}}, \tilde{\mathcal{D}})$ is a contraction. By (i), C is a contraction. We prove that C is a bicontraction by showing that it maps any maximal negative subspace $\mathcal{N}$ of $\mathcal{H}$ onto a maximal negative subspace of $\mathcal{K} \oplus \tilde{\mathcal{E}}$. Since $\tilde{\mathcal{E}}$ is a Hilbert space, any maximal uniformly negative subspace $\mathcal{K}_-$ of $\mathcal{K}$ is a maximal uniformly negative subspace of $\mathcal{K} \oplus \tilde{\mathcal{E}}$. We have

$$C\mathcal{N} = \left\{ \begin{pmatrix} Tf \\ \tilde{E}^* f \end{pmatrix} : f \in \mathcal{N} \right\},$$

and so by Theorem 1.3.6,

$$\mathrm{Pr}_{\mathcal{K}_-} C\mathcal{N} = \mathrm{Pr}_{\mathcal{K}_-} T\mathcal{N} = \mathcal{K}_-$$

because T is a bicontraction. Hence $C\mathcal{N}$ is maximal negative in $\mathcal{K} \oplus \tilde{\mathcal{E}}$, and another application of Theorem 1.3.6 shows that C is a bicontraction. ■

COROLLARY 2.2.2. *Let $\mathcal{H}$ and $\mathcal{K}$ be Kreĭn spaces, and let $T \in \mathbf{B}(\mathcal{H}, \mathcal{K})$ be a bicontraction. Let*

$$C = \begin{pmatrix} T \\ \tilde{E}^* \end{pmatrix} \in \mathbf{B}(\mathcal{H}, \mathcal{K} \oplus \tilde{\mathcal{E}})$$

be a column extension of T with $\tilde{\mathcal{E}}$ a Hilbert space. If C is a contraction, then C is a bicontraction.

Proof. Choose a defect operator for T as in Theorem 2.2.1. If C is a contraction, then by part (i) of the theorem, $\tilde{E} = \tilde{D}\tilde{G}$ where $\tilde{G} \in \mathbf{B}(\tilde{\mathcal{E}}, \tilde{\mathcal{D}})$ is a contraction. But then by part (ii) of the theorem, C is a bicontraction. ■

We note a norm estimate which will be useful in the commutant lifting theorem to be proved later.

COROLLARY 2.2.3. *Let $\mathcal{H}$ and $\mathcal{K}$ be Kreĭn spaces, and let $T \in \mathbf{B}(\mathcal{H}, \mathcal{K})$ be a contraction. Let*

$$C = \begin{pmatrix} T \\ \tilde{E}^* \end{pmatrix} \in \mathbf{B}(\mathcal{H}, \mathcal{K} \oplus \tilde{\mathcal{E}})$$

be a contractive column extension of T with $\tilde{\mathcal{E}}$ a Hilbert space. If norms are computed relative to fixed fundamental decompositions of $\mathcal{H}$ and $\mathcal{K}$ and the induced fundamental decomposition of $\mathcal{K} \oplus \tilde{\mathcal{E}}$, then

$$\|C\|^2 \le 1 + 2\|T\|^2.$$

Proof. Choose a defect operator for T and factor $\tilde{E} = \tilde{D}\tilde{G}$ as in Theorem 2.2.1 (i). Since T is a contraction, $\tilde{\mathcal{D}}$ is a Hilbert space by Corollary 1.4.2. For any $f \in \mathcal{H}$,

$$\left\|\tilde{D}^* f\right\|^2 = \left\langle \tilde{D}^* f, \tilde{D}^* f\right\rangle_{\tilde{\mathcal{D}}} = \left\langle \tilde{D}\tilde{D}^* f, f\right\rangle_{\mathcal{H}} = \langle (1 - T^*T)f, f\rangle_{\mathcal{H}}$$
$$\leq \left[1 + \|T\|^2\right]\|f\|^2.$$

Since $\tilde{G}$ is a Hilbert space contraction, for any $f \in \mathcal{H}$,

$$\|Cf\|^2 = \|Tf\|^2 + \|\tilde{E}^* f\|^2 = \|Tf\|^2 + \|\tilde{G}^*\tilde{D}^* f\|^2 \leq \left[1 + 2\|T\|^2\right]\|f\|^2,$$

as required. ■

2.3 Row Extensions

Let $T \in \mathbf{B}(\mathcal{H}, \mathcal{K})$, $\mathcal{H}$ and $\mathcal{K}$ Kreĭn spaces. By a *row extension* of T we mean an operator of the form

$$R = \begin{pmatrix} T & E \end{pmatrix} \in \mathbf{B}(\mathcal{H} \oplus \mathcal{E}, \mathcal{K}),$$

where $\mathcal{E}$ is a Kreĭn space and $E \in \mathbf{B}(\mathcal{E}, \mathcal{K})$. If T is a contraction, then there exist bicontractive row extensions of T. In fact, if $D \in \mathbf{B}(\mathcal{D}, \mathcal{K})$ is a defect operator for T^*, the operator

$$R = \begin{pmatrix} T & D \end{pmatrix} \in \mathbf{B}(\mathcal{H} \oplus \mathcal{D}, \mathcal{K})$$

is a bicontractive row extension of T by Corollary 1.4.2. We characterize all such extensions. It is convenient to begin with the special case of an extension by a Hilbert space.

Theorem 2.3.1. *Let $\mathcal{H}$ and $\mathcal{K}$ be Kreĭn spaces, $T \in \mathbf{B}(\mathcal{H}, \mathcal{K})$ a bicontraction, and $D \in \mathbf{B}(\mathcal{D}, \mathcal{K})$ a defect operator for T^*. Let*

$$R = \begin{pmatrix} T & E \end{pmatrix} \in \mathbf{B}(\mathcal{H} \oplus \mathcal{E}, \mathcal{K})$$

be a row extension of T with $\mathcal{E}$ a Hilbert space. Then R is a bicontraction if and only if $E = DG$ where $G \in \mathbf{B}(\mathcal{E}, \mathcal{D})$ is a contraction.

Note that $\mathcal{D}$ is a Hilbert space by the hypothesis that T is a bicontraction.

Proof. If R is a bicontraction, then the operators

$$R^* = \begin{pmatrix} T^* \\ E^* \end{pmatrix} \in \mathbf{B}(\mathcal{K}, \mathcal{H} \oplus \mathcal{E})$$

and T^* are contractions. By Theorem 2.2.1 (i) applied to T^*, $E = DG$ where $G \in \mathbf{B}(\mathcal{E}, \mathcal{D})$ is a contraction.

Conversely, assume that $E = DG$ where $G \in \mathbf{B}(\mathcal{E}, \mathcal{D})$ is a contraction. By Theorem 2.2.1 (ii) applied to T^*, the operator R^* is a bicontraction, hence R is a bicontraction. ■

A technical result is needed. It describes how bicontractive row extensions act with respect to the decompositions obtained in Theorem 2.1.2. This information will be used to determine the general form of any bicontractive row extension of a contraction.

THEOREM 2.3.2. *Let $\mathcal{H}$ and $\mathcal{K}$ be Kreĭn spaces, and let $T \in \mathbf{B}(\mathcal{H}, \mathcal{K})$ be a contraction. Let $\mathcal{K}_-$ be a maximal uniformly negative subspace of $\mathcal{K}$, and choose closed subspaces $\mathcal{L}_-$ and $\mathcal{L}_+$ of $\mathcal{K}_-$ such that*

$$\mathcal{K}_- = \mathcal{L}_- \dot{+} \mathcal{L}_+,$$

T^* maps $\mathcal{L}_-$ in a one-to-one way onto a maximal uniformly negative subspace of $\mathcal{H}$, and $T^*\mathcal{L}_+ \perp T^*\mathcal{L}_-$. Let

$$R = (\,T \quad E\,) \in \mathbf{B}(\mathcal{H} \oplus \mathcal{E}, \mathcal{K})$$

be any bicontractive row extension of T, where $\mathcal{E}$ a Kreĭn space and $E \in \mathbf{B}(\mathcal{E}, \mathcal{K})$. Then E^* maps $\mathcal{L}_+$ in a one-to-one way onto a maximal uniformly negative subspace of $\mathcal{E}$.

The existence of subspaces $\mathcal{L}_-$ and $\mathcal{L}_+$ with the required properties follows from Theorem 2.1.2.

Proof. Choose fundamental decompositions $\mathcal{E} = \mathcal{E}_+ \oplus \mathcal{E}_-$ and $\mathcal{H} = \mathcal{H}_+ \oplus \mathcal{H}_-$ with $\mathcal{H}_- = T^*\mathcal{L}_-$ and $\mathcal{H}_+ = (T^*\mathcal{L}_-)^\perp$. Let $\mathcal{K}_+ = \mathcal{K}_-^\perp$.

Since R^* is a bicontraction, $R^*\mathcal{K}_-$ is a maximal uniformly negative subspace of $\mathcal{H} \oplus \mathcal{E}$. If $b \in \mathcal{L}_+$, then $T^*b \in \mathcal{H}_+$ and for some positive number ϵ which does not depend on b,

$$\begin{aligned}\langle E^*b, E^*b\rangle_{\mathcal{E}} &\le \langle T^*b, T^*b\rangle_{\mathcal{H}} + \langle E^*b, E^*b\rangle_{\mathcal{E}} = \langle R^*b, R^*b\rangle_{\mathcal{H}\oplus\mathcal{E}} \\ &\le -\epsilon\|R^*b\|^2 = -\epsilon\|T^*b\|^2 - \epsilon\|E^*b\|^2 \\ &\le -\epsilon\|E^*b\|^2,\end{aligned}$$

where norms are computed in a suitable way. It follows that E^* maps $\mathcal{L}_+$ in a one-to-one way onto a uniformly negative subspace of $\mathcal{E}$.

We show that the projection of $E^*\mathcal{L}_+$ into $\mathcal{E}_-$ is all of $\mathcal{E}_-$. Since $R^*\mathcal{K}_-$ is maximal uniformly negative in $\mathcal{H} \oplus \mathcal{E}$,

$$\mathrm{Pr}_{\mathcal{H}_-\oplus\mathcal{E}_-} R^*\mathcal{K}_- = \mathcal{H}_- \oplus \mathcal{E}_-.$$

Given $e \in \mathcal{E}_-$, we may therefore choose $b \in \mathcal{K}_-$ such that

$$\mathrm{Pr}_{\mathcal{H}_-\oplus\mathcal{E}_-} R^*b = e.$$

Applying the projection onto $\mathcal{E}_-$, we obtain

$$\mathrm{Pr}_{\mathcal{E}_-} E^*b = \mathrm{Pr}_{\mathcal{E}_-} R^*b = e.$$

Applying instead the projection onto $\mathcal{H}_-$, we see that

$$\mathrm{Pr}_{\mathcal{H}_-} T^*b = \mathrm{Pr}_{\mathcal{H}_-} R^*b = 0,$$

that is, $T^*b \in \mathcal{H}_+$. Therefore if $b = u + v$, $u \in \mathcal{L}_-$, $v \in \mathcal{L}_+$, we have

$$T^*u = T^*b - T^*v \in \mathcal{H}_- \cap \mathcal{H}_+ = \{0\}.$$

Since $T^*|\mathcal{L}_-$ is one-to-one, $u = 0$ and $b = v \in \mathcal{L}_+$. We have shown that the projection of $E^*\mathcal{L}_+$ into $\mathcal{E}_-$ is all of $\mathcal{E}_-$, and so $E^*\mathcal{L}_+$ is maximal uniformly negative by the graph representation of negative subspaces. ■

The main result of this section follows.

THEOREM 2.3.3. *Let $\mathcal{H}$ and $\mathcal{K}$ be Kreĭn spaces, $T \in \mathbf{B}(\mathcal{H},\mathcal{K})$ a contraction, and $D \in \mathbf{B}(\mathcal{D},\mathcal{K})$ a defect operator for T^*. Let $E \in \mathbf{B}(\mathcal{E},\mathcal{K})$, where $\mathcal{E}$ is a Kreĭn space. Then the operator*

$$R = \begin{pmatrix} T & E \end{pmatrix} \in \mathbf{B}(\mathcal{H} \oplus \mathcal{E}, \mathcal{K})$$

is a bicontraction if and only if $E = DG$, where $G \in \mathbf{B}(\mathcal{E},\mathcal{D})$ is a bicontraction.

Proof of Sufficiency. Assume that $E = DG$ where $G \in \mathbf{B}(\mathcal{E},\mathcal{D})$ is a bicontraction. Since $1 - GG^* \geq 0$,

$$1 - RR^* = 1 - TT^* - DGG^*D^* = D(1 - GG^*)D^* \geq 0,$$

and R^* is a contraction. By Theorem 1.2.4, T has a Julia operator of the form

$$U = \begin{pmatrix} T & D \\ \tilde{D}^* & L \end{pmatrix} \in \mathbf{B}(\mathcal{H} \oplus D, \mathcal{K} \oplus \tilde{\mathcal{D}}).$$

Thus

$$\begin{aligned} 1 - R^*R &= 1 - \begin{pmatrix} T^* \\ G^*D^* \end{pmatrix} (T \quad DG) \\ &= \begin{pmatrix} 1 - T^*T & -T^*DG \\ -G^*D^*T & 1 - G^*D^*DG \end{pmatrix} \\ &= \begin{pmatrix} \tilde{D}\tilde{D}^* & \tilde{D}LG \\ G^*L^*\tilde{D}^* & G^*L^*LG \end{pmatrix} + \begin{pmatrix} 0 & 0 \\ 0 & 1 - G^*D^*DG - G^*L^*LG \end{pmatrix} \\ &= \begin{pmatrix} \tilde{D} \\ G^*L^* \end{pmatrix} (\tilde{D}^* \quad LG) + \begin{pmatrix} 0 & 0 \\ 0 & 1 - G^*G \end{pmatrix} \\ &\geq 0, \end{aligned}$$

because $\tilde{\mathcal{D}}$ is a Hilbert space by Theorem 1.4.1 and $1 - G^*G \geq 0$. Thus R is a bicontraction.

Proof of Necessity. Assume now that R is a bicontraction. Choose fundamental decompositions

$$\mathcal{E} = \mathcal{E}_+ \oplus \mathcal{E}_-, \quad \mathcal{D} = \mathcal{D}_+ \oplus \mathcal{D}_-, \quad \mathcal{K} = \mathcal{K}_+ \oplus \mathcal{K}_-.$$

Define $E_- = E|\mathcal{E}_- \in \mathbf{B}(\mathcal{E}_-, \mathcal{K})$. The operator

$$R_- = (T \quad E_-) \in \mathbf{B}(\mathcal{H} \oplus \mathcal{E}_-, \mathcal{K})$$

is a bicontraction by Lemma 2.1.3. Choose closed subspaces $\mathcal{L}_-$ and $\mathcal{L}_+$ of $\mathcal{K}_-$ as in Theorem 2.1.2. In particular, $\mathcal{K}_- = \mathcal{L}_- \dot{+} \mathcal{L}_+$.

By Theorem 2.3.2, E^* maps $\mathcal{L}_+$ in a one-to-one way onto a maximal uniformly negative subspace of $\mathcal{E}$. The projection of $\mathcal{E}$ onto $\mathcal{E}_-$ maps any maximal negative subspace of $\mathcal{E}$ onto $\mathcal{E}_-$ by the graph representation. Therefore E_-^* maps $\mathcal{L}_+$ in a one-to-one way onto $\mathcal{E}_-$.

We show that

$$\mathcal{K} = \mathcal{L}_+ \dot{+} \ker E_-^*.$$

If $k \in \mathcal{K}$, then $E_-^* k \in \mathcal{E}_-$ and there exists $u \in \mathcal{L}_+$ such that $E_-^* k = E_-^* u$. Then $k = u + v$ where $u \in \mathcal{L}_+$ and $v = k - u \in \ker E_-^*$. If $u \in \mathcal{L}_+ \cap \ker E_-^*$, then $u = 0$ because $E_-^*|\mathcal{L}_+$ is one-to-one and u is an element of $\mathcal{L}_+$, which is annihilated by E_-^*. Thus $\mathcal{K}$ is the direct sum of $\mathcal{L}_+$ and $\ker E_-^*$.

We next define a bicontractive operator $G_- \in \mathbf{B}(\mathcal{E}_-, \mathcal{D})$ such that $E_- = DG_-$. To this end first define a mapping X on the dense set $\operatorname{ran} D^*$ in $\mathcal{D}$ to $\mathcal{E}_-$ by

$$XD^*f = E_-^*f, \qquad f \in \mathcal{K}.$$

The mapping is well defined. For suppose $f \in \mathcal{K}$ and $D^*f = 0$. Write $f = u + v$, $u \in \mathcal{L}_+$, $v \in \ker E_-^*$. Then $D^*u = -D^*v$. By Corollary 1.4.2, $(T \quad D)$ is a bicontractive row extension of T. Applying Theorem 2.3.2 to $(T \quad D)$, we obtain

$$\langle D^*u, D^*u\rangle_{\mathcal{D}} \leq 0,$$

with equality only for $u = 0$. Since R_- is a bicontraction and $E_-^*v = 0$,

$$\begin{aligned} 0 &\leq \langle (1 - R_-R_-^*)v, v\rangle_{\mathcal{K}} = \langle (1 - TT^* - E_-E_-^*)v, v\rangle_{\mathcal{K}} \\ &= \langle D^*v, D^*v\rangle_{\mathcal{D}} = \langle D^*u, D^*u\rangle_{\mathcal{D}} \leq 0. \end{aligned}$$

Therefore equality holds throughout and u $= 0$. Thus $E_-^*f = E_-^*v = 0$, and X is well defined.

As far as it is defined, X satisfies

$$\langle Xg, Xg\rangle_{\mathcal{E}_-} \leq \langle g, g\rangle_{\mathcal{D}}, \qquad g \in \operatorname{dom} X.$$

For if $g = D^*f$, $f \in \mathcal{K}$,

$$\begin{aligned} 0 &\leq \langle (1 - R_-R_-^*)f, f\rangle_{\mathcal{K}} = \langle (1 - TT^* - E_-E_-^*)f, f\rangle_{\mathcal{K}} \\ &= \langle D^*f, D^*f\rangle_{\mathcal{D}} - \langle E_-^*f, E_-^*f\rangle_{\mathcal{E}_-} = \langle g, g\rangle_{\mathcal{D}} - \langle Xg, Xg\rangle_{\mathcal{E}_-}. \end{aligned}$$

Theorem 2.3.2 applied to $(T \quad D)$ implies that $D^*\mathcal{L}_+$ is a maximal uniformly negative subspace of $\mathcal{D}$. By the definition of X, $D^*\mathcal{L}_+ \subset \operatorname{dom} X$ and X maps $D^*\mathcal{L}_+$ onto the maximal uniformly negative subspace $E^*\mathcal{L}_+$ of $\mathcal{E}$. Therefore by Theorem 1.4.4, X has an extension by continuity to a bicontraction which we denote G_-^*. This completes the construction of a bicontractive operator $G_- \in \mathbf{B}(\mathcal{E}_-, \mathcal{D})$ such that $E_- = DG_-$.

Let $D_{R_-} \in \mathbf{B}(\mathcal{D}_{R_-}, \mathcal{K})$ and $D_{G_-} \in \mathbf{B}(\mathcal{D}_{G_-}, \mathcal{D})$ be defect operators for R_-^* and G_-^*. Then

$$\begin{aligned} D_{R_-}D_{R_-}^* &= 1 - R_-R_-^* = 1 - TT^* - E_-E_-^* \\ &= DD^* - DG_-G_-^*D^* = D(1 - G_-G_-^*)D^* \\ &= DD_{G_-}D_{G_-}^*D^*. \end{aligned}$$

Therefore there is a unitary operator $V \in \mathbf{B}(\mathcal{D}_{R_-}, \mathcal{D}_{G_-})$ such that

$$D_{R_-} = DD_{G_-}V.$$

Write $R = (T \quad E_- \quad E_+)$ where $E_+ = E|\mathcal{E}_+ \in \mathbf{B}(\mathcal{E}_+, \mathcal{K})$. The 1×3 row matrix acts on $\mathcal{H}\oplus\mathcal{E}_-\oplus\mathcal{E}_+$ to $\mathcal{K}$. Its restriction to $\mathcal{H}\oplus\mathcal{E}_-$ is the bicontraction R_-, and so by Theorem 2.3.1, $E_+ = D_{R_-}G_+$, where $G_+ \in \mathbf{B}(\mathcal{E}_+, \mathcal{D}_{R_-})$ is a contraction. Finally,

$$R = (\,T \quad DG_- \quad D_{R_-}G_+\,) = (\,T \quad DG_- \quad DD_{G_-}VG_+\,) = (\,T \quad DG\,),$$

where $G = \begin{pmatrix} G_- & D_{G_-}VG_+ \end{pmatrix} \in \mathbf{B}(\mathcal{E}_- \oplus \mathcal{E}_+, \mathcal{D}) = \mathbf{B}(\mathcal{E}, \mathcal{D})$ is a bicontraction by Theorem 2.3.1. ■

A similar theorem on contractive row extensions follows as a consequence.

THEOREM 2.3.4. *Let $\mathcal{H}$ and $\mathcal{K}$ be Kreĭn spaces, $T \in \mathbf{B}(\mathcal{H}, \mathcal{K})$ a contraction, and $D \in \mathbf{B}(\mathcal{D}, \mathcal{K})$ a defect operator for T^*. Let $\mathcal{E}$ be a Kreĭn space, and let $E \in \mathbf{B}(\mathcal{E}, \mathcal{K})$. Then the operator*

$$R = (\,T \quad E\,) \in \mathbf{B}(\mathcal{H} \oplus \mathcal{E}, \mathcal{K})$$

is a contraction if and only if $E = DG$ where $G \in \mathbf{B}(\mathcal{E}, \mathcal{D})$ is a contraction.

Proof of Sufficiency. If $E = DG$ where $G \in \mathbf{B}(\mathcal{E}, \mathcal{D})$ is a contraction, we show that $1 - R^*R \geq 0$ exactly as in the sufficiency part of the proof of Theorem 2.3.3.

Proof of Necessity. Assume that R is a contraction, and choose a defect operator $D_R \in \mathbf{B}(\mathcal{D}_R, \mathcal{K})$ for R^*. By Corollary 1.4.2, $\hat{R} = (R \quad D_R)$ is a bicontraction in $\mathbf{B}((\mathcal{H} \oplus \mathcal{E}) \oplus \mathcal{D}_R, \mathcal{K})$. In another view,

$$\hat{R} = (\,T \quad (\,E \quad D_R\,)\,) \in \mathbf{B}(\mathcal{H} \oplus (\mathcal{E} \oplus \mathcal{D}_R), \mathcal{K}),$$

is a bicontractive row extension of T, and we may apply the necessity part of Theorem 2.3.3. It follows that

$$(\,E \quad D_R\,) = D\Gamma,$$

where $\Gamma \in \mathbf{B}(\mathcal{E} \oplus \mathcal{D}_R, \mathcal{D})$ is a bicontraction. Thus

$$R = \hat{R}|(\mathcal{H} \oplus \mathcal{E}) = (\,T \quad DG\,)$$

where $G = \Gamma|\mathcal{E}$ is a contraction. ■

2.4 Two-by-Two Matrix Completions

Let $\mathcal{H}$ and $\mathcal{K}$ be Kreĭn spaces, and let $T \in \mathbf{B}(\mathcal{H}, \mathcal{K})$ be a contraction. We label all contractive two-by-two matrices with T in the upper left entry. The extension space for the range is required to be a Hilbert space.

Theorem 2.4.1. *Let $\mathcal{H}$ and $\mathcal{K}$ be Kreĭn spaces, and let $T \in \mathbf{B}(\mathcal{H}, \mathcal{K})$ be a contraction with Julia operator*

$$\begin{pmatrix} T & D \\ \tilde{D}^* & L \end{pmatrix} \in \mathbf{B}(\mathcal{H} \oplus \mathcal{D}, \mathcal{K} \oplus \tilde{\mathcal{D}}).$$

Let $\mathcal{E}$ be a Kreĭn space, and let $\tilde{\mathcal{E}}$ be a Hilbert space. If

$$Q = \begin{pmatrix} T & Q_1 \\ Q_2 & Q_3 \end{pmatrix} \in \mathbf{B}(\mathcal{H} \oplus \mathcal{E}, \mathcal{K} \oplus \tilde{\mathcal{E}})$$

is a contraction, then

$$Q = \begin{pmatrix} T & DG_1 \\ G_2^* \tilde{D}^* & G_2^* L G_1 + \tilde{D}_{G_2} G_3 \tilde{D}_{G_1}^* \end{pmatrix},$$

where

(i) $G_1 \in \mathbf{B}(\mathcal{E}, \mathcal{D})$ is a Kreĭn space contraction with defect operator $\tilde{D}_{G_1} \in \mathbf{B}(\tilde{\mathcal{D}}_{G_1}, \mathcal{E})$,

(ii) $G_2 \in \mathbf{B}(\tilde{\mathcal{E}}, \tilde{\mathcal{D}})$ is a Hilbert space contraction with defect operator $\tilde{D}_{G_2} \in \mathbf{B}(\tilde{\mathcal{D}}_{G_2}, \tilde{\mathcal{E}})$, and

(iii) $G_3 \in \mathbf{B}(\tilde{\mathcal{D}}_{G_1}, \tilde{\mathcal{D}}_{G_2})$ is a Hilbert space contraction.

Conversely, every operator $Q \in \mathbf{B}(\mathcal{H} \oplus \mathcal{E}, \mathcal{K} \oplus \tilde{\mathcal{E}})$ of this form is a contraction.

Proof. Assume that Q is a contraction. Then so is its restriction

$$\begin{pmatrix} T \\ Q_2 \end{pmatrix} \in \mathbf{B}(\mathcal{H}, \mathcal{K} \oplus \tilde{\mathcal{E}}).$$

Since T is a contraction and $\tilde{\mathcal{E}}$ is a Hilbert space, we can apply Theorem 2.2.1 (i) to write $Q_2 = G_2^* \tilde{D}^*$ where $G_2 \in \mathbf{B}(\tilde{\mathcal{E}}, \tilde{\mathcal{D}})$ is a Hilbert space contraction. Let $\tilde{D}_{G_2} \in \mathbf{B}(\tilde{\mathcal{D}}_{G_2}, \tilde{\mathcal{E}})$ be any defect operator for G_2.

Since $\tilde{\mathcal{E}}$ is a Hilbert space, $(T \quad Q_1) = \mathrm{Pr}_{\mathcal{K}} Q \in \mathbf{B}(\mathcal{H} \oplus \mathcal{E}, \mathcal{K})$ is a contraction. By Theorem 2.3.4, $Q_1 = DG_1$ where $G_1 \in \mathbf{B}(\mathcal{E}, \mathcal{D})$ is a Kreĭn space contraction. Let $\tilde{D}_{G_1} \in \mathbf{B}(\tilde{\mathcal{D}}_{G_1}, \mathcal{E})$ be any defect operator for G_1. Thus far we have

$$Q = \begin{pmatrix} T & DG_1 \\ G_2^* \tilde{D}^* & Q_3 \end{pmatrix}.$$

After a short calculation, the inequality $1 - Q^*Q \geq 0$ yields

$$\begin{aligned} 0 &\leq \begin{pmatrix} 1 - T^*T - \tilde{D}G_2G_2^*\tilde{D}^* & -T^*DG_1 - \tilde{D}G_2Q_3 \\ -G_1^*D^*T - Q_3^*G_2^*\tilde{D}^* & 1 - G_1^*D^*DG_1 - Q_3^*Q_3 \end{pmatrix} \\ &= \begin{pmatrix} \tilde{D}\tilde{D}^* - \tilde{D}G_2G_2^*\tilde{D}^* & \tilde{D}LG_1 - \tilde{D}G_2Q_3 \\ G_1^*L^*\tilde{D}^* - Q_3^*G_2^*\tilde{D}^* & 1 - G_1^*G_1 + G_1^*(1 - D^*D)G_1 - Q_3^*Q_3 \end{pmatrix} \\ &= \begin{pmatrix} \tilde{D} & 0 \\ 0 & 1 \end{pmatrix} \begin{pmatrix} 1 - G_2G_2^* & LG_1 - G_2Q_3 \\ G_1^*L^* - Q_3^*G_2^* & \tilde{D}_{G_1}\tilde{D}_{G_1}^* + G_1^*L^*LG_1 - Q_3^*Q_3 \end{pmatrix} \begin{pmatrix} \tilde{D}^* & 0 \\ 0 & 1 \end{pmatrix} \end{aligned}$$

Since $\tilde{D}$ is one-to-one, the range of $\tilde{D}^*$ is dense in $\tilde{\mathcal{D}}$. Therefore

$$\begin{pmatrix} 1 - G_2G_2^* & LG_1 - G_2Q_3 \\ G_1^*L^* - Q_3^*G_2^* & \tilde{D}_{G_1}\tilde{D}_{G_1}^* + G_1^*L^*LG_1 - Q_3^*Q_3 \end{pmatrix} \geq 0$$

as operators in $\mathbf{B}(\tilde{\mathcal{D}} \oplus \mathcal{E})$. A straightforward calculation brings this to the form $A^*A \geq B^*B$, where

$$A = \begin{pmatrix} 1 & LG_1 \\ 0 & \tilde{D}_{G_1}^* \end{pmatrix} \in \mathbf{B}(\tilde{\mathcal{D}} \oplus \mathcal{E}, \tilde{\mathcal{D}} \oplus \tilde{\mathcal{D}}_{G_1})$$

and

$$B = \begin{pmatrix} G_2^* & Q_3 \end{pmatrix} \in \mathbf{B}(\tilde{\mathcal{D}} \oplus \mathcal{E}, \tilde{\mathcal{E}}).$$

Since $\tilde{\mathcal{D}} \oplus \tilde{\mathcal{D}}_{G_1}$ and $\tilde{\mathcal{E}}$ are Hilbert spaces, there is a Hilbert space contraction

$$K = \begin{pmatrix} K_0 & K_1 \end{pmatrix} \in \mathbf{B}(\tilde{\mathcal{D}} \oplus \tilde{\mathcal{D}}_{G_1}, \tilde{\mathcal{E}})$$

such that $B = KA$, that is,

$$\begin{aligned} \begin{pmatrix} G_2^* & Q_3 \end{pmatrix} &= \begin{pmatrix} K_0 & K_1 \end{pmatrix} \begin{pmatrix} 1 & LG_1 \\ 0 & \tilde{D}_{G_1}^* \end{pmatrix} \\ &= \begin{pmatrix} K_0 & K_0LG_1 + K_1\tilde{D}_{G_1}^* \end{pmatrix}. \end{aligned}$$

Thus $K_0 = G_2^*$ and $Q_3 = K_0LG_1 + K_1\tilde{D}_{G_1}^*$. Applying Theorem 2.3.1 in the Hilbert space case, we obtain

$$\begin{pmatrix} K_0 & K_1 \end{pmatrix} = \begin{pmatrix} G_2^* & \tilde{D}_{G_2}G_3 \end{pmatrix}$$

where $G_3 \in \mathbf{B}(\tilde{\mathcal{D}}_{G_1}, \tilde{\mathcal{D}}_{G_2})$ is a Hilbert space contraction. Therefore

$$Q_3 = K_0LG_1 + K_1\tilde{D}_{G_1}^* = G_2^*LG_1 + \tilde{D}_{G_2}G_3\tilde{D}_{G_1}^*.$$

We have shown that Q has the required form if Q is a contraction. In the other direction, if Q has the stated form, we simply reverse the steps to show that $1 - Q^*Q \geq 0$. ■

A similar result characterizes bicontractive two-by-two matrices with T in the upper left entry when T is a contraction and the extension space for the range is a Hilbert space.

THEOREM 2.4.2. *Let $\mathcal{H}$ and $\mathcal{K}$ be Kreĭn spaces, and let $T \in \mathbf{B}(\mathcal{H}, \mathcal{K})$ be a contraction with Julia operator*

$$\begin{pmatrix} T & D \\ \tilde{D}^* & L \end{pmatrix} \in \mathbf{B}(\mathcal{H} \oplus \mathcal{D}, \mathcal{K} \oplus \tilde{\mathcal{D}}).$$

Let $\mathcal{E}$ be a Kreĭn space, and let $\tilde{\mathcal{E}}$ be a Hilbert space. If

$$Q = \begin{pmatrix} T & Q_1 \\ Q_2 & Q_3 \end{pmatrix} \in \mathbf{B}(\mathcal{H} \oplus \mathcal{E}, \mathcal{K} \oplus \tilde{\mathcal{E}})$$

is a bicontraction, then Q can be represented as in Theorem 2.4.1 with G_1 a bicontraction. Conversely, every $Q \in \mathbf{B}(\mathcal{H} \oplus \mathcal{E}, \mathcal{K} \oplus \tilde{\mathcal{E}})$ of the form described in Theorem 2.4.1 with G_1 a bicontraction is a bicontraction.

Proof. Assume first that Q has the form described in Theorem 2.4.1 with G_1 a bicontraction. By Theorem 2.3.3,

$$(T \quad Q_1) = (T \quad DG_1) \in \mathbf{B}(\mathcal{H} \oplus \mathcal{E}, \mathcal{K})$$

is a bicontraction. Now view

$$Q = \begin{pmatrix} (T & Q_1) \\ (Q_2 & Q_3) \end{pmatrix} \in \mathbf{B}(\mathcal{H} \oplus \mathcal{E}, \mathcal{K} \oplus \tilde{\mathcal{E}})$$

as a contractive column extension of the bicontraction $(T \quad Q_1)$ with extension space $\tilde{\mathcal{E}}$ a Hilbert space. By Corollary 2.2.2, Q is a bicontraction.

Conversely, let Q be a bicontraction. If $\mathcal{N}$ is a maximal negative subspace of $\mathcal{H} \oplus \mathcal{E}$, then $Q\mathcal{N}$ is a maximal negative subspace of $\mathcal{K} \oplus \tilde{\mathcal{E}}$. Since $\tilde{\mathcal{E}}$ is a Hilbert space, $\mathrm{Pr}_{\mathcal{K}} Q\mathcal{N} = (T \quad Q_1)\mathcal{N}$ is a maximal negative subspace of $\mathcal{K}$. As in the proof of Theorem 2.4.1, $(T \quad Q_1) \in \mathbf{B}(\mathcal{H} \oplus \mathcal{E}, \mathcal{K})$ is a contraction, and hence it is a bicontraction by Theorem 1.3.6. By Theorem 2.3.3, $Q_1 = DG_1'$ where $G_1' \in \mathbf{B}(\mathcal{E}, \tilde{\mathcal{D}})$ is a bicontraction. Since D has zero kernel, $G_1 = G_1'$ is a bicontraction. ■

COROLLARY 2.4.3. *Let $\mathcal{H}$ and $\mathcal{K}$ be Kreĭn spaces, and let $T \in \mathbf{B}(\mathcal{H}, \mathcal{K})$ be a bicontraction. Let $\mathcal{E}$ be a Kreĭn space, and let $\tilde{\mathcal{E}}$ be a Hilbert space. If*

$$Q = \begin{pmatrix} T & Q_1 \\ Q_2 & Q_3 \end{pmatrix} \in \mathbf{B}(\mathcal{H} \oplus \mathcal{E}, \mathcal{K} \oplus \tilde{\mathcal{E}})$$

is a contraction, then $\mathcal{E}$ is also a Hilbert space and Q is a bicontraction.

Proof. Choose a Julia operator

$$\begin{pmatrix} T & D \\ \tilde{D}^* & L \end{pmatrix} \in \mathbf{B}(\mathcal{H} \oplus \mathcal{D}, \mathcal{K} \oplus \tilde{\mathcal{D}}).$$

for T, and represent Q as in Theorem 2.4.1. Since T is a bicontraction, $\mathcal{D}$ is a Hilbert space by Corollary 1.4.3. Since $G_1 \in \mathbf{B}(\mathcal{E},\mathcal{D})$ is a contraction, $\mathcal{E}$ is a Hilbert space and G_1 is a bicontraction. So Q is a bicontraction by Theorem 2.4.2. ■

We recast the results of this section in a form that is convenient for application to the commutant lifting problem.

THEOREM 2.4.4. *Let $\mathcal{H}$, $\mathcal{K}$, $\mathcal{E}$ be Kreĭn spaces, $\tilde{\mathcal{E}}$ a Hilbert space. Assume that*

$$\begin{pmatrix} C_{11} & C_{12} \end{pmatrix} \in \mathbf{B}(\mathcal{H}\oplus\mathcal{E},\mathcal{K}) \qquad \text{and} \qquad \begin{pmatrix} C_{11} \\ C_{21} \end{pmatrix} \in \mathbf{B}(\mathcal{H},\mathcal{K}\oplus\tilde{\mathcal{E}})$$

are contraction operators. There exists an operator $C_{22} \in \mathbf{B}(\mathcal{E},\tilde{\mathcal{E}})$ such that

$$C = \begin{pmatrix} C_{11} & C_{12} \\ C_{21} & C_{22} \end{pmatrix} \in \mathbf{B}(\mathcal{H}\oplus\mathcal{E},\mathcal{K}\oplus\tilde{\mathcal{E}})$$

is a contraction. If $(C_{11} \quad C_{12})$ is a bicontraction, C_{22} may be chosen so that C is a bicontraction.

Proof. Since $(C_{11} \quad C_{12})$ is a contraction, so is its restriction C_{11}. Write $T = C_{11}$, and let

$$U = \begin{pmatrix} T & D \\ \tilde{D}^* & L \end{pmatrix} \in \mathbf{B}(\mathcal{H}\oplus\mathcal{D},\mathcal{K}\oplus\tilde{\mathcal{D}})$$

be a Julia operator for T. By Theorem 2.3.4, $C_{12} = DG_1$, where $G_1 \in \mathbf{B}(\mathcal{E},\mathcal{D})$ is a Kreĭn space contraction. By Theorem 2.2.1 (i), $C_{21} = G_2^*\tilde{D}^*$, where $G_2 \in \mathbf{B}(\tilde{\mathcal{E}},\tilde{\mathcal{D}})$ is a Hilbert space contraction. We may now invoke Theorem 2.4.1 with any choice of G_3, such as, $G_3 = 0$ (which is a contraction since the underlying spaces are Hilbert spaces), to produce an operator C_{22} such that C is a contraction.

If $(C_{11} \quad C_{12})$ is a bicontraction, then in the preceding argument we can use Theorem 2.3.3 in place of Theorem 2.3.4 to choose G_1 to be a bicontraction. In place of Theorem 2.4.1, we may use Theorem 2.4.2 to produce C_{22} such that C is a bicontraction. ■

To conclude this section, we use Theorem 2.4.1 to show that Julia operators for contractions, or operators whose adjoints are contractions, are essentially unique. See Appendix B for a more general result.

THEOREM 2.4.5. *Let $\mathcal{H}$ and $\mathcal{K}$ be Kreĭn spaces, and let $T \in \mathbf{B}(\mathcal{H},\mathcal{K})$ have two Julia operators*

$$\begin{pmatrix} T & D_j \\ \tilde{D}_j^* & L_j \end{pmatrix} \in \mathbf{B}(\mathcal{H} \oplus \mathcal{D}_j, \mathcal{K} \oplus \tilde{\mathcal{D}}_j), \qquad j = 1,2.$$

If either T or T^* is a contraction, then there exist unitary operators $V \in \mathbf{B}(\mathcal{D}_2, \mathcal{D}_1)$ and $\tilde{V} \in \mathbf{B}(\tilde{\mathcal{D}}_2, \tilde{\mathcal{D}}_1)$ such that

$$\begin{pmatrix} T & D_2 \\ \tilde{D}_2^* & L_2 \end{pmatrix} = \begin{pmatrix} 1 & 0 \\ 0 & \tilde{V}^* \end{pmatrix} \begin{pmatrix} T & D_1 \\ \tilde{D}_1^* & L_1 \end{pmatrix} \begin{pmatrix} 1 & 0 \\ 0 & V \end{pmatrix}.$$

Proof. Since the adjoint of a Julia operator for T is a Julia operator for T^*, it is sufficient to give the proof in the case in which T is a contraction. In this case, Theorem 2.4.1 is applicable and yields contractions $V \in \mathbf{B}(\mathcal{D}_2, \mathcal{D}_1)$ and $\tilde{V} \in \mathbf{B}(\tilde{\mathcal{D}}_2, \tilde{\mathcal{D}}_1)$ such that

$$\begin{pmatrix} T & D_2 \\ \tilde{D}_2^* & L_2 \end{pmatrix} = \begin{pmatrix} 1 & 0 \\ 0 & \tilde{V}^* \end{pmatrix} \begin{pmatrix} T & D_1 \\ \tilde{D}_1^* & L_1 \end{pmatrix} \begin{pmatrix} 1 & 0 \\ 0 & V \end{pmatrix} + \begin{pmatrix} 0 & 0 \\ 0 & \tilde{D}_{\tilde{V}} G \tilde{D}_V^* \end{pmatrix},$$

where $\tilde{D}_V \in \mathbf{B}(\tilde{\mathcal{D}}_V, \mathcal{D}_2)$ and $\tilde{D}_{\tilde{V}} \in \mathbf{B}(\tilde{\mathcal{D}}_{\tilde{V}}, \tilde{\mathcal{D}}_2)$ are defect operators for V and $\tilde{V}$ and $G \in \mathbf{B}(\tilde{\mathcal{D}}_V, \tilde{\mathcal{D}}_{\tilde{V}})$ is a contraction. In particular, $D_2 = D_1 V$ and

$$D_1 D_1^* - 1 - TT^* = D_2 D_2^* = D_1 V V^* D_1^*.$$

Since D_1 has zero kernel, VV^* is the identity on $\mathcal{D}_1$. Since D_2 has zero kernel, V is a unitary operator. In a similar way, $\tilde{D}_2 = \tilde{D}_1 \tilde{V}$ and

$$\tilde{D}_1 \tilde{D}_1^* = 1 - T^*T = \tilde{D}_2 \tilde{D}_2^* = \tilde{D}_1 \tilde{V} \tilde{V}^* \tilde{D}_1^*.$$

Since $\tilde{D}_1$ and $\tilde{D}_2$ have zero kernels, $\tilde{V}$ is a unitary operator. In particular, $\tilde{\mathcal{D}}_V$ and $\tilde{\mathcal{D}}_{\tilde{V}}$ each contain only the zero vector, and $\tilde{D}_{\tilde{V}} G \tilde{D}_V^* = 0$. ■

NOTES ON CHAPTER 2

The methods of Chapter 2 follow Dritschel [**30**]. Theorems 2.1.1 and 2.1.4 generalize a result of Kuzhel' [**48**], but the treatment here is different.

Theorems 2.3.3 and 2.3.4 first appear in Dritschel [**30**] in the present generality but were derived in special cases in earlier works. In fact, for most of the extension theorems in §2.2–2.4, there have been versions worked out by Arsene, Constantinescu, and Ghéondea [**8**]. The present work was done independently of the recent papers by Constantinescu and Ghéondea [**22**]. These papers contain a great amount of detailed information on related extension problems which we do not treat. The problems are generally of a more precise nature and require stronger hypotheses. While there is overlap with the present work in some cases, there is no inclusion of one theory in the other. In the bicontractive case, forerunners of the results in §2.2–2.4 also appear in Dritschel [**29**].

The method of proof of Theorem 2.4.1 is adapted from the account of the Hilbert space result in Pták and Vrbová [**56**]. In [**17**], de Branges uses complementation theory to prove a version of Theorem 2.4.1 when both extension spaces are Hilbert spaces. Two-by-two extension theorems in the Hilbert space case are due to Arsene and Ghéondea [**7**], Davis [**24**], Davis, Kahan, and Weinberger [**26**], and Shmul'yan and Yanovskaya [**65**]. Related papers are Davis [**25**] and Parrott [**54**].

The question of uniqueness of Julia operators is partially settled in Theorem 2.4.5 and Appendix B. We feel that uniqueness will fail in general, but we do not know an example.

Chapter 3: Commutant Lifting of Contraction Operators

3.1 Dilation Theory

The dilation properties of Kreĭn space operators are similar to the Hilbert space case. The main problem is to finesse the fact that a densely defined Kreĭn space isometry may fail to have a continuous extension: in some cases, this causes difficulty with uniqueness. It turns out that minimal isometric dilations of contractions are unique and have the familiar properties of the Hilbert space case. The same is true of minimal unitary dilations of bicontractions. For general operators, minimal isometric and minimal unitary dilations always exist with the properties of the Hilbert space case, but we cannot assert uniqueness nor that all minimal isometric and unitary dilations have desirable additional properties.

Definition 3.1.1. *Let $T \in \mathbf{B}(\mathcal{H})$, where $\mathcal{H}$ is a Kreĭn space. An* **isometric dilation** *of T is an isometry $U \in \mathbf{B}(\tilde{\mathcal{H}})$, where $\tilde{\mathcal{H}}$ is a Kreĭn space containing $\mathcal{H}$ isometrically as a regular subspace, such that*

$$T^n = \mathrm{Pr}_{\mathcal{H}} U^n | \mathcal{H}, \qquad n = 1, 2, 3, \ldots.$$

An isometric dilation $U \in \mathbf{B}(\tilde{\mathcal{H}})$ of T is **minimal** *if*

$$\bigvee_{n=0}^{\infty} U^n \mathcal{H} = \tilde{\mathcal{H}}.$$

Minimal isometric dilations always exist.

Theorem 3.1.2. *Let $T \in \mathbf{B}(\mathcal{H})$, $\mathcal{H}$ a Kreĭn space, and let $\tilde{D} \in \mathbf{B}(\tilde{\mathcal{D}}, \mathcal{H})$ be a defect operator for T. Define $\tilde{\mathcal{H}} = \mathcal{H} \oplus \tilde{\mathcal{D}} \oplus \tilde{\mathcal{D}} \oplus \cdots$ using fixed choices of fundamental decompositions for $\mathcal{H}$ and $\tilde{\mathcal{D}}$. Then the matrix*

$$U = \begin{pmatrix} T & 0 & 0 & 0 & \cdots \\ \tilde{D}^* & 0 & 0 & 0 & \cdots \\ 0 & 1 & 0 & 0 & \cdots \\ 0 & 0 & 1 & 0 & \cdots \\ & & \cdots & & \end{pmatrix}$$

acts as an everywhere defined and continuous operator on $\tilde{\mathcal{H}}$, and U is a minimal isometric dilation of T.

Proof. The infinite matrix determines an element U of $\mathbf{B}(\tilde{\mathcal{H}})$ because the shift operator is everywhere defined and continuous on $\tilde{\mathcal{D}} \oplus \tilde{\mathcal{D}} \oplus \cdots$. If $f = (f_0, f_1, \ldots)^t$ is in $\tilde{\mathcal{H}}$, then

$$\begin{aligned}\langle Uf, Uf\rangle_{\tilde{\mathcal{H}}} &= \left\langle (T^*T + \tilde{D}\tilde{D}^*)f_0, f_0\right\rangle_{\mathcal{H}} + \langle f_1, f_1\rangle_{\tilde{\mathcal{D}}} + \langle f_2, f_2\rangle_{\tilde{\mathcal{D}}} + \cdots \\ &= \langle f, f\rangle_{\tilde{\mathcal{H}}}\end{aligned}$$

because $T^*T + \tilde{D}\tilde{D}^* = 1$. Thus U is isometric. An inductive argument shows that for any element of $\tilde{\mathcal{H}}$ of the form $f = (f_0, 0, 0, \ldots)^t$ and any $n = 1, 2, 3, \ldots$,

$$U^n f = \begin{pmatrix} T^n f_0 \\ \tilde{D}^* T^{n-1} f_0 \\ \tilde{D}^* T^{n-2} f_0 \\ \vdots \\ \tilde{D}^* f_0 \\ 0 \\ 0 \\ \vdots \end{pmatrix}$$

Therefore $T^n f_0 = \mathrm{Pr}_{\mathcal{H}} U^n f_0$ for any $f_0 \in \mathcal{H}$. The same identity implies that for every positive integer n,

$$\bigvee_{j=0}^{n} U^j \mathcal{H} = \mathcal{H} \oplus \tilde{\mathcal{D}} \oplus \cdots \oplus \tilde{\mathcal{D}}$$

with n copies of $\tilde{\mathcal{D}}$. Hence U is a minimal isometric dilation of T. ■

The minimal isometric dilation of a Kreĭn space operator $T \in \mathbf{B}(\mathcal{H})$ constructed in Theorem 3.1.2 has special properties:

(a) $\mathcal{H}$ is invariant under U^* and $U^*|\mathcal{H} = T^*$;

(b) the closure $\mathcal{L}$ of $\{Uh - Th : h \in \mathcal{H}\}$ is a regular subspace of $\tilde{\mathcal{H}}$, and

$$\tilde{\mathcal{H}} = \mathcal{H} \oplus \tilde{\mathcal{L}} \oplus U\tilde{\mathcal{L}} \oplus U^2\tilde{\mathcal{L}} \oplus \cdots;$$

(c) the positive and negative indices of $\mathcal{L}$ coincide with $h_\pm(1 - T^*T)$.

Proofs are immediate from the construction in Theorem 3.1.2.

Let $T \in \mathbf{B}(\mathcal{H})$, $\mathcal{H}$ a Kreĭn space, and let $U \in \mathbf{B}(\tilde{\mathcal{H}})$ be a minimal isometric dilation of T. If $\tilde{\mathcal{H}}'$ is another Kreĭn space containing $\mathcal{H}$ as a regular subspace, and if $U' = WUW^{-1}$ where $W \in \mathbf{B}(\tilde{\mathcal{H}}, \tilde{\mathcal{H}}')$ is a unitary operator which coincides with the identity on $\mathcal{H}$, then the operator $U' \in \mathbf{B}(\tilde{\mathcal{H}}')$ is also a minimal isometric dilation of T. Two minimal isometric dilations of T related in this way are said to be *isomorphic*. Isomorphic minimal isometric dilations are abstractly indistinguishable. In particular, properties (a) - (c) are invariant under isomorphism.

The next result shows that minimal isometric dilations of contractions are unique up to isomorphism, and therefore all have the properties (a)–(c).

THEOREM 3.1.3. *Let $\mathcal{H}$ be a Kreĭn space, and assume that $T \in \mathbf{B}(\mathcal{H})$ is a contraction.*

(i) Any two minimal isometric dilations of T are isomorphic.

(ii) If $U \in \mathbf{B}(\tilde{\mathcal{H}})$ is a minimal isometric dilation of T, then $\tilde{\mathcal{H}} \ominus \mathcal{H}$ is uniformly positive, and hence any maximal negative subspace of $\mathcal{H}$ is maximal negative in $\tilde{\mathcal{H}}$.

(iii) The operator T is a bicontraction if and only if any minimal isometric dilation of T is a bicontraction.

Proof. To prove (i), it is sufficient to show that if $U \in \mathbf{B}(\tilde{\mathcal{H}})$ is a minimal isometric dilation of T of the form constructed in Theorem 3.1.2, and if $U' \in \mathbf{B}(\tilde{\mathcal{H}}')$ is any other minimal isometric dilation of T, then U and U' are isomorphic.

For any elements $f_0, \ldots, f_n$ and $g_0, \ldots, g_m$ of $\mathcal{H}$,

$$\left\langle \sum_{j=0}^{n} U^j f_j, \sum_{k=0}^{m} U^k g_k \right\rangle_{\tilde{\mathcal{H}}} = \left\langle \sum_{j=0}^{n} U'^j f_j, \sum_{k=0}^{m} U'^k g_k \right\rangle_{\tilde{\mathcal{H}}'}$$

by the definition of an isometric dilation. Since both dilations are minimal, in a routine way we obtain a well defined and densely defined isometry W_0 on $\tilde{\mathcal{H}}$, having dense range in $\tilde{\mathcal{H}}'$, such that

$$W_0 \left\{ \sum_{j=0}^{n} U^j f_j \right\} = \sum_{j=0}^{n} U'^j f_j$$

for any nonnegative integer n and vectors $f_0, \ldots, f_n$ in $\mathcal{H}$. Both the domain and range of W_0 include $\mathcal{H}$, and W_0 coincides with the identity on $\mathcal{H}$. The domain of W_0 is invariant under U, and $W_0 U g = U' W_0 g$ for all g in the domain of W_0.

Since T is a contraction, $\tilde{\mathcal{D}}$ is a Hilbert space by Corollary 1.4.2. Therefore $\tilde{\mathcal{H}} \ominus \mathcal{H}$ is uniformly positive. We show that $\tilde{\mathcal{H}}' \ominus \mathcal{H}$ is also uniformly positive. Since $\operatorname{ran} W_0$ is dense in $\tilde{\mathcal{H}}'$ and includes $\mathcal{H}$, the intersection of $\operatorname{ran} W_0$ and $\tilde{\mathcal{H}}' \ominus \mathcal{H}$ is dense in $\tilde{\mathcal{H}}' \ominus \mathcal{H}$. In a similar way, the intersection of $\operatorname{dom} W_0$ and $\tilde{\mathcal{H}} \ominus \mathcal{H}$ is dense in $\tilde{\mathcal{H}} \ominus \mathcal{H}$. Since $\tilde{\mathcal{H}} \ominus \mathcal{H}$ is positive and W_0 is isometric, $\tilde{\mathcal{H}}' \ominus \mathcal{H}$ is positive and therefore uniformly positive.

If $\mathcal{M}$ is a maximal negative subspace of $\mathcal{H}$, by what we just proved $\mathcal{M}$ is maximal negative in both $\tilde{\mathcal{H}}$ and $\tilde{\mathcal{H}}'$. Now $\mathcal{M}$ is in the domain of W_0 and $W_0 \mathcal{M} = \mathcal{M}$. Hence by Theorem 1.4.4, W_0 has an extension to an operator $W \in \mathbf{B}(\tilde{\mathcal{H}}, \tilde{\mathcal{H}}')$. It is clear that W is unitary, W coincides with the identity on $\mathcal{H}$, and $U' = WUW^{-1}$. Therefore U and U' are isomorphic, and (i) is proved. A proof of (ii) is obtained incidentally from the argument.

For (iii), consider any minimal isometric dilation $U \in \mathbf{B}(\tilde{\mathcal{H}})$ of T. By part (ii) of the theorem, $\tilde{\mathcal{H}} \ominus \mathcal{H}$ is uniformly positive. Hence any maximal uniformly negative subspace $\mathcal{H}_-$ of $\mathcal{H}$ is maximal negative in $\tilde{\mathcal{H}}$ as well. Both T and U are contractions, and $U\mathcal{H}_-$ and $T\mathcal{H}_-$ have the same projection onto $\mathcal{H}_-$ by the definition of an isometric dilation. By Theorem 1.3.6, U and T are simultaneously bicontractions or not. ■

Parallel results hold for unitary dilations.

DEFINITION 3.1.4. *Let $T \in \mathbf{B}(\mathcal{H})$, where $\mathcal{H}$ is a Kreĭn space. A* **unitary dilation** *of T is a unitary operator $U \in \mathbf{B}(\tilde{\mathcal{H}})$, where $\tilde{\mathcal{H}}$ is a Kreĭn space containing $\mathcal{H}$ isometrically as a regular subspace, such that*

$$T^n = \mathrm{Pr}_{\mathcal{H}} U^n|\mathcal{H} \qquad \text{and} \qquad T^{*n} = \mathrm{Pr}_{\mathcal{H}} U^{-n}|\mathcal{H}$$

for all $n = 1, 2, 3, \ldots$. A unitary dilation $U \in \mathbf{B}(\tilde{\mathcal{H}})$ of T is **minimal** *if*

$$\bigvee_{n=-\infty}^{\infty} U^n\mathcal{H} = \tilde{\mathcal{H}}.$$

We give two methods of constructing minimal unitary dilations. The first iterates the construction of minimal isometric dilations.

THEOREM 3.1.5. *Let $T \in \mathbf{B}(\mathcal{H})$, $\mathcal{H}$ a Kreĭn space. Let $V \in \mathbf{B}(\mathcal{H}')$ be any minimal isometric dilation of T, and let $U^* \in \mathbf{B}(\tilde{\mathcal{H}})$ be a minimal isometric dilation of V^* of the form constructed in Theorem 3.1.2. Then U is a minimal unitary dilation of T.*

Proof. By assumption, U^* has the form

$$U^* = \begin{pmatrix} V^* & 0 & 0 & 0 & \cdots \\ \tilde{E}^* & 0 & 0 & 0 & \cdots \\ 0 & 1 & 0 & 0 & \cdots \\ 0 & 0 & 1 & 0 & \cdots \\ & & \cdots & & \end{pmatrix} \in \mathbf{B}(\mathcal{H}' \oplus \tilde{\mathcal{E}} \oplus \tilde{\mathcal{E}} \oplus \cdots),$$

where $\tilde{E} \in \mathbf{B}(\tilde{\mathcal{E}}, \mathcal{H}')$ is a defect operator for $V^* \in \mathbf{B}(\mathcal{H}')$. Since V is an isometry, $\tilde{E}$ is an isometry with range equal to $\ker V^*$. To see this, note that $1 - VV^* = \tilde{E}\tilde{E}^*$ is a projection operator, and $\ker \tilde{E}^* = \ker \tilde{E}\tilde{E}^* = \ker(1 - VV^*) = V\mathcal{H}'$. Then apply Theorem 1.1.6 to obtain the assertion. It follows that

$$V^*\tilde{E} = 0 \qquad \text{and} \qquad \tilde{E}^*\tilde{E} = 1.$$

A matrix multiplication shows that $U^*U = 1$. Since U^* is an isometry by construction, U is unitary. For any $n = 1, 2, 3, \ldots$,

$$\mathrm{Pr}_{\mathcal{H}} V^n|\mathcal{H} = T^n \qquad \text{and} \qquad \mathrm{Pr}_{\mathcal{H}} V^{*n}|\mathcal{H} = T^{*n}$$

and

$$\mathrm{Pr}_{\mathcal{H}'} U^{*n}|\mathcal{H}' = V^{*n} \qquad \text{and} \qquad \mathrm{Pr}_{\mathcal{H}'} U^{n}|\mathcal{H}' = V^{n}$$

because V and U^* are isometric dilations of T and V^*. The second pair of relations can be further restricted to $\mathcal{H}$. Applying the projection onto $\mathcal{H}$ and using the first pair of relations, we then see that U is a unitary dilation of T.

It remains to show that U is minimal. From the matrix form of U, we have $U|\mathcal{H} = V|\mathcal{H}$, and so $\bigvee_0^n U^j\mathcal{H} = \bigvee_0^n V^j\mathcal{H}$. Since V is a minimal isometric dilation of T by assumption, $\bigvee_0^\infty U^j\mathcal{H} = \mathcal{H}'$. As in the proof of Theorem 3.1.2, an inductive argument shows that for any element of $\tilde{\mathcal{H}}$ of the form $f = (f_0, 0, 0, \ldots)^t$, $f_0 \in \mathcal{H}'$, and any $n = 1, 2, 3, \ldots$,

$$U^{*n} f = \begin{pmatrix} V^{*n} f_0 \\ \tilde{E}^* V^{*\,n-1} f_0 \\ \tilde{E}^* V^{*\,n-2} f_0 \\ \vdots \\ \tilde{E}^* f_0 \\ 0 \\ 0 \\ \vdots \end{pmatrix}.$$

The identity implies that $\bigvee_0^\infty U^{*j}\mathcal{H}' = \mathcal{H}' \oplus \tilde{\mathcal{E}} \oplus \cdots \oplus \tilde{\mathcal{E}}$ with n copies of $\tilde{\mathcal{E}}$. Since $\bigvee_0^\infty U^j\mathcal{H} = \mathcal{H}'$, this shows that U is a minimal unitary dilation of T. ∎

The preceding method of constructing unitary dilations is useful for applications, but it does not yield other desirable properties similar to the Hilbert space case. These properties are obtained from a different construction.

THEOREM 3.1.6. *Let $\mathcal{H}$ be a Kreĭn space, and let $T \in \mathbf{B}(\mathcal{H})$ have Julia operator*

$$\begin{pmatrix} T & D \\ \tilde{D}^* & L \end{pmatrix} \in \mathbf{B}(\mathcal{H} \oplus D, \mathcal{H} \oplus \tilde{D}),$$

Define $\tilde{\mathcal{H}} = \cdots \oplus \mathcal{D} \oplus \mathcal{D} \oplus \mathcal{H} \oplus \tilde{\mathcal{D}} \oplus \tilde{\mathcal{D}} \oplus \cdots$ using fixed choices of fundamental decompositions in the summands. Then the matrix

$$U = \begin{pmatrix} & & & & \cdots & & & & \\ \cdots & 1 & 0 & 0 & 0 & 0 & 0 & 0 & \cdots \\ \cdots & 0 & 1 & 0 & 0 & 0 & 0 & 0 & \cdots \\ \cdots & 0 & 0 & D & \boxed{T} & 0 & 0 & 0 & \cdots \\ \cdots & 0 & 0 & L & \tilde{D}^* & 0 & 0 & 0 & \cdots \\ \cdots & 0 & 0 & 0 & 0 & 1 & 0 & 0 & \cdots \\ \cdots & 0 & 0 & 0 & 0 & 0 & 1 & 0 & \cdots \\ & & & & \cdots & & & & \end{pmatrix}$$

acts as an everywhere defined and continuous operator on $\tilde{\mathcal{H}}$, and U is a minimal unitary dilation of T.

Proof. The continuity of shift operators implies that U exists as an element of $\mathbf{B}(\tilde{\mathcal{H}})$. For any element

$$h = (\ldots, d_2, d_1, f, \tilde{d}_1, \tilde{d}_2, \ldots)^t$$

of $\tilde{\mathcal{H}}$, with f as the element of $\mathcal{H}$, we have

$$Uh = \begin{pmatrix} \vdots \\ d_3 \\ d_2 \\ Dd_1 + Tf \\ Ld_1 + \tilde{D}^* f \\ \tilde{d}_1 \\ \tilde{d}_2 \\ \vdots \end{pmatrix} \quad \text{and} \quad U^* h = \begin{pmatrix} \vdots \\ d_2 \\ d_1 \\ D^* f + L^* \tilde{d}_1 \\ T^* f + \tilde{D}\tilde{d}_1 \\ \tilde{d}_2 \\ \tilde{d}_3 \\ \vdots \end{pmatrix}.$$

A straightforward calculation yields

$$\langle Uh, Uh \rangle_{\tilde{\mathcal{H}}} = \langle U^* h, U^* h \rangle_{\tilde{\mathcal{H}}} = \langle h, h \rangle_{\tilde{\mathcal{H}}},$$

and so U is unitary. If $h = (\ldots, 0, 0, f, 0, 0, \ldots)^t$, where $f \in \mathcal{H}$, then

$$U^n h = \begin{pmatrix} \vdots \\ 0 \\ 0 \\ T^n f \\ \tilde{D}^* T^{n-1} f \\ \tilde{D}^* T^{n-2} f \\ \vdots \\ \tilde{D}^* f \\ 0 \\ 0 \\ \vdots \end{pmatrix} \quad \text{and} \quad U^{*n} h = \begin{pmatrix} \vdots \\ 0 \\ 0 \\ D^* f \\ \vdots \\ D^* T^{*n-2} f \\ D^* T^{*n-1} f \\ T^{*n} f \\ 0 \\ 0 \\ \vdots \end{pmatrix}.$$

The identities yield

$$\bigvee_{j=0}^{n} U^j \mathcal{H} = \mathcal{H} \oplus \tilde{\mathcal{D}} \oplus \cdots \oplus \tilde{\mathcal{D}},$$

$$\bigvee_{j=0}^{n} U^{*j} \mathcal{H} = \mathcal{D} \oplus \cdots \oplus \mathcal{D} \oplus \mathcal{H},$$

with $n+1$ summands on the right side in each case. It follows that U is a minimal unitary dilation of T. ■

As in the case of isometric dilations, the construction of a minimal unitary dilation in Theorem 3.1.6 for a given operator $T \in \mathbf{B}(\mathcal{H})$ immediately yields additional properties:

(a′) the closures $\tilde{\mathcal{L}}$ and $\mathcal{L}$ of $\{Uh - Th : h \in \mathcal{H}\}$ and $\{U^*h - T^*h : h \in \mathcal{H}\}$ are regular subspaces of $\tilde{\mathcal{H}}$, and

$$\tilde{\mathcal{H}} = \cdots \oplus U^{*2}\mathcal{L} \oplus U^*\mathcal{L} \oplus \mathcal{L} \oplus \mathcal{H} \oplus \tilde{\mathcal{L}} \oplus U\tilde{\mathcal{L}} \oplus U^2\tilde{\mathcal{L}} \oplus \cdots;$$

(b′) the positive and negative indices of $\tilde{\mathcal{L}}$ are equal to $h_\pm(1-T^*T)$, and those of $\mathcal{L}$ are equal to $h_\pm(1-TT^*)$.

Let $T \in \mathbf{B}(\mathcal{H})$, $\mathcal{H}$ a Kreĭn space, and let $U \in \mathbf{B}(\tilde{\mathcal{H}})$ be a minimal unitary dilation of T. If $\tilde{\mathcal{H}}'$ is another Kreĭn space containing $\mathcal{H}$ as a regular subspace, and if $U' = WUW^{-1}$ where $W \in \mathbf{B}(\tilde{\mathcal{H}}, \tilde{\mathcal{H}}')$ is a unitary operator which coincides with the identity on $\mathcal{H}$, then the operator $U' \in \mathbf{B}(\tilde{\mathcal{H}}')$ is a minimal unitary dilation of T. In this situation, we say that U and U' are *isomorphic*. The properties (a′) and (b′) are invariant under isomorphism.

Minimal unitary dilations of bicontractions are unique up to isomorphism.

THEOREM 3.1.7. *Let $\mathcal{H}$ be a Kreĭn space, and let $T \in \mathbf{B}(\mathcal{H})$ be a bicontraction.*

(i) Any two minimal unitary dilations of T are isomorphic.

(ii) If $U \in \mathbf{B}(\tilde{\mathcal{H}})$ is a minimal unitary dilation of T, then $\tilde{\mathcal{H}} \ominus \mathcal{H}$ is uniformly positive, and hence any maximal negative subspace of $\mathcal{H}$ is maximal negative in $\tilde{\mathcal{H}}$.

Proof. The argument is similar to the proof of Theorem 3.1.3, parts (i) and (ii). In place of Theorem 3.1.2 we use Theorem 3.1.6. Since T is a bicontraction, the spaces $\mathcal{D}$ and $\tilde{\mathcal{D}}$ in Theorem 3.1.6 are Hilbert spaces by Corollary 1.4.3. Otherwise we proceed as before, except that in place of sums of the form $\sum_{j=0}^n U^j f_j$ with $f_0, \ldots, f_n \in \mathcal{H}$ we now use sums of the form $\sum_{j=-n}^n U^j f_j$ with $f_{-n}, \ldots, f_n \in \mathcal{H}$. ■

Theorem 3.1.3 (iii) has a variant form which does not require that T be a contraction. If $T \in \mathbf{B}(\mathcal{H})$, where $\mathcal{H}$ is a Kreĭn space, and if T^* is a contraction, then any minimal isometric dilation of T of the form constructed in Theorem 3.1.2 is a bicontraction. Conversely, if such a dilation is a bicontraction, then T^* is a contraction. For U is a bicontraction if and only if U^* is a contraction, and the condition for this is that $[T^* \quad \tilde{\mathcal{D}}]$ be a contraction. By Theorem 1.4.2, this holds if and only if T^* is a contraction.

We give an example. Let $\mathcal{H}$ be the anti-space of a Hilbert space, and let $T = -2V^*$, where $V \in \mathbf{B}(\mathcal{H})$ is an isometry whose range is not all of $\mathcal{H}$. Then T^* is a contraction, but T is not a contraction because its kernel is not uniformly positive. Let $\mathcal{D}$ and $\tilde{\mathcal{D}}$ be $\mathcal{H}$ as a vector space, with

$$\begin{aligned} \langle f,g\rangle_{\mathcal{D}} &= -\langle f,g\rangle_{\mathcal{H}}, \\ \langle f,g\rangle_{\tilde{\mathcal{D}}} &= -\langle Pf,Pg\rangle_{\mathcal{H}} + \langle (1-P)f,(1-P)g\rangle_{\mathcal{H}}, \end{aligned}$$

for all $f,g \in \mathcal{H}$, where P is the projection on $V\mathcal{H}$. A Julia operator for T is given by

$$\begin{pmatrix} T & \sqrt{3} \\ 1-P-\sqrt{3}P & 2V \end{pmatrix} \in \mathbf{B}(\mathcal{H}\oplus\mathcal{D},\mathcal{H}\oplus\tilde{\mathcal{D}}).$$

Its adjoint

$$\begin{pmatrix} T^* & 1-P+\sqrt{3}P \\ -\sqrt{3} & 2V^* \end{pmatrix} \in \mathbf{B}(\mathcal{H}\oplus\tilde{\mathcal{D}},\mathcal{H}\oplus\mathcal{D})$$

is a Julia operator for T^*. Since T^* is a contraction, the minimal isometric dilation of T constructed by Theorem 3.1.2 is a bicontraction by the previous remarks.

3.2 Commutant Lifting

The background is now in place for a commutant lifting theorem for contraction operators on Kreĭn spaces.

THEOREM 3.2.1. *Let $\mathcal{H}$ and $\mathcal{K}$ be Kreĭn spaces, and let $T_1 \in \mathbf{B}(\mathcal{H})$ and $T_2 \in \mathbf{B}(\mathcal{K})$ be contractions with minimal isometric dilations $U_1 \in \mathbf{B}(\tilde{\mathcal{H}})$ and $U_2 \in \mathbf{B}(\tilde{\mathcal{K}})$, respectively. Let $A \in \mathbf{B}(\mathcal{H},\mathcal{K})$ be a contraction satisfying*

$$AT_1 = T_2A.$$

Then there exists a contraction $\tilde{A} \in \mathbf{B}(\tilde{\mathcal{H}},\tilde{\mathcal{K}})$ such that

$$\tilde{A}U_1 = U_2\tilde{A}$$

and $A = \mathrm{Pr}_{\mathcal{K}}\,\tilde{A}|\mathcal{H}$.

Proof. Since any two minimal isometric dilations of a contraction are isomorphic by Theorem 3.1.3, we may assume that the minimal isometric dilations of T_1 and T_2 have the form constructed in Theorem 3.1.2. In other words, we may take

$$\tilde{\mathcal{H}} = \mathcal{H}\oplus\tilde{\mathcal{D}}_1\oplus\tilde{\mathcal{D}}_1\oplus\cdots \quad\text{and}\quad \tilde{\mathcal{K}} = \mathcal{K}\oplus\tilde{\mathcal{D}}_2\oplus\tilde{\mathcal{D}}_2\oplus\cdots$$

where $\tilde{\mathcal{D}}_1$, $\tilde{\mathcal{D}}_2$ are Hilbert spaces,

$$U_j = \begin{pmatrix} T_j & 0 & 0 & 0 & \cdots \\ \tilde{D}_j^* & 0 & 0 & 0 & \cdots \\ 0 & 1 & 0 & 0 & \cdots \\ 0 & 0 & 1 & 0 & \cdots \\ & & \cdots & & \end{pmatrix}, \qquad j = 1, 2,$$

and $\tilde{D}_1 \in \mathbf{B}(\tilde{\mathcal{D}}_1, \mathcal{H})$, $\tilde{D}_2 \in \mathbf{B}(\tilde{\mathcal{D}}_2, \mathcal{K})$ are defect operators for T_1, T_2. Let $\mathcal{H} = \mathcal{H}_+ \oplus \mathcal{H}_-$ and $\mathcal{K} = \mathcal{K}_+ \oplus \mathcal{K}_-$ be fundamental decompositions, used, for example, to induce Hilbert space norms on $\mathcal{H}$, $\tilde{\mathcal{H}}$, $\mathcal{K}$, $\tilde{\mathcal{K}}$. Set $\mathcal{K}_0 = \mathcal{K}$ and

$$\mathcal{K}_n = \mathcal{K} \oplus \tilde{\mathcal{D}}_2 \oplus \cdots \oplus \tilde{\mathcal{D}}_2, \qquad n = 1, 2, \ldots,$$

with n copies of $\tilde{\mathcal{D}}_2$ in the direct sum on the right. We construct contraction operators

$$\tilde{A}_n : \tilde{\mathcal{H}} \to \mathcal{K}_n, \qquad n = 0, 1, 2, \ldots,$$

such that $A = \Pr_{\mathcal{K}} \tilde{A}_n | \mathcal{H}$ for all nonnegative integers n and

(α) $\tilde{A}_n U_1 = U_2 \tilde{A}_{n-1}$,
(β) $\tilde{A}_{n-1} = \Pr_{\mathcal{K}_{n-1}} \tilde{A}_n$

if $n \geq 1$.

Let $\tilde{A}_0$ map any element $(f, g_1, g_2, \ldots)^t$ of $\tilde{\mathcal{H}}$ to the element Af of $\mathcal{K}_0$. Since A is a contraction and $\tilde{\mathcal{D}}_1$ is a Hilbert space, $\tilde{A}_0$ is a contraction. By construction, $A = \Pr_{\mathcal{K}} \tilde{A}_0 | \mathcal{H}$.

Assume that $\tilde{A}_0, \ldots, \tilde{A}_r$ have the required properties. We determine $\tilde{A}_{r+1}$ as a matrix

$$\tilde{A}_{r+1} = \begin{pmatrix} C_{11} & C_{12} \\ C_{21} & C_{22} \end{pmatrix},$$

relative to the decompositions

$$\tilde{\mathcal{H}} = U\tilde{\mathcal{H}} \oplus (\tilde{\mathcal{H}} \ominus U\mathcal{H}) \qquad \text{and} \qquad \mathcal{K}_{r+1} = \mathcal{K}_r \oplus \tilde{\mathcal{D}}_2.$$

The conditions (α) and (β) with $n = r + 1$ require that

$$\begin{pmatrix} C_{11} \\ C_{21} \end{pmatrix} U_1 = U_2 \tilde{A}_r \qquad \text{and} \qquad (\, C_{11} \quad C_{12} \,) = \tilde{A}_r$$

Define C_{11}, C_{12}, C_{21} by these equations. The two definitions of C_{11} are identical in the case $r = 0$, because

$$\tilde{A}_0 U_1 = \Pr_{\mathcal{K}_0} U_2 \tilde{A}_0.$$

In fact, this identity reduces to $AT_1 = T_2A$, which holds by hypothesis. In the case $r \geq 1$, the two definitions of C_{11} are identical because

$$\tilde{A}_r U_1 = U_2 \tilde{A}_{r-1} = U_2 \mathrm{Pr}_{\mathcal{K}_{r-1}} \tilde{A}_r = \mathrm{Pr}_{\mathcal{K}_r} U_2 \tilde{A}_r.$$

Here the first two equalities are by (α) and (β) with $n = r$, and the third is an elementary property of the minimal isometric dilation.

Since $\tilde{A}_r$ is a contraction and U_1 and U_2 are isometries, the operators

$$\begin{pmatrix} C_{11} \\ C_{21} \end{pmatrix} \quad \text{and} \quad \begin{pmatrix} C_{11} & C_{12} \end{pmatrix}$$

are contractions. By Theorem 2.4.4, there exists an operator C_{22} such that

$$\tilde{A}_{r+1} = \begin{pmatrix} C_{11} & C_{12} \\ C_{21} & C_{22} \end{pmatrix}$$

is a contraction. By construction, (α) and (β) hold with $n = r + 1$. Finally,

$$\mathrm{Pr}_{\mathcal{K}}\, \tilde{A}_{r+1}|\mathcal{H} = \mathrm{Pr}_{\mathcal{K}}\, \mathrm{Pr}_{\mathcal{K}_r} \tilde{A}_{r+1}|\mathcal{H} = \mathrm{Pr}_{\mathcal{K}}\, \tilde{A}_r|\mathcal{H} = A.$$

This completes the inductive construction of operators $\tilde{A}_0, \tilde{A}_1, \ldots$ with the stated properties.

We show that $\|\tilde{A}_n\|^2 \leq 1 + 2\|A\|^2$ for all n. This is true for $n = 0$ because $\|\tilde{A}_0\|^2 = \|A\|^2$ by the definition of $\tilde{A}_0$. For $n \geq 1$, we obtain $\tilde{A}_0 = \mathrm{Pr}_{\mathcal{K}}\, \tilde{A}_n$ by repeated application of (β). Hence we may write

$$\tilde{A}_n = \begin{pmatrix} \tilde{A}_0 \\ X \end{pmatrix} \in \mathbf{B}(\tilde{\mathcal{H}}, \mathcal{K} \oplus (\mathcal{K}_n \ominus \mathcal{K})),$$

where $X \in \mathbf{B}(\tilde{\mathcal{H}}, \mathcal{K}_n \ominus \mathcal{K})$. Since $\tilde{A}_0$ and $\tilde{A}_n$ are contractions and $\mathcal{K}_n \ominus \mathcal{K}$ is a Hilbert space, Corollary 2.2.3 yields

$$\|\tilde{A}_n\|^2 \leq 1 + 2\|\tilde{A}_0\|^2 = 1 + 2\|A\|^2,$$

which proves the assertion.

By (β), $\tilde{A}_{n-1}^* = \tilde{A}_n^*|\mathcal{K}_{n-1}$ for all $n \geq 1$. Hence there is an operator $\tilde{A} \in \mathbf{B}(\tilde{\mathcal{H}}, \tilde{\mathcal{K}})$ such that $\|\tilde{A}\|^2 \leq 1 + 2\|A\|^2$ and $\tilde{A}_n^* = \tilde{A}^*|\mathcal{K}_n$, equivalently, $\tilde{A}_n = \mathrm{Pr}_{\mathcal{K}_n} \tilde{A}$ for every nonnegative integer n. It follows that $\lim_{n\to\infty} \tilde{A}_n f = \tilde{A} f, \quad f \in \tilde{\mathcal{H}}$, and from this we verify without difficulty that $\tilde{A}$ has the required properties. ■

The norm estimate obtained in the proof of Theorem 3.2.1, namely,

$$\|\tilde{A}\|^2 \leq 1 + 2\|A\|^2,$$

holds for any operator $\tilde{A}$ having the properties stated in the theorem (it is not special to the construction in the proof). Moreover, arbitrary fundamental decompositions of $\mathcal{H}$ and $\mathcal{K}$ may be used in the computation of norms. To see this, recall that by Theorem 3.1.3 (ii),

$$\tilde{\mathcal{H}} = \mathcal{H} \oplus (\tilde{\mathcal{H}} \ominus \mathcal{H}) \qquad \text{and} \qquad \tilde{\mathcal{K}} = \mathcal{K} \oplus (\tilde{\mathcal{K}} \ominus \mathcal{K}),$$

where $\tilde{\mathcal{H}} \ominus \mathcal{H}$ and $\tilde{\mathcal{K}} \ominus \mathcal{K}$ are Hilbert spaces. Thus fundamental decompositions of $\mathcal{H}$ and $\mathcal{K}$ determine fundamental decompositions of $\tilde{\mathcal{H}}$ and $\tilde{\mathcal{K}}$. The norm estimate follows from Corollary 2.2.3 by an argument used in the proof of Theorem 3.2.1.

Theorem 3.2.1 has a companion for bicontractions, which is deduced as an immediate consequence.

THEOREM 3.2.2. *Let $\mathcal{H}$ and $\mathcal{K}$ be Kreĭn spaces, and let $T_1 \in \mathbf{B}(\mathcal{H})$ and $T_2 \in \mathbf{B}(\mathcal{K})$ be contractions with minimal isometric dilations $U_1 \in \mathbf{B}(\tilde{\mathcal{H}})$ and $U_2 \in \mathbf{B}(\tilde{\mathcal{K}})$, respectively. Let $A \in \mathbf{B}(\mathcal{H}, \mathcal{K})$ be a bicontraction satisfying*

$$AT_1 = T_2 A.$$

Then there exists a bicontraction $\tilde{A} \in \mathbf{B}(\tilde{\mathcal{H}}, \tilde{\mathcal{K}})$ such that

$$\tilde{A}U_1 = U_2 \tilde{A}$$

and $A = \mathrm{Pr}_{\mathcal{K}} \tilde{A}|\mathcal{H}$.

Proof. Let $\mathcal{N}_1$ be a maximal negative subspace of $\mathcal{H}$, and let $\mathcal{N}_2$ be a maximal uniformly negative subspace of $\mathcal{K}$. Since A is a bicontraction, $A\mathcal{N}_1$ is a maximal negative subspace of $\mathcal{K}$ by Theorem 1.3.6, and so

$$\mathrm{Pr}_{\mathcal{N}_2} A\mathcal{N}_1 = \mathcal{N}_2.$$

By Theorem 3.1.3 (ii), $\mathcal{N}_1$ is a maximal negative subspace of $\tilde{\mathcal{H}}$ and $\mathcal{N}_2$ is a maximal uniformly negative subspace of $\tilde{\mathcal{K}}$. Since $\tilde{A}$ is a contraction, $\tilde{A}\mathcal{N}_1$ is a negative subspace of $\tilde{\mathcal{K}}$. It is maximal negative in $\tilde{\mathcal{K}}$ because

$$\mathrm{Pr}_{\mathcal{N}_2} \tilde{A}\mathcal{N}_1 = \mathrm{Pr}_{\mathcal{N}_2} \mathrm{Pr}_{\mathcal{K}} \tilde{A}\mathcal{N}_1 = \mathrm{Pr}_{\mathcal{N}_2} A\mathcal{N}_1 = \mathcal{N}_2.$$

Therefore $\tilde{A}$ is a bicontraction by Theorem 1.3.6. ■

With stronger hypotheses, it is also possible to lift an intertwining relation to minimal unitary dilations.

THEOREM 3.2.3. *Let $\mathcal{H}$ and $\mathcal{K}$ be Kreĭn spaces, and let $T_1 \in \mathbf{B}(\mathcal{H})$ and $T_2 \in \mathbf{B}(\mathcal{K})$ be bicontractions with minimal unitary dilations $U_1 \in \mathbf{B}(\tilde{\mathcal{H}})$ and $U_2 \in \mathbf{B}(\tilde{\mathcal{K}})$, respectively. Let $A \in \mathbf{B}(\mathcal{H}, \mathcal{K})$ be a bicontraction satisfying*

$$AT_1 = T_2 A.$$

Then there exists a bicontraction $\tilde{A} \in \mathbf{B}(\tilde{\mathcal{H}}, \tilde{\mathcal{K}})$ such that

$$\tilde{A}U_1 = U_2 \tilde{A}$$

and $A = \mathrm{Pr}_{\mathcal{K}} \tilde{A}|\mathcal{H}$.

Proof. By Theorem 3.1.7, minimal unitary dilations of bicontractions are unique up to isomorphism. Therefore we may assume that U_1, U_2 are constructed as in Theorem 3.1.5. Let $V_1 \in \mathbf{B}(\mathcal{H}')$, $V_2 \in \mathbf{B}(\mathcal{K}')$ be minimal isometric dilations of T_1, T_2. We take U_1^*, U_2^* to be minimal isometric dilations of V_1^*, V_2^*.

By Theorem 3.2.2, there exists a bicontraction $A' \in \mathbf{B}(\mathcal{H}', \mathcal{K}')$ such that

$$A'V_1 = V_2 A'$$

and $A = \mathrm{Pr}_{\mathcal{K}} A'\mathcal{H}$. The operators

$$T_1' = V_2^* \in \mathbf{B}(\mathcal{K}') \qquad \text{and} \qquad T_2' = V_1^* \in \mathbf{B}(\mathcal{H}')$$

are contractions by Theorem 3.1.3 (iii), and $A'^* \in \mathbf{B}(\mathcal{K}', \mathcal{H}')$ is a bicontraction satisfying

$$A'^* T_1' = T_2' A'^*.$$

Since U_2^* and U_1^* are minimal isometric dilations of T_1' and T_2', respectively, a second application of Theorem 3.2.2 produces a bicontraction $\tilde{A}^* \in \mathbf{B}(\tilde{\mathcal{K}}, \tilde{\mathcal{H}})$ such that

$$\tilde{A}^* U_2^* = U_1^* \tilde{A}^*$$

and $A'^* = \mathrm{Pr}_{\mathcal{H}'} \tilde{A}^*|\mathcal{K}'$. Thus $\tilde{A} \in \mathbf{B}(\tilde{\mathcal{H}}, \tilde{\mathcal{K}})$ is a bicontraction satisfying $\tilde{A}U_1 = U_2\tilde{A}$. Finally, we have $A' = \mathrm{Pr}_{\mathcal{K}'} \tilde{A}|\mathcal{H}'$, and restricting this relation further to $\mathcal{H}$, we obtain

$$A = \mathrm{Pr}_{\mathcal{K}} A'|\mathcal{H} = \mathrm{Pr}_{\mathcal{K}} \tilde{A}|\mathcal{H}.$$

Thus A has the required properties. ■

3.3 Characterization of Extensions

We do not solve the problem of labeling all operators produced by the commutant lifting theorem, but we give some information in this direction.

Let $\mathcal{H}$ and $\mathcal{K}$ be Kreĭn spaces, and let $T_1 \in \mathbf{B}(\mathcal{H})$ and $T_2 \in \mathbf{B}(\mathcal{K})$ be contractions with minimal isometric dilations $U_1 \in \mathbf{B}(\tilde{\mathcal{H}})$ and $U_2 \in \mathbf{B}(\tilde{\mathcal{K}})$. Assume that $A \in \mathbf{B}(\mathcal{H}, \mathcal{K})$ is a contraction such that

$$AT_1 = T_2 A.$$

By Theorem 3.1.3, $\mathcal{H}^\perp = \tilde{\mathcal{H}} \ominus \mathcal{H}$ and $\mathcal{K}^\perp = \tilde{\mathcal{K}} \ominus \mathcal{K}$ are uniformly positive in the Kreĭn spaces $\tilde{\mathcal{H}}$ and $\tilde{\mathcal{K}}$.

Lemma 3.3.1. *If $\tilde{A} \in \mathbf{B}(\tilde{\mathcal{H}}, \tilde{\mathcal{K}})$, $\tilde{A}U_1 = U_2\tilde{A}$, and $\mathrm{Pr}_{\mathcal{K}}\, \tilde{A}|\mathcal{H} = A$, then $\tilde{A}\mathcal{H}^\perp \subset \mathcal{K}^\perp$ and therefore $\tilde{A}^*|\mathcal{K} = A^*$.*

Proof. By Theorem 3.1.3, we may choose U_1 and U_2 as in Theorem 3.1.2. Thus

$$\tilde{\mathcal{H}} = \mathcal{H} \oplus \tilde{\mathcal{D}}_1 \oplus \tilde{\mathcal{D}}_1 \oplus \cdots \quad \text{and} \quad \tilde{\mathcal{K}} = \mathcal{K} \oplus \tilde{\mathcal{D}}_2 \oplus \tilde{\mathcal{D}}_2 \oplus \cdots$$

where $\tilde{\mathcal{D}}_1$, $\tilde{\mathcal{D}}_2$ are Hilbert spaces,

$$U_j = \begin{pmatrix} T_j & 0 & 0 & 0 & \cdots \\ \tilde{D}_j^* & 0 & 0 & 0 & \cdots \\ 0 & 1 & 0 & 0 & \cdots \\ 0 & 0 & 1 & 0 & \cdots \\ & & \cdots & & \end{pmatrix}, \qquad j = 1, 2,$$

and $\tilde{D}_1 \in \mathbf{B}(\tilde{\mathcal{D}}_1, \mathcal{H})$, $\tilde{D}_2 \in \mathbf{B}(\tilde{\mathcal{D}}_2, \mathcal{K})$ are defect operators for T_1, T_2. Writing $\tilde{A} = (A_{jk})_{j,k=1}^\infty$ relative to the same decompositions and comparing entries in the relation $\tilde{A}U_1 = U_2\tilde{A}$, we find that $A_{11} = A$ and $A_{1n} = 0$ for $n \geq 1$. In particular, $\tilde{A}\mathcal{H}^\perp \subset \mathcal{K}^\perp$. ■

Let $\mathcal{L}$ be the direct sum of the anti-space of $\mathcal{H}$ together with $\mathcal{K}$, so $\mathcal{L}$ is the space of pairs $\binom{f}{g}$ with f in $\mathcal{H}$ and g in $\mathcal{K}$, and

$$\left\langle \begin{pmatrix} f \\ g \end{pmatrix}, \begin{pmatrix} f \\ g \end{pmatrix} \right\rangle_{\mathcal{L}} = -\langle f, f\rangle_{\mathcal{H}} + \langle g, g\rangle_{\mathcal{K}}, \qquad f \in \mathcal{H}, g \in \mathcal{K}.$$

Since $A \in \mathbf{B}(\mathcal{H}, \mathcal{K})$ is a contraction, its graph

$$\mathcal{G}(A) = \left\{ \begin{pmatrix} f \\ Af \end{pmatrix} : f \in \mathcal{H} \right\}$$

is a closed negative subspace of $\mathcal{L}$.

Given a contraction $\tilde{A} \in \mathbf{B}(\tilde{\mathcal{H}}, \tilde{\mathcal{K}})$, we define $\tilde{\mathcal{L}}$ and $\mathcal{G}(\tilde{A})$ in a parallel way, so $\tilde{\mathcal{L}}$ is the space of pairs $\binom{f}{g}$ with f in $\tilde{\mathcal{H}}$ and g in $\tilde{\mathcal{K}}$, and

$$\left\langle \begin{pmatrix} f \\ g \end{pmatrix}, \begin{pmatrix} f \\ g \end{pmatrix} \right\rangle_{\tilde{\mathcal{L}}} = -\langle f, f\rangle_{\tilde{\mathcal{H}}} + \langle g, g\rangle_{\tilde{\mathcal{K}}}, \qquad f \in \tilde{\mathcal{H}}, g \in \tilde{\mathcal{K}}.$$

The graph

$$\mathcal{G}(\tilde{A}) = \left\{ \begin{pmatrix} f \\ \tilde{A}f \end{pmatrix} : f \in \tilde{\mathcal{H}} \right\}$$

of $\tilde{A}$ is a closed negative subspace of $\tilde{\mathcal{L}}$. If

$$\tilde{U} = \begin{pmatrix} U_1 & 0 \\ 0 & U_2 \end{pmatrix} \in \mathbf{B}(\tilde{\mathcal{L}}),$$

the relation $\tilde{A}U_1 = U_2\tilde{A}$ holds if and only if $\tilde{U}\mathcal{G}(\tilde{A}) \subset \mathcal{G}(\tilde{A})$.

Choose fundamental decompositions $\mathcal{H} = \mathcal{H}_+ \oplus \mathcal{H}_-$, $\mathcal{K} = \mathcal{K}_+ \oplus \mathcal{K}_-$, $\mathcal{L} = \mathcal{L}_+ \oplus \mathcal{L}_-$, $\tilde{\mathcal{L}} = \tilde{\mathcal{L}}_+ \oplus \tilde{\mathcal{L}}_-$ in a consistent way such that

$$\mathcal{L}_+ = \mathcal{K}_+ \oplus \mathcal{H}_-, \qquad \mathcal{L}_- = \mathcal{H}_+ \oplus \mathcal{K}_-,$$

and

$$\begin{aligned} \tilde{\mathcal{L}}_+ &= \mathcal{L}_+ \oplus \mathcal{K}^\perp = \mathcal{K}_+ \oplus \mathcal{H}_- \oplus \mathcal{K}^\perp, \\ \tilde{\mathcal{L}}_- &= \mathcal{L}_- \oplus \mathcal{H}^\perp = \mathcal{H}_+ \oplus \mathcal{K}_- \oplus \mathcal{H}^\perp. \end{aligned}$$

Notice that $\mathcal{L}^\perp = \tilde{\mathcal{L}} \ominus \mathcal{L}$ is given by $\mathcal{L}^\perp = \mathcal{H}^\perp \oplus \mathcal{K}^\perp$.

THEOREM 3.3.2. *In the preceding situation, the relation*

$$\mathcal{M} = \mathcal{G}(\tilde{A})$$

establishes a one-to-one correspondence between all contractions $\tilde{A} \in \mathbf{B}(\tilde{\mathcal{H}}, \tilde{\mathcal{K}})$ such that $\tilde{A}U_1 = U_2\tilde{A}$ and $\Pr_{\mathcal{K}} \tilde{A}|\mathcal{H} = A$ and all closed negative $\tilde{U}$-invariant subspaces $\mathcal{M}$ of $\tilde{\mathcal{L}}$ satisfying

(i) $\mathcal{M} \subset \mathcal{G}(A) + \mathcal{L}^\perp$, and
(ii) $\Pr_{\tilde{\mathcal{L}}_-} \mathcal{M} = \Pr_{\tilde{\mathcal{L}}_-} \left[\mathcal{G}(A) + \mathcal{L}^\perp\right]$.

Proof. Assume that $\tilde{A} \in \mathbf{B}(\tilde{\mathcal{H}}, \tilde{\mathcal{K}})$ is a contraction, $\tilde{A}U_1 = U_2\tilde{A}$, and $\Pr_{\mathcal{K}} \tilde{A}|\mathcal{H} = A$, and define $\mathcal{M} = \mathcal{G}(\tilde{A})$. Clearly $\mathcal{M}$ is a closed negative $\tilde{U}$-invariant subspace of $\tilde{\mathcal{L}}$. For any $f \in \tilde{\mathcal{H}}$,

$$\begin{pmatrix} f \\ \tilde{A}f \end{pmatrix} = \begin{pmatrix} (\Pr_{\mathcal{H}} + \Pr_{\mathcal{H}^\perp})f \\ (\Pr_{\mathcal{K}} + \Pr_{\mathcal{K}^\perp})\tilde{A}(\Pr_{\mathcal{H}} + \Pr_{\mathcal{H}^\perp})f \end{pmatrix} = \begin{pmatrix} \Pr_{\mathcal{H}} f \\ A\Pr_{\mathcal{H}} f \end{pmatrix} + \begin{pmatrix} \Pr_{\mathcal{H}^\perp} f \\ \Pr_{\mathcal{K}^\perp} \tilde{A} f \end{pmatrix}$$

by Lemma 3.3.1, and (i) holds.

Let $P_\pm$ be the projections of $\mathcal{H}$ onto $\mathcal{H}_\pm$, $Q_\pm$ the projections of $\mathcal{K}$ onto $\mathcal{K}_\pm$. For any $f \in \tilde{\mathcal{H}}$,

$$\mathrm{Pr}_{\tilde{\mathcal{L}}_-} \begin{pmatrix} f \\ \tilde{A}f \end{pmatrix} = \begin{pmatrix} P_+ \mathrm{Pr}_{\mathcal{H}} f + \mathrm{Pr}_{\mathcal{H}^\perp} f \\ Q_- A \mathrm{Pr}_{\mathcal{H}} f \end{pmatrix} = (P_+ + Q_- A)\mathrm{Pr}_{\mathcal{H}} f + \mathrm{Pr}_{\mathcal{H}^\perp} f$$

by Lemma 3.3.1. Therefore

$$\mathrm{Pr}_{\tilde{\mathcal{L}}_-} \mathcal{M} = \mathrm{ran}\,(P_+ + Q_- A) + \mathcal{H}^\perp.$$

In a similar way,

$$\begin{aligned} \mathrm{Pr}_{\tilde{\mathcal{L}}_-} [\mathcal{G}(A) + \mathcal{L}^\perp] &= \mathrm{Pr}_{\mathcal{H}_+ \oplus \mathcal{K}_-} \left\{ \begin{pmatrix} f \\ Af \end{pmatrix} : f \in \mathcal{H} \right\} + \mathcal{H}^\perp \\ &= \mathrm{ran}\,(P_+ + Q_- A) + \mathcal{H}^\perp, \end{aligned}$$

and (ii) follows.

Conversely, let $\mathcal{M}$ be a closed negative $\tilde{U}$-invariant subspace of $\tilde{\mathcal{L}}$ satisfying (i) and (ii). If $g \in \tilde{\mathcal{K}}$ and $\binom{0}{g} \in \mathcal{M}$, then by (i) we can write

$$\begin{pmatrix} 0 \\ g \end{pmatrix} = \begin{pmatrix} u \\ Au \end{pmatrix} + h + k$$

with $u \in \mathcal{H}$, $h \in \mathcal{H}^\perp$, $k \in \mathcal{K}^\perp$. Projecting onto $\mathcal{H}$ and $\mathcal{H}^\perp$, we find that $u = 0$ and $h = 0$. Therefore $g = k \in \mathcal{K}^\perp$. But $\mathcal{M}$ is a negative subspace of $\tilde{\mathcal{L}}$, and $\mathcal{K}^\perp$ is uniformly positive in $\tilde{\mathcal{L}}$, so $g = 0$. It follows that $\mathcal{M} = \mathcal{G}(\tilde{A})$ is the graph of a linear transformation $\tilde{A}$ with domain in $\tilde{\mathcal{H}}$ and range in $\tilde{\mathcal{K}}$.

We use conditions (i) and (ii) to show that $\mathrm{dom}\,\tilde{A} = \tilde{\mathcal{H}}$. For any $f \in \mathrm{dom}\,\tilde{A}$, by (i) there exist $u \in \mathcal{H}$, $h \in \mathcal{H}^\perp$, $k \in \mathcal{K}^\perp$ such that

$$\begin{pmatrix} f \\ \tilde{A}f \end{pmatrix} = \begin{pmatrix} u \\ Au \end{pmatrix} + h + k.$$

Projecting onto $\mathcal{H}$, $\mathcal{H}^\perp$, $\mathcal{K}^\perp$, we find that $u = \mathrm{Pr}_{\mathcal{H}} f$, $h = \mathrm{Pr}_{\mathcal{H}^\perp} f$, $k = \mathrm{Pr}_{\mathcal{K}^\perp} \tilde{A}f$, and so

$$\begin{pmatrix} f \\ \tilde{A}f \end{pmatrix} = \begin{pmatrix} \mathrm{Pr}_{\mathcal{H}} f \\ A\mathrm{Pr}_{\mathcal{H}} f \end{pmatrix} + \mathrm{Pr}_{\mathcal{H}^\perp} f + \mathrm{Pr}_{\mathcal{K}^\perp} \tilde{A}f \tag{3.3.1}$$

and

$$\begin{aligned} \mathrm{Pr}_{\tilde{\mathcal{L}}_-} \begin{pmatrix} f \\ \tilde{A}f \end{pmatrix} &= \mathrm{Pr}_{\mathcal{H}_+ \oplus \mathcal{K}_- \oplus \mathcal{H}^\perp} \left\{ \begin{pmatrix} \mathrm{Pr}_{\mathcal{H}} f \\ A\mathrm{Pr}_{\mathcal{H}} f \end{pmatrix} + \mathrm{Pr}_{\mathcal{H}^\perp} f + \mathrm{Pr}_{\mathcal{K}^\perp} \tilde{A}f \right\} \\ &= (P_+ + Q_- A)\mathrm{Pr}_{\mathcal{H}} f + \mathrm{Pr}_{\mathcal{H}^\perp} f. \end{aligned}$$

But since

$$\mathrm{Pr}_{\tilde{\mathcal{L}}_-}\mathcal{M} = \mathrm{Pr}_{\tilde{\mathcal{L}}_-}\left[\mathcal{G}(A) + \mathcal{L}^{\perp}\right] = \mathrm{ran}\,(P_+ + Q_- A) + \mathcal{H}^{\perp}$$

by (ii), it follows that $\mathrm{dom}\,\tilde{A} = \tilde{\mathcal{H}}$.

Since $\mathcal{M}$ is closed, $\tilde{A}$ is continuous by the closed graph theorem. That is, $\tilde{A} \in \mathbf{B}(\tilde{\mathcal{H}}, \tilde{\mathcal{K}})$. Since $\mathcal{M}$ is negative, $\tilde{A}$ is a contraction. The invariance of $\mathcal{M}$ under $\tilde{A}$ implies that $\tilde{A}U_1 = U_2\tilde{A}$. The identity (3.3.1) implies that $\mathrm{Pr}_{\mathcal{K}}\,\tilde{A}|\mathcal{H} = A$.

The correspondence between operators and their graphs is obviously one-to-one, and so the result is proved. ■

The result takes a simpler form when A is a bicontraction. As above, $T_1 \in \mathbf{B}(\mathcal{H})$ and $T_2 \in \mathbf{B}(\mathcal{K})$ are contractions with minimal isometric dilations $U_1 \in \mathbf{B}(\tilde{\mathcal{H}})$ and $U_2 \in \mathbf{B}(\tilde{\mathcal{K}})$.

THEOREM 3.3.3. *Assume that A is a bicontraction. Then the relation*

$$\mathcal{M} = \mathcal{G}(\tilde{A})$$

establishes a one-to-one correspondence between all bicontractions $\tilde{A} \in \mathbf{B}(\tilde{\mathcal{H}}, \tilde{\mathcal{K}})$ such that $\tilde{A}U_1 = U_2\tilde{A}$ and $\mathrm{Pr}_{\mathcal{K}}\,\tilde{A}|\mathcal{H} = A$ and all maximal negative $\tilde{U}$-invariant subspaces $\mathcal{M}$ of $\tilde{\mathcal{L}}$ satisfying $\mathcal{M} \subset \mathcal{G}(A) + \mathcal{L}^{\perp}$.

Proof. Let $\tilde{A} \in \mathbf{B}(\tilde{\mathcal{H}}, \tilde{\mathcal{K}})$ be a bicontraction such that $\tilde{A}U_1 = U_2\tilde{A}$ and $\mathrm{Pr}_{\mathcal{K}}\,\tilde{A}|\mathcal{H} = A$. By Theorem 3.3.2, $\mathcal{G}(\tilde{A})$ is a negative $\tilde{U}$-invariant subspace of $\mathcal{L}$ satisfying $\mathcal{M} \subset \mathcal{G}(A) + \mathcal{L}^{\perp}$. In addition,

$$\mathrm{Pr}_{\tilde{\mathcal{L}}_-}\mathcal{M} = \mathrm{Pr}_{\tilde{\mathcal{L}}_-}\left[\mathcal{G}(A) + \mathcal{L}^{\perp}\right] = \mathrm{ran}\,(P_+ + Q_- A) + \mathcal{H}^{\perp},$$

where the last equality follows as in the proof of Theorem 3.3.2. Since A is a bicontraction, $\mathrm{ran}\,(P_+ + Q_- A) = \mathcal{L}_-$ by Theorems 1.3.3 and 1.3.4. Therefore $\mathrm{Pr}_{\tilde{\mathcal{L}}_-}\mathcal{M} = \mathcal{L}_- + \mathcal{H}^{\perp} = \tilde{\mathcal{L}}_-$, and $\mathcal{M}$ is maximal negative.

Conversely, let $\mathcal{M}$ be a maximal negative $\tilde{U}$-invariant subspace of $\tilde{\mathcal{L}}$ satisfying $\mathcal{M} \subset \mathcal{G}(A) + \mathcal{L}^{\perp}$. Then $\mathcal{M}$ is closed and

$$\mathrm{Pr}_{\tilde{\mathcal{L}}_-}\mathcal{M} = \tilde{\mathcal{L}}_- = \mathrm{ran}\,(P_+ + Q_- A) + \mathcal{H}^{\perp} = \mathrm{Pr}_{\tilde{\mathcal{L}}_-}\left[\mathcal{G}(A) + \mathcal{L}^{\perp}\right].$$

By Theorem 3.3.2, $\mathcal{M} = \mathcal{G}(\tilde{A})$ is the graph of a contraction $\tilde{A} \in \mathbf{B}(\tilde{\mathcal{H}}, \tilde{\mathcal{K}})$ such that $\tilde{A}U_1 = U_2\tilde{A}$ and $\mathrm{Pr}_{\mathcal{K}}\,\tilde{A}|\mathcal{H} = A$. But A is a bicontraction, and so

$$\mathrm{Pr}_{\mathcal{K}_-}\tilde{A}\mathcal{H}_- = \mathrm{Pr}_{\mathcal{K}_-}A\mathcal{H}_- = \mathcal{K}_-.$$

Since $\mathcal{K}_-$ is maximal uniformly negative in $\tilde{\mathcal{K}}$, the operator $\tilde{A}$ is a bicontraction by Theorem 1.3.6. ■

3.4 Abstract Leech Theorem

We use a theorem of Shmul'yan on the factorization of bicontractions.

Theorem 3.4.1. *If $\mathcal{A}$, $\mathcal{B}$, $\mathcal{C}$ are Kreĭn spaces and $A \in \mathbf{B}(\mathcal{A},\mathcal{C})$ and $B \in \mathbf{B}(\mathcal{B},\mathcal{C})$ are bicontractions, then $A = BC$ for some bicontraction $C \in \mathbf{B}(\mathcal{A},\mathcal{B})$ if and only if $AA^* \le BB^*$.*

Proof. The condition is necessary: if $A = BC$ with C a bicontraction, then $AA^* = BCC^*B^* \le BB^*$.

Conversely, assume $AA^* \le BB^*$. We show that $\ker B^* \subset \ker A^*$. Suppose $f \in \ker B^*$ and $f \ne 0$. Let $\mathcal{C}_-$ be a maximal uniformly negative subspace of $\mathcal{C}$. Then $f \notin \mathcal{C}_-$ because $\ker B^*$ is a uniformly positive subspace of $\mathcal{C}$ by Theorem 1.3.1. Let $\tilde{\mathcal{C}}_-$ be the span of f and $\mathcal{C}_-$, so $\mathcal{C}_-$ is properly contained in $\tilde{\mathcal{C}}_-$. Note that $A^*\tilde{\mathcal{C}}_-$ is a negative subspace of $\mathcal{A}$. For if $g = \alpha f + h$ where α is a complex number and h is in $\mathcal{C}_-$, then since $AA^* \le BB^*$ by assumption,

$$\langle A^*g, A^*g\rangle_{\mathcal{A}} \le \langle B^*g, B^*g\rangle_{\mathcal{B}} = \langle B^*h, B^*h\rangle_{\mathcal{B}} \le 0. \tag{3.4.1}$$

Since $A^*\tilde{\mathcal{C}}_- \supset A^*\mathcal{C}_-$ and $A^*\mathcal{C}_-$ is maximal negative by Theorem 1.3.6, $A^*\tilde{\mathcal{C}}_- = A^*\mathcal{C}_-$. Therefore A^* annihilates some nonzero element $g = \alpha f + h$ of $\tilde{\mathcal{C}}_-$, where α is a complex number and $h \in \mathcal{C}_-$. For such g, equality holds in (3.4.1), and so $h = 0$ and $\alpha \ne 0$. Hence $f \in \ker A^*$. It follows that $\ker B^* \subset \ker A^*$.

The inclusion on kernels allows us to construct a linear transformation X_0 on $\operatorname{ran} B^* \oplus \ker B$ into $\mathcal{A}$ such that $A^* = X_0 B^*$ and X_0 annihilates $\ker B$. Since $AA^* \le BB^*$ and $\ker B$ is uniformly positive by Theorem 1.3.1, we obtain

$$\langle X_0u, X_0u\rangle_{\mathcal{A}} \le \langle u, u\rangle_{\mathcal{B}}, \quad u \in \operatorname{dom} X_0.$$

The domain of X_0 is dense in $\mathcal{B}$ and contains $B^*\mathcal{C}_-$, which is maximal uniformly negative in $\mathcal{B}$ by Theorem 1.3.6. Moreover, X_0 maps $B^*\mathcal{C}_-$ onto $A^*\mathcal{C}_-$, which is maximal uniformly negative in $\mathcal{A}$. By Theorem 1.4.4, X_0 extends to a bicontraction $C^* \in \mathbf{B}(\mathcal{B},\mathcal{A})$. Then $C \in \mathbf{B}(\mathcal{A},\mathcal{B})$ is a bicontraction and $A = BC$. ■

Leech's theorem is a structured form of Shmul'yan's theorem.

Theorem 3.4.2. *Let $\mathcal{A}$, $\mathcal{B}$, $\mathcal{C}$ be Kreĭn spaces, $S_{\mathcal{A}} \in \mathbf{B}(\mathcal{A})$, $S_{\mathcal{B}} \in \mathbf{B}(\mathcal{B})$, $S_{\mathcal{C}} \in \mathbf{B}(\mathcal{C})$ isometries, and $A \in \mathbf{B}(\mathcal{A},\mathcal{C})$, $B \in \mathbf{B}(\mathcal{B},\mathcal{C})$ bicontractions. Assume that*

$$AS_{\mathcal{A}} = S_{\mathcal{C}}A \qquad \text{and} \qquad BS_{\mathcal{B}} = S_{\mathcal{C}}B.$$

Then $A = BC$ for some bicontraction $C \in \mathbf{B}(\mathcal{A},\mathcal{B})$ satisfying

$$CS_{\mathcal{A}} = S_{\mathcal{B}}C$$

if and only if $AA^* \le BB^*$.

The case in which $S_{\mathcal{A}}$, $S_{\mathcal{B}}$, $S_{\mathcal{C}}$ coincide with the identity operators is Shmul'yan's theorem.

Proof. Necessity follows from Theorem 3.4.1.

Conversely, assume $AA^* \leq BB^*$. Set $\mathcal{H} = \overline{A^*\mathcal{C}} \subset \mathcal{A}$ and $\mathcal{K} = \overline{B^*\mathcal{C}} \subset \mathcal{B}$. Since A and B are contractions, $\mathcal{H}^\perp = \ker A$ and $\mathcal{K}^\perp = \ker B$ are uniformly positive by Theorem 1.3.1. In particular, $\mathcal{H}$ and $\mathcal{K}$ are regular subspaces of $\mathcal{A}$ and $\mathcal{B}$ and may be viewed as Kreĭn spaces in the scalar products of $\mathcal{A}$ and $\mathcal{B}$. By Shmul'yan's theorem there is a bicontraction $X \in \mathbf{B}(\mathcal{A}, \mathcal{B})$ such that $A = BX$. Define

$$T_1 = \mathrm{Pr}_{\mathcal{H}}\, S_{\mathcal{A}}|\mathcal{H} \in \mathbf{B}(\mathcal{H}), \qquad T_2 = \mathrm{Pr}_{\mathcal{K}}\, S_{\mathcal{B}}|\mathcal{K} \in \mathbf{B}(\mathcal{K}),$$

and

$$Y = \mathrm{Pr}_{\mathcal{K}}\, X|\mathcal{H} \in \mathbf{B}(\mathcal{H}, \mathcal{K}).$$

Then T_1 and T_2 are contractions, and Y is a bicontraction. Since $S_{\mathcal{A}}^*\mathcal{H} \subset \mathcal{H}$ and $S_{\mathcal{B}}^*\mathcal{K} \subset \mathcal{K}$, we have $T_1^* = S_{\mathcal{A}}^*|\mathcal{H}$ and $T_2^* = S_{\mathcal{B}}^*|\mathcal{K}$. We also have $X^*|\mathcal{K} = Y^*$ because $X^*B^* = A^*$, and so

$$\begin{aligned} Y^*T_2^*B^* &= X^*S_{\mathcal{B}}^*B^* = X^*B^*S_{\mathcal{C}}^* \\ &= A^*S_{\mathcal{C}}^* = S_{\mathcal{A}}^*A^* = S_{\mathcal{A}}^*X^*B^* = T_1^*Y^*B^*. \end{aligned}$$

Therefore $YT_1 = T_2Y$.

Next note that

$$\tilde{\mathcal{H}} = \bigvee_0^\infty S_{\mathcal{A}}^n\mathcal{H} \qquad \text{and} \qquad \tilde{\mathcal{K}} = \bigvee_0^\infty S_{\mathcal{B}}^n\mathcal{K}$$

are regular subspaces of $\mathcal{A}$ and $\mathcal{B}$ and hence Kreĭn spaces in the scalar products of $\mathcal{A}$ and $\mathcal{B}$. For example, we have

$$\mathcal{H} \subset \mathcal{H} \vee S_{\mathcal{A}}\mathcal{H} \subset \mathcal{H} \vee S_{\mathcal{A}}\mathcal{H} \vee S_{\mathcal{A}}^2\mathcal{H} \subset \cdots,$$

where at each stage the extension is obtained by forming a direct sum with a Hilbert space. Therefore each subspace in the chain is regular. By Lemma 1.1.9, $\tilde{\mathcal{H}}$ is a regular subspace of $\mathcal{A}$. In a similar way, $\tilde{\mathcal{K}}$ is a regular subspace of $\mathcal{B}$. Note that $\tilde{\mathcal{H}}$ is invariant under $S_{\mathcal{A}}$ and $S_{\mathcal{A}}^*$, and $\tilde{\mathcal{K}}$ is invariant under $S_{\mathcal{B}}$ and $S_{\mathcal{B}}^*$. Therefore $U_1 = S_{\mathcal{A}}|\tilde{\mathcal{H}}$ and $U_2 = S_{\mathcal{B}}|\tilde{\mathcal{K}}$ are minimal isometric dilations of T_1 and T_2.

By Theorem 3.2.2, there is a bicontraction $\tilde{Y} \in \mathbf{B}(\tilde{\mathcal{H}}, \tilde{\mathcal{K}})$ such that $\tilde{Y}U_1 = U_2\tilde{Y}$ and $\mathrm{Pr}_{\mathcal{K}}\, \tilde{Y}|\mathcal{H} = Y$. By Lemma 3.3.1, $\tilde{Y}^*|\mathcal{K} = Y^*$. Set

$$C = \tilde{Y}\mathrm{Pr}_{\tilde{\mathcal{H}}} \in \mathbf{B}(\mathcal{A}, \mathcal{B}).$$

Then $C = \mathrm{Pr}_{\tilde{\mathcal{K}}}\tilde{Y}\mathrm{Pr}_{\tilde{\mathcal{H}}}$, $C^* = \mathrm{Pr}_{\tilde{\mathcal{H}}}\tilde{Y}^*\mathrm{Pr}_{\tilde{\mathcal{K}}}$, and

$$X^*|\mathcal{K} = Y^* = \tilde{Y}^*|\mathcal{K} = C^*|\mathcal{K}.$$

So from $A^* = X^*B^*$ we obtain $A^* = C^*B^*$, hence $A = BC$. Since $\tilde{Y}$ is a bicontraction and the orthogonal complements of $\tilde{\mathcal{H}}$ and $\tilde{\mathcal{K}}$ in $\mathcal{A}$ and $\mathcal{B}$ are uniformly positive, C is a bicontraction. Finally,

$$CS_{\mathcal{A}}|\tilde{\mathcal{H}} = \tilde{Y}U_1 = U_2\tilde{Y} = S_{\mathcal{B}}C|\tilde{\mathcal{H}},$$

whereas

$$CS_{\mathcal{A}}|\tilde{\mathcal{H}}^{\perp} = 0 = S_{\mathcal{B}}C|\tilde{\mathcal{H}}^{\perp}.$$

Thus $CS_{\mathcal{A}} = S_{\mathcal{B}}C$. ■

Notes on Chapter 3

The dilation theory in §3.1 follows the methods of the Hilbert space case as given in Sz.-Nagy and Foiaş **[66]**, with some modifications. An extension of dilation theory to arbitrary operators was first made by Davis **[23]**; see Bognár **[12]**. In the indefinite setting, dilation theory is used by Bruinsma, Dijksma, and de Snoo **[20]**, Constantinescu and Ghéondea **[21,22]**, and Dritschel **[29]**.

The commutant lifting theorem in the Hilbert space case is given in Sz.-Nagy and Foiaş **[66]**. It was inspired by applications in interpolation theory due to Sarason **[60]**. There have been numerous accounts and applications; references are given in Rosenblum and Rovnyak **[59]**. In the indefinite case, extensions of the commutant lifting theorem are proved in Alpay **[1]**, de Branges **[14,16,17]**, Constantinescu and Ghéondea **[21,22]**, and Dritschel **[29,30]** in various settings. The general case of Theorem 3.2.1 first appears in Dritschel **[30]**. The corresponding result on bicontractions, Theorem 3.2.2, is due to Dritschel **[29]**; different proofs and extensions are given in Constantinescu and Ghéondea **[22]** and de Branges **[17]**. Theorem 3.2.3 is new.

The use of matrix completions in commutant lifting problems is a known way to organize calculations in the Hilbert space case. The idea seems to be due to Parrott **[54]**. It has been adopted by other authors, including Frazho **[35]** and Pták and Vrbová **[56]**. The method is also used in Dritschel **[29,30]** and Constantinescu and Ghéondea **[21,22]**.

Theorems 3.3.2 and 3.3.3 appear in Dritschel **[30]** and **[29]**, respectively. The graph approach is due to Ball and Helton **[11]**. Much has been written on the labeling problem in the Hilbert space case. See Arsene, Ceauşescu, and Foiaş **[6]**, Foiaş and Frazho **[34]**, and Helton et al. **[39]**. A comparison of methods is given in Frazho **[35]**.

Leech's original theorem **[51]** concerns the factorization of power series with operator coefficients in the form $A(z) = B(z)C(z)$. Helton **[39, p. 52]** gives a version for bounded measurable matrix valued functions. Rosenblum **[58]** proves an abstract form of Leech's theorem using the commutant lifting theorem (see also **[59]**). Leech's theorem was generalized to Kreĭn spaces by de Branges **[14,16]**. The version in Theorem 3.4.2 appears in Dritschel **[30]** and extends the abstract result in Rosenblum **[58]**.

APPENDIX A: COMPLEMENTATION THEORY

The operator methods in this paper are related to a theory of complementation in Kreĭn spaces due to de Branges [**13,15**]. The Kreĭn space version of complementation extends Hilbert space notions which appear in de Branges and Rovnyak [**18,19**]. We state without proof some results on an operator approach to complementation, with a full account to appear in a later paper.

A Kreĭn space $\mathcal{P}$ is *contained continuously*, *contractively*, or *isometrically* in a Kreĭn space $\mathcal{H}$ if $\mathcal{P}$ is a linear subspace of $\mathcal{H}$ and the inclusion mapping is continuous, contractive, or isometric, respectively. Let $\mathcal{P}$ be a Kreĭn space which is contained continuously in a Kreĭn space $\mathcal{H}$, and let A be the inclusion mapping. We associate with $\mathcal{P}$ the selfadjoint operator $P \in \mathbf{B}(\mathcal{H})$ given by

$$P = AA^*.$$

The operator P plays the role of a generalized projection for $\mathcal{P}$. Viewed as mappings on $\mathcal{H}$, the operators P and A^* have the same action. In the terminology of de Branges, P is the selfadjoint operator on $\mathcal{H}$ which coincides with the adjoint of the inclusion of $\mathcal{P}$ in $\mathcal{H}$. The positive and negative indices of the Kreĭn space $\mathcal{P}$ coincide with the hermitian indices $h_\pm(P)$ of the operator P by Theorem 1.2.1. The range $\mathcal{P}_0$ of P is a scalar product space with scalar product defined by

$$\langle Pf, Pg\rangle_{\mathcal{P}_0} = \langle Pf, g\rangle_{\mathcal{H}}, \qquad f, g \in \mathcal{H}.$$

The space $\mathcal{P}_0$ is contained in $\mathcal{P}$ as a dense subspace, and the $\mathcal{P}_0$ and $\mathcal{P}$ scalar products coincide on $\mathcal{P}_0$. Every selfadjoint operator P arises in this way.

THEOREM A1. *Let $\mathcal{H}$ be a Kreĭn space, and let $P \in \mathbf{B}(\mathcal{H})$ be a selfadjoint operator. Write P in any way in the form*

$$P = EE^*,$$

where $E \in \mathbf{B}(\mathcal{E}, \mathcal{H})$ for some Kreĭn space $\mathcal{E}$ and E has zero kernel. Let $\mathcal{P}_E$ be the range of E viewed as a Kreĭn space in the scalar product which makes E a Kreĭn space isomorphism of $\mathcal{E}$ onto $\mathcal{P}_E$. Then $\mathcal{P}_E$ is contained continuously in $\mathcal{H}$, and the adjoint of the inclusion of $\mathcal{P}_E$ in $\mathcal{H}$ coincides with P.

Special properties hold whenever the operators P and Q for two spaces $\mathcal{P}$ and $\mathcal{Q}$ satisfy $P+Q = 1$. In the special case of isometric inclusion, these properties reflect the fact that $\mathcal{P}$ and $\mathcal{Q}$ are regular subspaces of the larger Kreĭn space $\mathcal{H}$, and $\mathcal{H}$ decomposes into the orthogonal direct sum of $\mathcal{P}$ and $\mathcal{Q}$.

THEOREM A2. *Let $\mathcal{P}$ and $\mathcal{Q}$ be Kreĭn spaces which are contained continuously in a Kreĭn space $\mathcal{H}$, and let P and Q be the selfadjoint operators on $\mathcal{H}$ which coincide with the adjoints of the inclusions of $\mathcal{P}$ and $\mathcal{Q}$ in $\mathcal{H}$. Assume that $P + Q = 1$.*

(i) *The mapping $U : (f, g) \to f + g$ is a partial isometry from $\mathcal{P} \times \mathcal{Q}$ onto $\mathcal{H}$ with adjoint $U^* : h \to (Ph, Qh)$.*

(ii) *The intersection $\mathcal{L}$ of $\mathcal{P}$ and $\mathcal{Q}$ is a Kreĭn space in the scalar product defined by*

$$\langle f, g\rangle_{\mathcal{L}} = \langle f, g\rangle_{\mathcal{P}} + \langle f, g\rangle_{\mathcal{Q}}, \qquad f, g \in \mathcal{L}.$$

The Kreĭn space $\mathcal{L}$ is called the **overlapping space** *for $\mathcal{P}$ and $\mathcal{Q}$. It is contained continuously in $\mathcal{H}$, and the adjoint of the inclusion coincides with PQ.*

(iii) *The following conditions are equivalent: (a) $\mathcal{P}$ is contained contractively in $\mathcal{H}$, (b) $\mathcal{Q}$ is contained contractively in $\mathcal{H}$, (c) $P^2 \leq P$, (d) $Q^2 \leq Q$, (e) U is a contraction, and (f) the overlapping space $\mathcal{L}$ is a Hilbert space.*

(iv) *The following conditions are equivalent: (a) $\mathcal{P}$ and $\mathcal{Q}$ are contained isometrically in $\mathcal{H}$ as regular subspaces with $\mathcal{Q} = \mathcal{P}^{\perp}$, (b) $P^2 = P$, (c) $Q^2 = Q$, (d) U is an isometry, and (e) the overlapping space $\mathcal{L}$ contains no nonzero element.*

The ideas in Theorem A2 go back to Schwartz [**63**], who created a theory of operator ranges both in the Hilbert space and Kreĭn space settings. The authors thank Daniel Alpay for calling their attention to Schwartz's paper; a related work is Alpay [**2**]. It is of interest to know when a unique Kreĭn space $\mathcal{P}$ is associated with a given selfadjoint operator P. The following uniqueness condition, using an extra hypothesis, is similar to one given by Schwartz [**63**].

THEOREM A3. *Let $\mathcal{H}$ be a Kreĭn space, and let $P \in \mathbf{B}(\mathcal{H})$ be a selfadjoint operator. Let $\mathcal{P}_1$ and $\mathcal{P}_2$ be Kreĭn spaces which are contained continuously in $\mathcal{H}$ such that the adjoints of the inclusions each coincide with P. If $\mathcal{P}_1$ is contained continuously in $\mathcal{P}_2$, then $\mathcal{P}_1$ and $\mathcal{P}_2$ are equal isometrically.*

Uniqueness holds under a condition of a different nature.

THEOREM A4. *Let $\mathcal{H}$ be a Kreĭn space, and let $P \in \mathbf{B}(\mathcal{H})$ be a selfadjoint operator. Let $\mathcal{P}_1$ be a Kreĭn space which is contained continuously in $\mathcal{H}$ such that the adjoint of the inclusion of $\mathcal{P}_1$ in $\mathcal{H}$ coincides with P. Assume that the range of P contains a subspace $\mathcal{M}$ which is maximal uniformly definite in $\mathcal{P}_1$. Then if $\mathcal{P}_2$ is any Kreĭn space which is contained continuously in $\mathcal{H}$ such that the adjoint of the inclusion of $\mathcal{P}_2$ in $\mathcal{H}$ coincides with P, $\mathcal{P}_1$ and $\mathcal{P}_2$ are equal isometrically.*

COROLLARY A5. *Let $\mathcal{P}$ be a Kreĭn space which is contained continuously in a Kreĭn space $\mathcal{H}$, and let $P \in \mathbf{B}(\mathcal{H})$ be the selfadjoint operator which coincides with the adjoint of the inclusion of $\mathcal{P}$ in $\mathcal{H}$. Assume that the range of P contains a subspace $\mathcal{M}$ which is maximal uniformly definite in $\mathcal{P}$. If*

$$P = EE^*,$$

where $E \in \mathbf{B}(\mathcal{E}, \mathcal{H})$ for some Kreĭn space $\mathcal{E}$ and E has zero kernel, then E is a Kreĭn space isomorphism of $\mathcal{E}$ onto $\mathcal{P}$.

The condition for uniqueness in Theorem A4 is always satisfied in the important special case of contractive inclusion. This yields a new derivation of a result of de Branges [**13**].

THEOREM A6. *Let $\mathcal{P}_1$ and $\mathcal{P}_2$ be Kreĭn spaces which are contained continuously and contractively in a Kreĭn space $\mathcal{H}$ such that the adjoints of the inclusions each coincide with the selfadjoint operator P on $\mathcal{H}$. Then $\mathcal{P}_1$ and $\mathcal{P}_2$ are equal isometrically.*

Combining Theorem A6 and Corollary 5, we see that the Kreĭn spaces in de Branges' theory of complementation [**13**] are operator ranges. In the indefinite case, this was first shown by Heinz Langer (private communication, 1988) using his theory of definitizable operators and spectral functions [**49**]. In Hilbert spaces, this has been known for a long time. It was shown, for example, in seminar lectures by Marvin Rosenblum at the University of Virginia in the 1960's. In Hilbert spaces, Sarason has used the operator view in applications to function theory in a series of papers including [**61,62**].

The choice of a contraction operator leads to an example of complementation Kreĭn spaces.

THEOREM A7. *Let $\mathcal{H}$ and $\mathcal{K}$ be Kreĭn spaces, and let $T \in \mathbf{B}(\mathcal{H}, \mathcal{K})$ be a contraction. Define $\mathcal{M}(T)$ to be the range of T in the scalar product which makes T a partial isometry of $\mathcal{H}$ onto $\mathcal{M}(T)$. Let $D \in \mathbf{B}(\mathcal{D}, \mathcal{K})$ be any defect operator for T^*. Define $\mathcal{H}(T)$ to be the range of D in the scalar product which makes D a Kreĭn space isomorphism of $\mathcal{D}$ onto $\mathcal{H}(T)$. Then $\mathcal{M}(T)$ and $\mathcal{H}(T)$ are Kreĭn spaces which are contained continuously and contractively in $\mathcal{K}$, and the adjoints of the inclusions coincide with TT^* and $1 - TT^*$, respectively. The definition of $\mathcal{H}(T)$ is independent of the choice of defect operator D for T^*. An element g of $\mathcal{K}$ belongs to $\mathcal{H}(T)$ if and only if*

$$\sup_{u \in \mathcal{H}} \left[\langle g + Tu, g + Tu \rangle_{\mathcal{K}} - \langle u, u \rangle_{\mathcal{H}} \right] < \infty,$$

in which case the value of the supremum is $\langle g, g \rangle_{\mathcal{H}(T)}$.

Appendix B: More on Julia Operators

The existence of a Julia operator was proved in §1.2 by an argument based on the factorization of any selfadjoint operator in the form AA^*, where A is an operator having zero kernel. We present an alternative proof following the original method due to Arsene, Constantinescu, and Ghéondea [**8**]. The method gives the additional information that a particular choice of Julia operator satisfies norm estimates (Theorem B3). We also show that uniqueness holds under weaker assumptions than that of Theorem 2.4.5 (Theorem B4).

Two preliminary Hilbert space results are needed. As usual, we write $*$ for Kreĭn space adjoint and $\times$ for Hilbert space adjoint. If $\mathcal{H}$ and $\mathcal{K}$ are Hilbert spaces and T is a continuous everywhere defined operator on $\mathcal{H}$ to $\mathcal{K}$, then by the *polar factorization* of T we mean the representation $T = RU$, where U is a partial isometry on $\mathcal{H}$ to $\mathcal{K}$ with kernel equal to the kernel of T and R is a nonnegative operator on $\mathcal{K}$ which is zero on the orthogonal complement of the range of U.

Lemma B1. *Let $\mathcal{H}$ be a Kreĭn space, and let $H \in \mathbf{B}(\mathcal{H})$ be a selfadjoint operator. Assume that $\mathcal{H} = \mathcal{H}_+ \oplus \mathcal{H}_-$ is a fundamental decomposition. Then there is a Kreĭn space $\mathcal{A}$ with fundamental decomposition $\mathcal{A} = \mathcal{A}_+ \oplus \mathcal{A}_-$ and an operator $A \in \mathbf{B}(\mathcal{A}, \mathcal{H})$ with these properties:*

(i) A has zero kernel and $H = AA^*$;

(ii) if $A = RW$ is the polar factorization of A as an operator on $\mathcal{A}_+ \oplus |\mathcal{A}_-|$ to $\mathcal{H}_+ \oplus |\mathcal{H}_-|$, then $AA^* = R^2WW^*$ and $R^2 = AA^*(WW^*)^\times$.

Proof. The operator A constructed in the proof of Theorem 1.2.2 satisfies (i). We show that A also satisfies (ii).

The representation $A = RW$ in the proof of Theorem 1.2.2 is the polar factorization of A as an operator on $\mathcal{A}_+ \oplus |\mathcal{A}_-|$ to $\mathcal{H}_+ \oplus |\mathcal{H}_-|$. The operator $WJ_{\mathcal{A}}W^\times$ is 1 on $\mathcal{M}_+$, -1 on $\mathcal{M}_-$, and 0 on the orthogonal complement of $\mathcal{M}_+ + \mathcal{M}_-$ in $\mathcal{H}_+ \oplus |\mathcal{H}_-|$. Hence $R(WJ_{\mathcal{A}}W^\times)R = R^2(WJ_{\mathcal{A}}W^\times)$, and

$$AA^* = RWW^*R^* = R(WJ_{\mathcal{A}}W^\times)RJ_{\mathcal{H}} = R^2(WJ_{\mathcal{A}}W^\times)J_{\mathcal{H}} = R^2WW^*.$$

Since $WW^*(WW^*)^\times = WW^\times$ is the projection of $\mathcal{H}_+ \oplus |\mathcal{H}_-|$ onto $\overline{\operatorname{ran} R}$,

$$AA^*(WW^*)^\times = R^2WW^*(WW^*)^\times = R^2,$$

and so (ii) holds. ■

We use a result due to Kreĭn [**41**] and Reid [**57**] and rediscovered by Lax [**50**] and Dieudonné [**27**]. We include Reid's proof for completeness.

LEMMA B2. *Let $\mathcal{H}$ be a Hilbert space and $A, X \in \mathbf{B}(\mathcal{H})$. If A is nonnegative and AX is selfadjoint, then for all $f \in \mathcal{H}$,*

$$|\langle AXf, f\rangle| \leq \|X\|\langle Af, f\rangle.$$

Proof. It is enough to give the proof when $\|X\| = 1$. Since A is nonnegative, for any f and g in $\mathcal{H}$,

$$|\langle Af, g\rangle| \leq \langle Af, f\rangle^{1/2}\langle Ag, g\rangle^{1/2} \leq \frac{1}{2}\left[\langle Af, f\rangle + \langle Ag, g\rangle\right].$$

Since $AX = X^*A$, for any positive integer n,

$$|\langle AX^n f, f\rangle| \leq \frac{1}{2}\left[\langle AX^n f, X^n f\rangle + \langle Af, f\rangle\right] = \frac{1}{2}\left[\langle AX^{2n} f, f\rangle + \langle Af, f\rangle\right].$$

Iteration of this inequality yields

$$|\langle AXf, f\rangle| \leq 2^{-n}\left\langle AX^{2^n} f, f\right\rangle + \left(\frac{1}{2} + \frac{1}{4} + \cdots + \frac{1}{2^n}\right)\langle Af, f\rangle,$$

and we obtain the result on letting n tend to ∞. ■

THEOREM B3. *Let $\mathcal{H}$ and $\mathcal{K}$ be Kreĭn spaces with fundamental decompositions $\mathcal{H} = \mathcal{H}_+ \oplus \mathcal{H}_-$ and $\mathcal{K} = \mathcal{K}_+ \oplus \mathcal{K}_-$. For any $T \in \mathbf{B}(\mathcal{H}, \mathcal{K})$, there exist Kreĭn spaces $\mathcal{D}$ and $\tilde{\mathcal{D}}$ with fundamental decompositions $\mathcal{D} = \mathcal{D}_+ \oplus \mathcal{D}_-$ and $\tilde{\mathcal{D}} = \tilde{\mathcal{D}}_+ \oplus \tilde{\mathcal{D}}_-$ and a Julia operator*

$$U = \begin{pmatrix} T & D \\ \tilde{D}^* & L \end{pmatrix} \in \mathbf{B}(\mathcal{H} \oplus \mathcal{D}, \mathcal{K} \oplus \tilde{\mathcal{D}}).$$

such that

$$\|L\| \leq \|T\|$$

and

$$\max\left(\|D\|, \|\tilde{D}\|\right) \leq \left[1 + \|T\|^2\right]^{1/2}.$$

Proof. By Lemma B1, there exist Kreĭn spaces $\mathcal{D}$ and $\tilde{\mathcal{D}}$ with fundamental decompositions $\mathcal{D} = \mathcal{D}_+ \oplus \mathcal{D}_-$ and $\tilde{\mathcal{D}} = \tilde{\mathcal{D}}_+ \oplus \tilde{\mathcal{D}}_-$, and operators $D \in \mathbf{B}(\mathcal{D}, \mathcal{K})$ and $\tilde{D} \in \mathbf{B}(\tilde{\mathcal{D}}, \mathcal{H})$ with zero kernels, such that

$$1 - TT^* = DD^*, \qquad 1 - T^*T = \tilde{D}\tilde{D}^*, \tag{B-1}$$

and the polar representations $D = RW$ and $\tilde{D} = \tilde{R}\tilde{W}$ as Hilbert space operators satisfy

$$1 - TT^* = R^2WW^*, \qquad \tilde{R}^2 = (1 - T^*T)(\tilde{W}\tilde{W}^*)^\times.$$

In particular, $\|D\|$ and $\|\tilde{D}\|$ are bounded by $\left[1 + \|T\|^2\right]^{1/2}$.

We construct $L \in \mathbf{B}(\mathcal{D}, \tilde{\mathcal{D}})$ with the aid of Lemma B2, applied to $A = R^2$ and $X = (WW^*)T(\tilde{W}\tilde{W}^*)^\times T^\times$ viewed as operators on $\mathcal{K}_+ \oplus |\mathcal{K}_-|$. Clearly A is nonnegative, and

$$\begin{aligned} AX &= R^2(WW^*)T(\tilde{W}\tilde{W}^*)^\times T^\times = (1 - TT^*)T(\tilde{W}\tilde{W}^*)^\times T^\times \\ &= T(1 - T^*T)(\tilde{W}\tilde{W}^*)^\times T^\times = T\tilde{R}^2T^\times \end{aligned}$$

is selfadjoint. Since $\|X\| \leq \|T\|^2$, Lemma B2 yields

$$\left|\langle R^2Xf, f\rangle_{\mathcal{K}_+\oplus|\mathcal{K}_-|}\right| \leq \|T\|^2\langle R^2f, f\rangle_{\mathcal{K}_+\oplus|\mathcal{K}_-|}, \qquad f \in \mathcal{K}.$$

In other words, $T\tilde{R}^2T^\times \leq \|T\|^2R^2$ as operators on $\mathcal{K}_+\oplus|\mathcal{K}_-|$. Therefore $T\tilde{R} = RC$ where $C \in \mathbf{B}(\mathcal{H}, \mathcal{K})$ and $\|C\| \leq \|T\|$. We can choose C so that $\operatorname{ran} C \subset \overline{\operatorname{ran} R}$, and then $T\tilde{R}\tilde{W} = RWW^\times C\tilde{W}$. We obtain

$$T\tilde{D} = -DL^*, \tag{B - 2}$$

where $L = -(W^\times C\tilde{W})^* \in \mathbf{B}(\mathcal{D}, \tilde{\mathcal{D}})$ and $\|L\| \leq \|T\|$.

It remains to prove the six identities in (1.2.1) and (1.2.2). The relations (1.2.1a) and (1.2.2a,b) hold by (B-1) and (B-2). By (1.2.2b),

$$\tilde{D}LD^* = -\tilde{D}\tilde{D}^*T^* = -(1 - T^*T)T^* = -T^*(1 - TT^*) = -T^*DD^*.$$

Since $\ker D = \{0\}$, (1.2.1b) follows. Similarly,

$$\begin{aligned} DL^*L &= -T\tilde{D}L = TT^*D = (1 - DD^*)D = D(1 - D^*D), \\ \tilde{D}LL^* &= -T^*DL^* = T^*T\tilde{D} = (1 - \tilde{D}\tilde{D}^*)\tilde{D} = \tilde{D}(1 - \tilde{D}^*\tilde{D}), \end{aligned}$$

and (1.2.1c) and (1.2.2c) hold. ∎

We proved uniqueness of Julia operators in Theorem 2.4.5 under the assumption that either T or T^* is a contraction. The conclusion can be obtained with a weaker hypothesis, namely, that one of the operators $1 - T^*T$ or $1 - TT^*$ has at most a finite number of negative squares.

THEOREM B4. *Let $\mathcal{H}$ and $\mathcal{K}$ be Kreĭn spaces, and let $T \in \mathbf{B}(\mathcal{H},\mathcal{K})$ have two Julia operators*

$$\begin{pmatrix} T & D_j \\ \tilde{D}_j^* & L_j \end{pmatrix} \in \mathbf{B}(\mathcal{H}\oplus\mathcal{D}_j, \mathcal{K}\oplus\tilde{\mathcal{D}}_j), \qquad j=1,2.$$

If one of the indices $h_-(1-T^*T)$, $h_-(1-TT^*)$, $h_+(1-T^*T)$, or $h_+(1-TT^*)$ is finite, then there exist unitary operators $V \in \mathbf{B}(\mathcal{D}_2,\mathcal{D}_1)$ and $\tilde{V} \in \mathbf{B}(\tilde{\mathcal{D}}_2,\tilde{\mathcal{D}}_1)$ such that

$$\begin{pmatrix} T & D_2 \\ \tilde{D}_2^* & L_2 \end{pmatrix} = \begin{pmatrix} 1 & 0 \\ 0 & \tilde{V}^* \end{pmatrix}\begin{pmatrix} T & D_1 \\ \tilde{D}_1^* & L_1 \end{pmatrix}\begin{pmatrix} 1 & 0 \\ 0 & V \end{pmatrix}.$$

LEMMA B5. *Any dense subspace $\mathcal{M}$ of a Pontryagin space $\mathcal{H}$ contains a maximal uniformly negative subspace.*

Proof of Lemma B5. We prove the assertion by showing that if a uniformly negative subspace $\mathcal{N}$ of $\mathcal{M}$ is not maximal, then there exists a uniformly negative subspace $\mathcal{N}'$ of $\mathcal{M}$ which properly contains $\mathcal{N}$. Granting this, then starting with $\mathcal{N}=\{0\}$ one can construct in a finite number of steps a maximal uniformly negative subspace of $\mathcal{M}$.

Assume that $\mathcal{N}$ is a uniformly negative subspace which is contained in $\mathcal{M}$ and that $\mathcal{N}$ is not maximal. Choose a fundamental decomposition $\mathcal{H}=\mathcal{H}_+\oplus\mathcal{H}_-$ such that $\mathcal{N}\subset\mathcal{H}_-$. Norms are to be computed in the associated Hilbert space $\mathcal{H}_+\oplus|\mathcal{H}_-|$. Note that $\dim\mathcal{H}_- < \infty$ because $\mathcal{H}$ is a Pontryagin space.

Since $\mathcal{N}$ is not maximal, there is a unit vector $e\in\mathcal{H}_-$ which is orthogonal to $\mathcal{N}$. Since $\mathcal{M}$ is dense in $\mathcal{H}$, we can choose $\phi\in\mathcal{M}$ such that $\|e-\phi\|\le 1/3$. Then $\|\phi\|\ge 2/3$. We show that the span $\mathcal{N}'$ of $\mathcal{N}$ and ϕ is uniformly negative. For any $g\in\mathcal{N}'$ we can write $g=f+\alpha\phi$ where $f\in\mathcal{N}$ and α is a scalar. Then

$$\begin{aligned}\langle g,g\rangle_{\mathcal{H}} &= \langle f,f\rangle_{\mathcal{H}} + |\alpha|^2\langle\phi,\phi\rangle_{\mathcal{H}} + 2\,\mathrm{Re}\,[\overline{\alpha}\langle f,\phi\rangle_{\mathcal{H}}] \\ &= -\|f\|^2 - |\alpha|^2\|\phi\|^2 + 2\,\mathrm{Re}\,[\overline{\alpha}\langle f,\phi-e\rangle_{\mathcal{H}}]\end{aligned}$$

because $e\perp f$. Thus

$$\begin{aligned}\langle g,g\rangle_{\mathcal{H}} &\le -\|f\|^2 - |\alpha|^2\|\phi\|^2 + 2|\alpha|\|f\|\|\phi-e\| \\ &\le -\|f\|^2 - \frac{4}{9}|\alpha|^2 + 2\cdot\frac{1}{3}|\alpha|\|f\| \\ &= -\left[\|f\| - \frac{1}{3}|\alpha|\right]^2 - \frac{1}{3}|\alpha|^2 \\ &\le 0\end{aligned}$$

with equality only if $g=0$. Thus the form $\langle g,g\rangle_{\mathcal{H}}$ is strictly negative on the unit sphere of $\mathcal{N}'$, which is a compact set. Hence $\langle g,g\rangle_{\mathcal{H}}\le -\delta\|g\|^2$ for all $g\in\mathcal{N}'$ and some $\delta > 0$. Therefore $\mathcal{N}'$ is a subspace of $\mathcal{M}$ which is uniformly negative and properly contains $\mathcal{N}$. ■

Proof of Theorem B4. We assume that $h_-(1 - T^*T) < \infty$. For the case $h_-(1-TT^*) < \infty$, we obtain the conclusion by replacing T by T^*. The other cases are handled by easy modifications as noted below.

The hypothesis $h_-(1-T^*T) < \infty$ implies that $\tilde{\mathcal{D}}_1$ and $\tilde{\mathcal{D}}_2$ are Pontryagin spaces each with finite negative index equal to $h_-(1 - T^*T)$ (see the remarks following Definition 1.2.3).

Since $\tilde{D}_1\tilde{D}_1^* = 1 - T^*T = \tilde{D}_2\tilde{D}_2^*$, for any $f, g \in \mathcal{H}$,

$$\left\langle \tilde{D}_1^* f, \tilde{D}_1^* g \right\rangle_{\tilde{\mathcal{D}}_1} = \left\langle \tilde{D}_2^* f, \tilde{D}_2^* g \right\rangle_{\tilde{\mathcal{D}}_2}.$$

It follows that there is a well defined and densely defined isometry X from $\tilde{\mathcal{D}}_1$ to $\tilde{\mathcal{D}}_2$ with $\operatorname{dom} X = \operatorname{ran} \tilde{D}_1^*$ and $\operatorname{ran} X = \operatorname{ran} \tilde{D}_2^*$ such that

$$X\tilde{D}_1^* f = \tilde{D}_2^* f, \qquad f \in \mathcal{H}.$$

Since $\tilde{\mathcal{D}}_1$ is a Pontryagin space, Lemma B5 implies that $\operatorname{dom} X$ contains a maximal uniformly negative subspace $\tilde{\mathcal{D}}_1^-$. But X is an isometry and $\tilde{\mathcal{D}}_2$ is a Pontryagin space with the same negative index as that of $\tilde{\mathcal{D}}_1$. So $X\tilde{\mathcal{D}}_1^-$ is a maximal uniformly negative subspace $\tilde{\mathcal{D}}_2^-$ of $\tilde{\mathcal{D}}_2$. Therefore by Theorem 1.4.4, X has an extension by continuity to a unitary operator $\tilde{V}^* \in \mathbf{B}(\tilde{\mathcal{D}}_1, \tilde{\mathcal{D}}_2)$. By construction, $\tilde{D}_2^* = \tilde{V}^*\tilde{D}_1^*$.

If we had assumed instead that one of the positive hermitian indices is finite, then the same conclusion could be drawn. We simply replace the defect spaces by their anti-spaces and note that the densely defined isometry X is still an isometry on the new spaces. Then we apply the argument given above to conclude that X may be extended to a unitary operator.

We show next that

$$\begin{pmatrix} D_1 \\ \tilde{V}^* L_1 \end{pmatrix} \in \mathbf{B}(\mathcal{D}_1, \mathcal{K} \oplus \tilde{\mathcal{D}}_2)$$

is a defect operator for the adjoint of $\binom{T}{\tilde{D}_2^*} \in \mathbf{B}(\mathcal{H}, \mathcal{K} \oplus \tilde{\mathcal{D}}_2)$. The operator has zero kernel because D_1 has zero kernel. We have

$$\begin{aligned}
1 - \begin{pmatrix} T \\ \tilde{D}_2^* \end{pmatrix} \begin{pmatrix} T^* & \tilde{D}_2 \end{pmatrix} &= 1 - \begin{pmatrix} T \\ \tilde{V}^*\tilde{D}_1^* \end{pmatrix} \begin{pmatrix} T^* & \tilde{D}_1\tilde{V} \end{pmatrix} \\
&= \begin{pmatrix} 1 - TT^* & -T\tilde{D}_1\tilde{V} \\ -\tilde{V}^*\tilde{D}_1^*T^* & 1 - \tilde{V}^*\tilde{D}_1^*\tilde{D}_1\tilde{V} \end{pmatrix} \\
&= \begin{pmatrix} D_1D_1^* & D_1L_1^*\tilde{V} \\ \tilde{V}^*L_1D_1^* & \tilde{V}^*L_1L_1^*\tilde{V} \end{pmatrix} \\
&= \begin{pmatrix} D_1 \\ \tilde{V}^*L_1 \end{pmatrix} \begin{pmatrix} D_1^* & L_1^*\tilde{V} \end{pmatrix},
\end{aligned}$$

which proves the assertion.

Now to complete the proof, think of $\begin{pmatrix} T & D_2 \\ \tilde{D}_2^* & L_2 \end{pmatrix}$ as $(A \quad B)$, where $A = \begin{pmatrix} T \\ \tilde{D}_2^* \end{pmatrix} \in \mathbf{B}(\mathcal{H}, \mathcal{K} \oplus \tilde{\mathcal{D}}_2)$ and $B = \begin{pmatrix} D_2 \\ L_2 \end{pmatrix} \in \mathbf{B}(\mathcal{D}_2, \mathcal{K} \oplus \tilde{\mathcal{D}}_2)$. The operator A is an isometry and hence a contraction. The operator A^* has defect operator

$$\begin{pmatrix} D_1 \\ \tilde{V}^* L_1 \end{pmatrix} \in \mathbf{B}(\mathcal{D}_1, \mathcal{K} \oplus \tilde{\mathcal{D}}_2)$$

by what was shown above. Since $(A \quad B)$ is unitary and hence a bicontraction, Theorem 2.3.3 yields a bicontraction $V \in \mathbf{B}(\mathcal{D}_2, \mathcal{D}_1)$ such that

$$\begin{pmatrix} D_2 \\ L_2 \end{pmatrix} = \begin{pmatrix} D_1 \\ \tilde{V}^* L_1 \end{pmatrix} V.$$

We see easily that V is unitary. In fact,

$$D_1 V V^* D_1^* = D_2 D_2^* = 1 - TT^* = D_1 D_1^*,$$

and since D_1 has zero kernel, this implies that $VV^* = 1$, that is, V^* is an isometry. But $D_2 = D_1 V$ and therefore V has zero kernel. By elementary properties of isometries and partial isometries described in §1.1(C), V is unitary. By construction,

$$\begin{pmatrix} T & D_2 \\ \tilde{D}_2^* & L_2 \end{pmatrix} = \begin{pmatrix} T & D_1 V \\ \tilde{V}^* \tilde{D}_1^* & \tilde{V}^* L_1 V \end{pmatrix},$$

as required. ■

After this paper was completed, the authors obtained a copy of Agnes Yang's thesis [**68**]. This includes a stronger form of Kreĭn 's theorem, Lemma B2, which is due to Dijksma, Langer, and de Snoo ("Unitary colligations in Krein spaces and their role in the extension theory of isometries and symmetric linear relations in Hilbert spaces," Functional Analysis, II (Dubrovnic 1985), pp. 1–42, Lecture Notes in Math., Vol. 1242, Springer, Berlin-New York, 1987; MR 89a:47055). The stronger form of Kreĭn 's theorem is essentially no more difficult to prove, and by appropriate choices of operators one obtains a somewhat more direct proof of the construction of Julia operators in Theorem B3.

References

[1] D. Alpay, *Dilatations des commutants d'operateurs pour des espaces de Krein de fonctions analytiques,* Annales de l'Institut Fourier, to appear.

[2] D. Alpay, *Some remarks on reproducing kernel Krein spaces,* Rocky Mountain J. Math., to appear.

[3] D. Alpay, *The Schur Algorithm, System Theory and Reproducing Kernel Hilbert Spaces of Analytic Functions,* lecture notes from Groningen University, 1989.

[4] T. Andô, *Linear Operators on Kreĭn Spaces,* Hokkaido University, Research Institute of Applied Electricity, Division of Applied Mathematics, Sapporo, 1979. MR 81e:47032.

[5] D. Z. Arov, *Darlington realization of matrix-valued functions,* Izv. Akad. Nauk SSSR Ser. Mat. **37** (1973), 1299–1331; Math. USSR-Izv. **7** (1973), 1295–1326. MR 50#10287.

[6] Gr. Arsene, Z. Ceauşescu, and C. Foiaş, *On intertwining dilations. VIII,* J. Operator Theory **4** (1980), 55–91. MR 82d:47013f.

[7] Gr. Arsene and A. Ghéondea, *Completing matrix contractions,* J. Operator Theory **7** (1982), 179–189. MR 83i:4710.

[8] Gr. Arsene, T. Constantinescu, and A. Ghéondea, *Lifting of operators and prescribed numbers of negative squares,* Michigan J. Math. **34** (1987), 201–216. MR 88j:47008.

[9] T. Azizov and I. S. Iokhvidov, *Linear operators in spaces with an indefinite metric and their applications,* Mathematical Analysis, Vol. 17, pp. 113–205, 272, Akad. Nauk SSSR, Vsesojuz. Inst. Naučn. i Tehn. Informacii, Moscow, 1979; J. Soviet Math. **15** (1981), 438–490. MR 81m:47052.

[10] T. Azizov and I. S. Iokhvidov, *Foundations of the Theory of Linear Operators in Spaces with Indefinite Metric,* "Nauka", Moscow, 1986; English transl. *Linear Operators in Spaces with Indefinite Metric,* Wiley, New York, 1989. MR 88g:47070.

[11] J. A. Ball and J. W. Helton, *A Beurling-Lax theorem for the Lie group $U(m,n)$ which contains most interpolation theory,* J. Operator Theory **9** (1983), 107–142. MR 84m:47046.

[12] J. Bognár, *Indefinite Inner Product Spaces,* Springer, Berlin-New York, 1974. MR 57#7125.

[13] L. de Branges, *Complementation in Krein spaces,* Trans. Amer. Math. Soc. **305** (1988), 277–291. MR 89c:46034.

[14] L. de Branges, *Krein spaces of analytic functions,* J. Funct. Anal. **81** (1988), 219–259.

[15] L. de Branges, *Square Summable Power Series,* in preparation.

[16] L. de Branges, *A construction of Krein spaces of analytic functions,* preprint, 1989.

[17] L. de Branges, *Commutant lifting in Krein spaces,* in preparation.

[18] L. de Branges and J. Rovnyak, *Canonical models in quantum scattering theory,* Perturbation Theory and its Applications in Quantum Mechanics, (Proc. Adv. Sem. Math. Res. Center, U. S. Army, Theoret. Chem. Inst., Univ. of Wisconsin, Madison, Wis., 1965), C. H. Wilcox, ed., pp. 295–392, Wiley, New York, 1966. MR 39#6109.

[19] L. de Branges and J. Rovnyak, *Square Summable Power Series,* Holt, Rinehart and Winston, New York, 1966. MR 35#5909.

[20] P. Bruinsma, A. Dijksma, H. de Snoo, *Unitary dilations of contractions in* Π_κ*-spaces,* Oper. Theory: Adv. Appl., Vol. 28, pp. 27–42, Birkhäuser, Basel-Boston, 1988. MR 90b:47062.

[21] T. Constantinescu and A. Ghéondea, *On unitary dilations and characteristic functions in indefinite product spaces,* Oper. Theory: Adv. Appl., Vol. 24, pp. 87–102, Birkhäuser, Basel-Boston, 1987. MR 88j:47049.

[22] T. Constantinescu and A. Ghéondea, *Minimal signature in lifting of operators. I,II,* INCREST, Bucharest, preprints; Part I: No. 56/1988; Part II: March 1989.

[23] Ch. Davis, *J-unitary dilation of a general operator,* Acta Sci. Math. (Szeged) **31** (1970), 75–86. MR 41#9032.

[24] Ch. Davis, *An extremal problem for extensions of a sesquilinear form,* Linear Algebra and Appl. **13** (1976), 91–102. MR 52#15068.

[25] Ch. Davis, *A factorization of an arbitrary* $m \times n$ *contractive operator-matrix,* Toeplitz Centennial (Tel Aviv, 1981), Oper. Theory: Adv. Appl., Vol. 4, pp. 217–232, Birkhäuser, Basel-Boston, 1982. MR 84c:47011

[26] Ch. Davis, W. M. Kahan, and H. F. Weinberger, *Norm-preserving dilations and their applications to optimal error bounds,* SIAM J. Numer. Anal. **19** (1982), 445–469. MR 84b:47010.

[27] J. Dieudonné, *Quasi-hermitian operators,* Proc. Internat. Sympos. Linear Spaces (Jerusalem, 1960), pp. 115–122, Jerusalem Academic Press, Jerusalem; Pergamon, Oxford, 1961. MR 32#4540.

[28] R. G. Douglas, *On majorization, factorization, and range inclusion of operators on Hilbert spaces,* Proc. Amer. Math. Soc. 17 (1966), 413–415. MR 34#3315.

[29] M. Dritschel, *A lifting theorem for bicontractions,* J. Funct. Anal. **88** (1990), 61–89.

[30] M. Dritschel, *Extension Theorems for Operators on Kreĭn Spaces,* Dissertation, University of Virginia, 1989.

[31] N. Dunford and J. T. Schwartz, *Linear Operators,* Vol. I, Interscience, New York, 1958. MR22#8302.

[32] H. Dym, *J Contractive Matrix Functions, Reproducing Kernel Hilbert Spaces, and Interpolation,* CBMS Regional Conference Series in Mathematics, Vol. 71, Amer. Math. Soc., Providence, 1989.

[33] C. Foiaş, *Contractive intertwining dilations and waves in layered media,* Proceedings of the International Congress of Mathematicians (Helsinki, 1978), pp. 605–613, Acad. Sci. Fennica, Helsinki, 1980. MR 82a:47002.

[34] C. Foiaş and A. Frazho, *On the Schur representation in the commutant lifting problem. I,II,* Part I: I. Schur Methods in Operator Theory and Signal Processing, Oper. Theory: Adv. Appl., Vol. 18, pp. 207–217, Birkhäuser, Basel-Boston, 1986; Part II: Oper. Theory: Adv. Appl., Vol. 29, pp. 171–179, Birkhäuser, Basel-Boston, 1988. MR 89a:47011.

[35] A. Frazho, *Three inverse scattering algorithms for the lifting theorem,* I. Schur Methods in Operator Theory and Signal Processing, Oper. Theory: Adv. Appl., Vol. 18, pp. 219–248, Birkhäuser, Basel-Boston, 1986. MR 89c:47012.

[36] A. Ghéondea, *Canonical forms of unbounded unitary operators in Kreĭn spaces,* Publ. Res. Inst. Math. Sci. **24** (1988), no. 2, 205–224. MR 90a:47087.

[37] Yu. P. Ginsburg, *J-nonexpansive analytic operator-functions,* Dissertation, Odessa, 1958.

[38] I. Gohberg, P. Lancaster, and L. Rodman, *Matrices and Indefinite Scalar Products,* Birkhäuser, Basel-Boston, 1983. MR 87j:15001.

[39] J. W. Helton, with the assistance of J. A. Ball, C. R. Johnson, and J. N. Palmer, *Operator Theory, Analytic Functions, Matrices, and Electrical Engioneering,* CBMS Regional Conference Series in Mathematics, Vol. 68, Amer. Math. Soc., Providence, 1987. MR 89f:47001.

[40] I. S. Iokhvidov, M. G. Kreĭn, and H. Langer, *Introduction to the Spectral Theory of Operators in Spaces with an Indefinite Metric,* Akademie-Verlag, Berlin, 1982. MR 85g:47050.

[41] M. G. Kreĭn, *Completely continuous linear operators in function spaces with two norms,* Akad. Nauk Ukrain. RSR. Zbirnik Prac' Inst. Mat., No. 9, (1947), 104–129.

[42] M. G. Kreĭn, *On an application of a fixed point principle in the theory of linear transformations of spaces with an indefinite metric,* Uspekhi Mat. Nauk **5**, No. 2, (1950), 180–190. MR 14,56.

[43] M. G. Kreĭn, *Introduction to the geometry of indefinite J-spaces and to the theory of operators in those spaces,* Second Math. Summer School, Part I, pp. 15–92. Naukova Dumka, Kiev, 1965; Amer. Math. Soc. Transl. (2) **93** (1970), 103–176. MR 33#574.

[44] M. G. Kreĭn and H. Langer, *Continuation of Hermitian positive definite functions and related questions,* preprint.

[45] M. G. Kreĭn and Yu. L. Shmul'yan, *Plus operators in a space with indefinite metric,* Mat. Issled. **1** (1966), 131–161; Amer. Math. Soc. Transl. (2) **85** (1969), 93–113. MR 34#4923.

[46] M. G. Kreĭn and Yu. L. Shmul'yan, *J-polar representation of plus-operators,* Mat. Issled. **1** (1966), 172–210; Amer. Math. Soc. Transl. (2) **85** (1969), 115–143. MR 34#8183.

[47] M. G. Kreĭn and Yu. L. Shmul'yan, *On linear-fractional transformations with operator coefficients,* Mat. Issled. **2** (1967), 64–96; Amer. Math. Soc. Transl. (2) **103** (1974), 125–152. MR 39#813.

[48] S. A. Kuzhel', *J-nonexpansive operators,* Teor. Funktsional. Anal. i. Prilozhen., No. 45, (1986), 63–68, ii. MR 87m:47086.

[49] H. Langer, *Spectral functions of definitizable operators in Kreĭn spaces,* Functional analysis (Dubrovnik, 1981), pp. 1 - 46, Lecture Notes in Math., Vol. 948, Springer, Berlin-New York, 1982. MR 84g:47034.

[50] P. D. Lax, *Symmetrizable linear transformations,* Comm. Pure Appl. Math. **7** (1954), 633–647. MR 16,832.

[51] R. B. Leech, *Factorization of analytic functions and operator inequalities,* unpublished manuscript (early 1970's?).

[52] B. McEnnis, *Shifts on indefinite inner product spaces,* Pacific J. Math. **81** (1979), 113–130; part II, ibid. **100** (1982), 177–183. MR 81c:47040; 84b:47043.

[53] B. McEnnis, *Shifts on Krein Spaces,*, Proceedings of Symposia in Pure Mathematics: Operator Theory, Operator Algebras, and Applications, Amer. Math. Soc., to appear.

[54] S. Parrott, *On the quotient norm and the Sz.-Nagy–Foiaş lifting theorem,* J. Funct. Anal. **30** (1978), 311–328. MR 81h:47006.

[55] V. P. Potapov, *The multiplicative structure of J-contractive matrix functions,* Trudy Moskov. Mat. Obshch. **4** (1955), 125–236; Amer. Math. Soc. Transl. (2) **15** (1960), 131–243. MR 17,958.

[56] V. Pták and P. Vrbová, *Lifting intertwining relations,* Integral Equations and Operator Theory **11** (1988), 128–147.

[57] W. T. Reid, *Symmetrizable completely continuous linear transformations in Hilbert space,* Duke Math. J. **18** (1951), 41–56. MR 13,564.

[58] M. Rosenblum, *A corona theorem for countably many functions,* Integral Equations and Operator Theory **3** (1980), 125–137. MR 81e:46034.

[59] M. Rosenblum and J. Rovnyak, *Hardy Classes and Operator Theory,* Oxford Math. Monographs, Oxford University Press, New York, 1985. MR 87e:47001.

[60] D. Sarason, *Generalized interpolation in H^∞,* Trans. Amer. Math. Soc. **127** (1967), 179–203. MR 34#8193.

[61] D. Sarason, *Shift-invariant subspaces from the Brangesian point of view,* The Bieberbach Conjecture (West Lafayette, Ind., 1985), 153–166, Math. Surveys Monographs, Vol. 21, Amer. Math. Soc., Providence, 1986. MR 88d:47014a.

[62] D. Sarason, *Doubly shift-invariant spaces in H^2,* J. Operator Theory **16** (1986), 75–97. MR 88d:47014b.

[63] L. Schwartz, *Sous-espaces hilbertiens d'espaces vectoriels topologiques et noyaux associés (noyaux reproduisants),* J. Analyse Math. **13** (1964), 115–256. MR 31#3835.

[64] Yu. L. Shmul'yan, *Division in the class of J-expansive operators,* Mat. Sb. (N.S.) **74** (116) (1967), 516–525; Math. USSR-Sbornik **3** (1967), 471–479. MR 37#784.

[65] Yu. L. Shmul'yan and R. N. Yanovskaya, *Blocks of a contractive operator matrix,* Izv. Vyssh. Uchebn. Zaved. Mat. 1981, No. 7, 72–75; Soviet Math. (Iz. VUZ) **25** (1981), No. 7, 82–86. MR 83e:47007.

[66] B. Sz.-Nagy and C. Foiaş, *Harmonic Analysis of Operators on Hilbert Space,* North Holland, New York, 1970. MR 43#947.

[**67**] Shao Zong Yan, *The contraction operators on space* Π , Chinese Ann. Math. Ser. B **7** (1986), 75–89. MR 87m:47088.

[**68**] Agnes Yang, *A Construction of Krein spaces of Analytic Functions,* Dissertation, Purdue University, 1990.

Department of Mathematics
Mathematical Sciences Building
Purdue University
West Lafayette, Indiana 47907

Department of Mathematics
Mathematics-Astronomy Building
University of Virginia
Charlottesville, Virginia 22903